AF775371

AIR, WATER AND SOIL POLLUTION SCIENCE AND TECHNOLOGY

MYCORRHIZAL FUNGI: SOIL, AGRICULTURE AND ENVIRONMENTAL IMPLICATIONS

AIR, WATER AND SOIL POLLUTION SCIENCE AND TECHNOLOGY

Additional books in this series can be found on Nova's website under the Series tab.

Additional E-books in this series can be found on Nova's website under the E-books tab.

AGRICULTURE ISSUES AND POLICIES

Additional books in this series can be found on Nova's website under the Series tab.

Additional E-books in this series can be found on Nova's website under the E-books tab.

AIR, WATER AND SOIL POLLUTION SCIENCE AND TECHNOLOGY

MYCORRHIZAL FUNGI: SOIL, AGRICULTURE AND ENVIRONMENTAL IMPLICATIONS

SUSANNE M. FULTON
EDITOR

Nova Science Publishers, Inc.
New York

Copyright © 2011 by Nova Science Publishers, Inc.

All rights reserved. No part of this book may be reproduced, stored in a retrieval system or transmitted in any form or by any means: electronic, electrostatic, magnetic, tape, mechanical photocopying, recording or otherwise without the written permission of the Publisher.

For permission to use material from this book please contact us:
Telephone 631-231-7269; Fax 631-231-8175
Web Site: http://www.novapublishers.com

NOTICE TO THE READER

The Publisher has taken reasonable care in the preparation of this book, but makes no expressed or implied warranty of any kind and assumes no responsibility for any errors or omissions. No liability is assumed for incidental or consequential damages in connection with or arising out of information contained in this book. The Publisher shall not be liable for any special, consequential, or exemplary damages resulting, in whole or in part, from the readers' use of, or reliance upon, this material. Any parts of this book based on government reports are so indicated and copyright is claimed for those parts to the extent applicable to compilations of such works.

Independent verification should be sought for any data, advice or recommendations contained in this book. In addition, no responsibility is assumed by the publisher for any injury and/or damage to persons or property arising from any methods, products, instructions, ideas or otherwise contained in this publication.

This publication is designed to provide accurate and authoritative information with regard to the subject matter covered herein. It is sold with the clear understanding that the Publisher is not engaged in rendering legal or any other professional services. If legal or any other expert assistance is required, the services of a competent person should be sought. FROM A DECLARATION OF PARTICIPANTS JOINTLY ADOPTED BY A COMMITTEE OF THE AMERICAN BAR ASSOCIATION AND A COMMITTEE OF PUBLISHERS.

Additional color graphics may be available in the e-book version of this book.

LIBRARY OF CONGRESS CATALOGING-IN-PUBLICATION DATA

Mycorrhizal fungi : soil, agriculture, and environmental implications /
editor, Susanne M. Fulton.
p. cm.
Includes index.
ISBN 978-1-61122-659-1 (hardcover)
1. Mycorrhizal fungi. 2. Mycorrhizas in agriculture. I. Fulton, Susanne M.
QK604.2.M92M964 2010
631.4'6--dc22
2010041375

Published by Nova Science Publishers, Inc. † New York

CONTENTS

PREFACE

Abuscular mycorrhizal (AM) fungi are ubiquitous in soils around the world where they form symbiotic associations with the majority of plant species. The beneficial effects of AM fungi are well known. They include improved plant nutrient uptake, enhanced N2-fixation in legumes, increased plant tolerance to biotic and abiotic stress and improved soil structural quality. They are seen as the cornerstone of sustainability in agro-ecosystems. This book presents current research in the study of mycorrhizal fungi.

Chapter 1 - Global environmental change caused by human activities is one of the greatest challenges facing our society. Many soils have become nutrient deficient due to years of intensive farming, over use of chemical fertilizers and from the effects of industrialization and urbanization. The world is losing between 5 to 10 hectares of agricultural land annually due to severe degradation. To achieve this agriculturist are following alternate technology, where microbes are the key players especially for protecting and regaining the soil health. Amongst, the key players, arbuscular mycorrhizal fungi (AMF) are one of the important organisms for improving and sustaining soil qualities. Varied responses of arbuscular mycorrhizal fungi (AMF) to these aspects are well documented. The studies conducted with these effective fungi have shown 10 to 40 percent increase in yield/biomass. However, its association varies widely in structure and function. The beneficial effects of AMF depend on the combined efficacy of plants and the fungus/fungi involved, and also on their ecological interactions within the system (soil, water, sediment, etc.) in which the contaminant is present. The symbiosis, they form, is potentially valuable not only for development programme like low input agriculture, but also act as a complex experimental model in challenging environment. Although relatively few specific plant-fungus combinations have been studied for their efficacy and application in remediation and resource conservation, the existing data on the benefits for AMF are promising. AMF improve plant growth, help in contaminant removal, reduce the need for fertilizer application in commercial plant production, and improve the soil structure and health. In order to gain the optimum benefit from these fungi, care should be taken to use the appropriate management strategies to encourage their survival. In present article, the authors are focusing on some recent progress in improving soil under challenged conditions.

Chapter 2 - This review was done to explore the current information on the benefit of arbuscular mycorrhizal symbioses in agricultural systems, drawing on results of research on agroecosystems in Argentina and Brazil. The increasing recognition that agricultural intensification and use of agrochemicals can adversely affect soil quality, modifying the number, diversity and activity of the soil microbiota, including symbiotic fungal populations, have improved the research aimed at crop yield improvement and sustainability. Thus, microbial-based approaches have been proposed and quantitative relationships between soil or plant phosphorus (P) levels and arbuscular mycorrhizas (AM) are being investigated in order to adjust the fertilizer dose to obtain acceptable yield production without depressing the symbiosis. Biodiversity of AM is high in agricultural soils from Argentina both under tillage or non-tillage management. Soils cultivated with wheat, one of the most important crops in Argentina, contain indigenous AM that colonize this crop; however the selection of effective fungus, as well as the study of the mycorrhizal populations in soils supplied with P may allow the attainment of acceptable yield levels with low fertilizer inputs, while also reducing costs and environmental pollution risk. Fertilization with synthetic P sources as superphosphate decreases both mycorrhizal potential and indigenous arbuscular mycorrhizal colonization (AMC) of grain and grasses crops. However, fertilization with rock phosphate doesn't negatively affect both mycorrhizal potential and indigenous AMC. Additionally, the positive effects of mycorrhizal inoculation are shown with the rock phosphate fertilization, while when the fertilizer was superphosphate mycorrhizal benefit was only evident at moderate/low doses of P. Alternatively, fertilizer placement also affects indigenous AMC. Applications with banded superphosphate depress the AM formation of wheat roots as compared with broadcasted applications both under no-tillage and under conventional tillage. In Brazil, reports for maize, peanut, soybean and sorghum as well as for green manure species have pointed their mycotrophy and AM dependency. However, more research is needed in relation to germplasm banks of AM for increase the production of agricultural plants as well as to optimize the use of renewable P sources, which is essential to the Brazilian agribusiness. This chapter discusses the AM parameters, such as root colonization and spore number in agricultural systems and the relationship with some soil characteristics, such as organic matter and glomalin contents. Relevant findings are emphasized, such as inoculation and management benefits for farmers by increasing soil fertility without loss of yield or added costs. Research directions that are needed to increase understanding of mycorrhizal associations in agriculture are indicated.

Chapter 3 - The acceleration of species' lost is such that understanding relationships between biodiversity and ecosystem function is a crucial aspect of ecology. Plant community composition and distribution has been thought, for a long time, to be a matter of aboveground interactions such as plant – plant feedbacks, plant – herbivores and/or parasites relationships, disturbance creating new patches for plant colonization, … Due to relevant advances in our understanding of plant and soil ecology, particularly in the investigation of rhizosphere (that is the soil biologically active compartment where root – root and root – microbes communications occurred), several findings provide evidence that plant – soil microbes feedbacks, particularly those with mycorrhizal fungi, are of great importance in determining plant species competitive performance and, could thus contribute to regulate plant community structure. Indeed, mycorrhizal symbioses are involved in plant nutrient mobilization and acquisition strategies, in plant health promotion, in interactions between plant and other biotic and/or abiotic factors, ... As plants differ in their specificity with mycorrhizal symbionts, plant host – mycorrhizal symbiont feedback would result in plant species' differential fitness and abilities when colonizing soil patches. Moreover, mycorrhizal hyphal network is known to interconnect conspecific as well as interspecific plants, and nutrient exchange may occur *via* this route to support subordinates. Consequently, mycorrhizal associations may be considered as an important biological mediator for plant coexistence and diversity.

This chapter reviews some of the recent research work implemented on the involvement of soil microbial community in regulating plant community composition and distribution. The authors particularly focus on the implication of mycorrhizal symbioses in promoting multi-species assemblages in plant communities.

Chapter 4 - Arbuscular mycorrhizal fungi are soil fungi developing symbiotic association with most terrestrial plants as their usually non-specific hosts. In their symbioses the fungi provide their host plant with water and nutrients in the exchange for carbon. The beneficiary effects of mycorrhizal fungi on the ecosystem have been previously indicated by numerous number of research work. Their contribution to the ecosystem is by enhancing the efficiency of environment including soil under different conditions including stress, which eventually increases the production of crop yield. In greater details such implications include: 1) enhanced water and nutrient uptake by plant, 2) alleviation of different soil stresses on plant growth including drought, salinity, compaction, and pathogens, 3) their interactions with other soil microbes, and 4) their likely use as bio-fertilizers. In this article the most recent advances regarding such implications, which eventually contribute to the enhanced ecosystem efficiency, are reviewed. Some likely approaches, which can improve the efficiency of mycorrhizal fungi under different conditions, are also highlighted.

Chapter 5 - The numerous benefits contributed by arbuscular mycorrhizal (AM) fungi has lead to recognition of their benefits in sustainable agriculture and various natural environmental processes. Application of anthropogenic and other compounds intentionally in the form of pesticides, fungicides, fertilizers or unintentionally like polycyclic aromatic hydrocarbons (PAH) which affect soil fungi have shown varying effects on AM association. Not all organic pollutants are detrimental to AM fungi. Some fungicides such as Fludioxonil have a stimulatory effect on mycorrhizal formation and function. Rapid decrease in the metabolic activity of AM fungi, a decline in arbuscule production, which is the main site of nutrient exchange between the host plant and the fungus, reduction in the total root length colonized and slow colonization rates are the negative effects shown by certain fungicides. Pesticides generally inhibit AM colonization and spore numbers. Slow release mineral fertilizers though benefit AM fungi, their excessive use can result in suppression of the AM fungi. Certain PAH especially anthracene reduce the development of extra radical mycelium, sporulation, root colonization and spore germination. Different AM fungal species show differential sensitivity even within a particular class of organic pollutant. Caution should also be exercised while irrigating soils with effluents as these could affect AM fungi. A reduction in our dependency on conventional agricultural practices especially organic pesticides by using AM fungi as a biological tool which is possible only by maintaining low levels of organic pollutants in the soil. Minimal use of organic pollutants especially pesticides would enable the maintenance of a healthy AM fungal populations essential for sustainable plant production systems.

Chapter 6 - Arbuscular mycorrhizal (AM) fungi are ubiquitous in soils around the world where they form symbiotic associations with the majority of plant species. The beneficial effects of AM fungi are well known. They include improved plant nutrient uptake, enhanced N_2-fixation in legumes, increased plant tolerance to biotic and abiotic stress and improved soil structural quality, and are seen as the cornerstone of sustainability in agro-ecosystems. In soil, AM fungi form extensive mycelial networks connected to plant roots. These AM mycelial networks, which increase the surface for nutrient and water absorption, also offer large surfaces for interactions with other microorganisms. These fungi often interact positively with biofertilizers and biocontrol agents. The presence of AM fungi in plant roots and soils was repeatedly shown to reduce the incidence of diseases caused by pathogenic nematodes and fungi and to improve plant health. AM fungi exert profound effects on soil microorganisms through competition for nutrients, and modification of root exudation, plant physiology and signaling. Better understanding of the interactions between AM fungi and other microorganisms is necessary for the development of sustainable management of soil fertility and crop production. The purpose of this review is to outline the current knowledge on the mechanisms by which the symbiotic association with mycorrhizal fungi alters the existing mycorrhizosphere microbial community and discusses the possible applications these mechanisms may have in sustainable agriculture.

Chapter 7 - Sustainable agriculture, conservation-oriented production, is the optimal target of all countries worldwide with the increased concerns on food security and resources & environmental protection. More and more biological measures and recycle technologies were attempted to apply in crop production to reduce the chemicals' input, such as fertilizer and pesticides. Many studies have been carried out to determine the efficacy of arbuscular mycorrhizal fungi (AMF) in promoting plant fitness, productivity and produce quality over the past two decades in China except for the biodiversity and distribution investigations. Besides field crops, many industrial crops (fibre crops, sugar crops, tea plants, offcinals), horticultural crops (vegetables and fruits) as well as micropropagated seedlings, were proved to benefit from the arbuscular mycorrhizal symbiosis with respect to stress resistance, yield and quality enhancement. In this paper, the data were reviewed the beneficial functions of AMF in agricultural production, highlighting the environment-friendly and non-pollution advantages of AMF application in agriculture, and the application strategies were also discussed. In particular, the authors hope that our review will make some of the recent work published in the Chinese literature more accessible to a wider international audience.

Chapter 8 - The fungal symbiont, Arbuscular mycorrhiza (AM) may benefit greenhouse gas reduction at the same time as increasing plant nutrients, moisture availability and soil carbon. The lack of increased mineralization observed in response to higher carbon availability in AM colonized soils is addressed through a literature review of studies from fields with a range of soil carbon, soil texture from sand to clay and over various time frames. Many studies pointed to highest carbon solubility in organic systems with carbon amendments and on undisturbed prairie, both of which host abundant AM colonization. Published data from several large field studies with wide variation in carbon levels, provided evidence of significantly decreased nitrogen mineralization in high soil carbon conditions, and more so in sand than clay soils. Multiple explanations were proposed in literature, including increased nitrogen immobilization with abundant soluble carbon, changes in carbon storage form and turnover time and the delayed effect of AM colonization on soil water repellency and aggregation. Our review is contrary to ideology found in some computer modeling programs, where nitrate levels and nitrogen mineralization were increased in high soil carbon conditions, and decreased with low soil carbon in response to nitrogen level. Arbuscular mycorrhizal fungal colonization may provide a unique system to retain carbon and nitrogen in the environment in association with plant growth. Further research in this area could provide necessary detail to understand soil carbon feedback from plant growth that may alter carbon and nitrogen mineralization with increased atmospheric CO_2 concentrations.

Chapter 9 - Persistent chemical usage in conventional agriculture aggravated the environmental contamination and paved the way to the conception of several approaches to integrated pest management. In this context, microbe-microbe interactions such as the mycorrhizal relationships with other soil micro biota in the rhizosphere like the rhizosphere bacteria and saprotrophic fungi are particularly important. A good understanding of the interactions between beneficial microbial groups is indispensable to the improvement of their management as biocontrol agents or biofertilizers. In recent years, our knowledge of the interaction effects between arbuscular mycorrhizal fungi (AMF) and rhizosphere bacteria or saprotrophic fungi in relation to plant growth and induction of plant resistance against diseases and nematodes have greatly increased. This chapter seeks to review the current research results on the effects of combined inoculation of the AMF and rhizosphere bacteria on plant growth, root colonization and induction of systemic resistance and compare it with those of AMF and saprotrophic fungi and to discuss their mechanisms of action and implications to biological control of plant diseases.

Chapter 10 - There are spacious studies concerning nutritional and physiological effects of arbuscular mycorrhizal fungi (AMF) on the plant and microbial communities of natural and agricultural ecosystems, but there are less information about importance and tolerance mechanisms of AMF in the heavy metals (HMs) polluted soils that all limited to two last decades. Totally, potentially toxic metals influence on the abundance, species richness and soil microbial and plant diversity. HM pollution does not affect on all organisms in an equal manner. Many researchers were reported the existence of AMF and colonization of root plants in metals polluted sites. It was described the identified and unidentified species of one or more AMF genus in the studies. *Glomus* was a common genus in the polluted soils. This chapter reviewed and discussed the results of some scientists regarding the presence and diversity of AMF and their role in imparting metal tolerance to plants in the HMs polluted soils.

Chapter 11 - Since dry land becoming arid or semi-arid is rising in the last decades, more agricultural practices in accordance with the principles of sustainable agriculture are needed. One practice could be the use of arbuscular mycorhizal (AM) fungi as inoculants to render the plants more tolerant to drought stress. In the present chapter is described how AM symbiosis improves three aspects of plant drought tolerance. The three aspects described here are root water uptake capacity, water use efficiency (WUE), and antioxidant systems. Regarding water uptake capacity it is clear that AM plants are able to take up more water than non AM plants under drought conditions. This higher capacity of water extraction is linked in several cases to a more apoplastic water flow through roots and the water uptake by the mycelium itself. The precise role of aquaporins on this water uptake capacity enhancement has not been solved yet. WUE it is usually enhanced in AM plants under drought conditions when it is calculated as the ratio between biomass or yield and the water consumed. This fact means that AM plants can produce more biomass with the same amount of water consumed, that is, they use water more efficiently. These higher WUE could be linked to a better photosystem II efficiency. But how AM symbiosis affects dark respiration or CO_2 assimilatory enzymes activities under drought conditions is far to be known. Finally, although general speaking, AM symbiosis diminishes oxidative damage caused by drought in terms of lipid peroxidation, the question if this is a consequence of a better water status or if it is directly caused by a better antioxidant system is still a matter of debate.

Chapter 12 - High temperature in summer often restricts rooting, root growth and induces browning of roots and shoots during nursery production resulting in withering and causing great losses of plants in strawberry cultivation. Then, it is necessary to establish stable technique for nursery production in strawberry cultivation during the high temperature period in summer including the problems of global warming. Arbuscular mycorrhizal fungus (AMF) has the effect of promoting host plant growth mainly by enhancing phosphorus uptake through symbiosis. On the other hand, promotion of active oxygen synthesis in plants occurs under several stress conditions including temperature stress, and antioxidative ability has relation with the resolution. In this study, the changes in activities of antioxidative ability through symbiosis with AMF in strawberry plants under natural high temperature stress were investigated. The strawberry (*Fragaria×ananassa* Duch., cv. Nohime) runner plants were treated with split root system method, and plants were raised in greenhouse during summer period. Dry weight of shoots and roots became greater in AMF plants than in non-AMF, resulting that growth promotion appeared in mycorrhizal plants under high temperature stress in summer period. The severity of browning in roots caused by high temperature stress was eased in AMF plants than in non-AMF. In this case, the alleviation of root browning appeared in both AMF- and AMF+ root system, so that induced tolerance to high temperature stress was recognized in mycorrhizal plants. AMF plots showed higher activities in SOD and APX than non-AMF; the effect differs with plant portions and investigated periods. In addition, some AMF plots showed higher values in followed items; chlorophyll content, ascorbic acid content and DPPH radical-scavenging activity. From these findings, tolerance to high temperature stress and the increase in antioxidative ability occurred in mycorrhizal strawberry plants, so that the activation of antioxidative ability is supposed to be closely associated with the high temperature stress tolerance in summer.

In: Mycorrhizal Fungi
Editor: Susanne M. Fulton
ISBN 978-1-61122-659-1
© 2011 Nova Science Publishers, Inc.

Chapter 1

ARBUSCULAR MYCORRHIZAL FUNGI IN CHALLENGING ENVIRONMENT– A PROSPECTIVE

Kajal Srivastava and A. K. Sharma*
Department of Biological Sciences, CBSH
G. B. Pant University of Agriculture and Technology
Pantnagar – 263145, UK, India

ABSTRACT

Global environmental change caused by human activities is one of the greatest challenges facing our society. Many soils have become nutrient deficient due to years of intensive farming, over use of chemical fertilizers and from the effects of industrialization and urbanization. The world is losing between 5 to 10 hectares of agricultural land annually due to severe degradation. To achieve this agriculturist are following alternate technology, where microbes are the key players especially for protecting and regaining the soil health. Amongst, the key players, arbuscular mycorrhizal fungi (AMF) are one of the important organisms for improving and sustaining soil qualities. Varied responses of arbuscular mycorrhizal fungi (AMF) to these aspects are well documented. The studies conducted with these effective fungi have shown 10 to 40 percent increase in yield/biomass. However, its association varies widely in structure and function. The beneficial effects of AMF depend on the combined efficacy of plants and the fungus/fungi involved, and also on their ecological interactions within the system (soil, water, sediment, etc.) in which the contaminant is present. The symbiosis, they form, is potentially valuable not only for development programme like low input agriculture, but also act as a complex experimental model in challenging environment. Although relatively few specific plant-fungus combinations have been studied for their efficacy and application in remediation and resource conservation, the existing data on the benefits for AMF are promising. AMF improve plant growth, help in

* Email: anilksharma_99@yahoo.com

contaminant removal, reduce the need for fertilizer application in commercial plant production, and improve the soil structure and health. In order to gain the optimum benefit from these fungi, care should be taken to use the appropriate management strategies to encourage their survival. In present article, we are focusing on some recent progress in improving soil under challenged conditions.

Keywords: Arbuscular mycorrhizal fungi, agriculture, association, environment, potential

INTRODUCTION

Worldwide increasing environmental, ecological, and economic concerns have been providing momentum for maintenance of soil quality and remediation strategies for management of soil contaminated with pollutants and other sources. Arbuscular mycorrhizal fungi (AMF) are recognized as indirect contributors through their influence on soil aggregation, plant physiology, and community composition. In present review, we attempt to show the development of research on arbuscular mycorrhizae and their role in mitigating environmental challenges, which have been arisen due to man made activities.

Arbuscular mycorrhizal fungi (AMF), comprising fungi in the order Glomales (Zygomycota), are ubiquitous symbionts of the majority of higher plants. Most plant species in their undisturbed natural environments form a beneficial association with mycorrhizal fungi. The resulting structure is called a mycorrhiza, or fungus-root. They are the most common type of endomycorrhiza, and known to have originated 460 Million Years Ago. These fungi are obligate symbionts occurring with most herbaceous and woody plants throughout the temperate deciduous forests, grasslands, and sub-tropical and tropical regions (Leake, 2007). The arbuscule, a highly branched hyphal structure that forms within the cortical root cells, is of central importance in AM because nutrient exchange occurs across the interface between the arbuscule and the cellular contents.

AMF have numerous benefits to host plants, including improved plant growth and mineral nutrition (Marschner and Dell, 1994), tolerance to disease (Matsubara et al., 2000, 2001) and abiotic stresses such as drought (Subramanian and Charest, 1995), chilling (El-Tohamy et al., 1999), and salinity (Ho, 1987). AMF can improve fertilizer utilization, rooting depth, the speed of establishment, disease and drought resistance of turf. They are present in soil as spores, as hyphae (filaments) or as colonized roots. Hyphae of mycorrhiza penetrate into and between the outer cells of the root. Inside the root, the fungus forms special coiled hyphae (arbuscules) that provide increased surface area for exchanges of food to the fungus and nutrients for the plant. The mycorrhizal fungi once established on the turf root system radiate out from the roots to form a dense network of filaments called extraradical hyphae. These filaments form an extensive system of hyphae that grow into the surrounding soil and provide a variety of benefits for the plant. This network of filaments obtains major macro and micro nutrients and water and transport these materials back to the root system. They are especially important for uptake of nutrients that do not readily move through the soil such as phosphorous and many of the micro-nutrients. The network also binds soil particles together, improves soil porosity and the movement of air and water within the soil.

New scientific advancements in the cost effective growing of certain mycorrhizal species beneficial to crops are rapidly bringing mycorrhizal products market place. They can help lower costs over the long run. The mycorrhizal condition is esseantial for most plants because up to 80% of the flowering plants (angiosperms) and up to 95% of all plants form mycorrhizal relationships (Read, 1992). Photosynthetic plants support the fungi by providing fixed carbon (up to 20% of the photosynthate may be allocated to roots to support mycorrhiza) and nutrients; the fungi in return provide the main plant-growth-limiting nutrients, nitrogen and phosphorus The fungal partner(s) in mycorrhiza may account between 80% and 100% of phosphorus (P) taken up by a plant (Hodge et al., 2001). Plants with roots colonized by mycorrhizal fungi are potentially more effective at nutrient and water acquisition, less susceptible to disease, and can be more productive under certain stressful environmental growing conditions than plants without mycorrhiza. Some of the important practical applications of mycorrhiza are: a) in soil/substrate (including transplanting media) that are constantly fumigated or receiving high rates of fungicides to eliminate/reduce soil borne pathogens, such as in horticultural crops and fruits; b) in re-vegetation of eroded or mined areas (extreme pH; metal toxicity; low organic matter content, natural AM inoculum, and overall fertility); and c) in arid and semi-arid regions. Their contribution to biological, chemical, and physical soil quality has been acknowledged, although many questions remain how to optimally manage these beneficial fungi. More fundamental and strategic studies in this field are therefore needed.

Importance of AMF for Environmental Remediation and Resource Conservation

Environmental remediation is defined as the removal of pollution or contaminants from environmental media such as soil, groundwater, sediment, or surface water for the general protection of human health and the environment. In the plant, rhizosphere (the zone of soil under the direct influence of a plant root), biodegradation or transformation of pollutants by root-associated bacteria and fungi under the influence of plant species occurs. Plants can increase the total numbers of beneficial fungi and bacteria in contaminated soil from a general rhizosphere effect. This is substantiated by the observation of higher microbial biomass and activity in the rhizosphere (Olson et al., 2003). AMF have been suggested to improve biodegradation of persistent organic pollutants because of the immense size and very high surface interface with soil. AM fungi are of importance as they play a vital role in metal tolerance (del Val et al., 1999) and accumulation (Zhu et al., 2001; Jamal et al., 2002). External mycelium of AM fungi provides a wider exploration of soil volumes by spreading beyond the root exploration zone (Khan et al., 2000; Malcova et al., 2003), thus providing access to greater volume of heavy metals present in the rhizosphere. These fungi have enzymes which are known to metabolize and degrade compounds such as polychlorinated biphenyls (PCBs) and polycyclic aromatic hydrocarbons (PAHs) (Olson et al., 2003). They also improve the remediation potential of plants by producing plant-growth-stimulating substances and by encouraging mineral nutrition, better general growth, and high biomass necessary for plant-based remediation. Currently, numerous studies are being conducted on the isolation of mycorrhizal strains, especially their ecotypes that are the most effective in counteracting the toxicity of metals in soils. In most studies showing the beneficial role of AM fungi the protective mechanisms involve the prevention of translocation of metals into the host (Krupa, 1999). On naturally contaminated sites, lack of essential nutrients together with metal concentration have been found to be the main difficulties for re-vegetation (Vangronsveld et al., 1996). AMF contribute and improve host plant P nutrition, which might help the plant to tolerate metal toxicity, through the resulting increase in plant growth. This has long been accepted as the main beneficial effect of AMF association on host plants. In addition, in view of the poor fertility and unbalanced mineral nutrient supply of many contaminated sites, application of arbuscular mycorrhiza would be even more useful. They got associated with metal-tolerant plants, e.g. metallophytes, may contribute to the accumulation of metals in plant roots in a non-toxic form. Occurrence of AM fungi has been reported in rhizosphere of metallophytes such as *Viola calaminaria* (Tonin et al., 2001) and *Berkheyacoddii* (Turnau et al., 2003) in Zn/Pb and Ni-rich soils, respectively. Hence, the prospect of these fungi existing in metal-contaminated soils has important implications for phytoremediation and it can further serve as a filtration barrier against transfer of metals to plant shoots. The protection and enhanced capability of uptake of minerals result in greater biomass production, a prerequisite for successful remediation.

Because mycorrhizal plants have greater access to nutrients and water in the substrate, the benefits of this symbiotic association in resource conservation in agriculture are important to consider. The network of fungal hyphae vastly extends the area available for absorption of substances required for plant growth, and thereby can help in reducing the inputs of fertilizers and water in agricultural and horticultural systems. In addition, they improve plant growth, help in contaminant removal, reduce the need for fertilizer application in commercial plant production, and improve the soil structure and health. Although relatively few specific plant-fungus combinations have been studied for their efficacy and application in remediation and resource conservation, the existing data on the benefits for mycorrhiza are promising.

Transgenic Crops

The development and use of genetically modified plants (GMPs) has undergone a sustained increase over the past few years. The advent of agricultural biotechnology and genetic engineering of plants has created new transgenic crop plants with superior yields and improved traits for resistance to insect pests and pathogens, tolerance to herbicides, and improved ability to endure environmental stress (Sharma and Srivastava, 2008). Along with modification of agronomic traits, genetically modified plants are now also being developed to produce industrial and pharmaceutical compounds. This new technology has the potential to provide enormous economic and agronomic benefits in the future and can also provide environmental benefits such as reducing the need for pesticide usage (Huang et al., 2003). Nevertheless, the release of transgenic plants is still highly controversial in many countries due to concern over the potential detrimental effects of genetically modified plants on human health and the environment. Among the major concerns are the possibility of creating invasive plant species, the unintended consequences of transgene flow to indigenous plants and microorganisms, development of super pests, and the effects of transgenic plants on non-target organisms including soil microbial communities (Wolfenbarger and Phifer, 2000). These are all serious issues and require that the effects of transgenic plants on the environment be carefully evaluated. Since, transgenic crops reflect, potentially unwanted effects on the environment and human health (Liu et al., 2005), therefore ecological risk of these crops should not be ruled out.

AMF are ubiquitous soil microorganisms and help in improvement in mineral nutrition status, which mainly depends on the nutrient uptake by the extended hyphae of AMF beyond the depleted zone in the soil (Li et al., 1991). But recent development and use of transgenic crops has drawn ecological risks on soil microorganisms. AMF, important in regulating aboveground and underground processes in ecosystems, are the most crucial soil microbial community worthy of being monitored in ecological risks assessment of transgenic crops for their sensitivity to environmental alterations (plant, soil, climatic factor etc.), for example, Bt insecticidal proteins constitutive expression and rhizosphere release during cultivation of BtPs may damage some critical steps of the AMF symbiotic development (Liu, 2009). Study carried out by Turrini et al. (2004) revealed the significant negative effects of BtPs on root colonization and the development of AMF. On contrary, fragmentary information is available regarding to no adverse affect of transgenic crops on AMF development (Powell et al., 2007).

Generally, symbiotic AMF consist of an internal phase inside the root and an external phase, which forms an extensive network within the plant–soil system. Spore, hyphae and infected root segments are effective propogules for the survival of AMF. The entire environmental conditions for AMF life changed when the conventional lines were replaced by transgenic ones (Liu, 2009). Taking BtPs for example, in the rhizosphere, Bt toxins might accumulate to a high level that directly impacts extraradical mycelium growth, development, and sporulation. Moreover, Bt toxins are constitutively expressed in the roots at a high level (at least for some cultivars), to which the important organs (e.g., the arbuscular and hyphae of AMF) are exposed, which may directly influence their intraradical development. In summary, AMF are thought to be exposed to Bt toxins both for their intraradical and extraradical tissues (Liu, 2009). Transgenic crops determine the release and exudation of inserted genetic products, which is toxin in nature. When this product reaches to soil it alter the structure and function of soil microbial communities, and finally the activity of soil enzymes and related nutrient cycling and transformation processes. One ecological consequence of reducing mycorrhizal colonization was a decrease in the soil's mycorrhizal propagule reserve, which diminished the next crop's production, especially under low-input cropping practices (Miller 1993, Liu and Du 2008). AMF are important non-target soil microorganisms for monitoring the ecological risks of transgenic crop. Therefore, the cumulative effects of long-term transgenic crops on AMF are important topics and need more attention.

Heavy Metals

The metal content of soils is derived partly from the chemical nature of the parent materials. In some areas, it may be derived from dry and wet atmospheric deposition as dusts and water droplets (Dosskey and Adriano, 1992). Sources of heavy metals in dusts and droplets include wind and water movement of polluted soils, acid rain and fogs, and volcanic eruptions. Anthropogenic atmospheric sources of metals include mining, smelting, industrial and agricultural activities, burning of fossil fuels, land clearing, and incineration of municipal wastes. Direct additions of municipal sludges to soils also are a source of metals. Availability and toxicity of metals to plants and mycorrhizal fungi varies, depending on the actual concentrations and oxidation states of the metals; soil and rhizosphere pH; and soil cation exchange capacity (CEC), texture, organic matter content, and redox potential (Dosskey and Adriano, 1992; Meharg and Cairney, 2000; Liu et al., 2000;Cairney and Meharg, 2000). In roots, metals such as aluminum can impair cell division, increase cell wall rigidity, alter root respiration, precipitate nucleic acids, and interfere with the uptake and transport of Ca, Mg, P, and Fe (Foy, 1983). Fungal hyphae sequester heavy metals, which may serve to reduce movement into, and toxicity to, the host (stress tolerance) (Weissenhorn et al., 1994; Weissenhorn and Leyval, 1996; Leyval et al., 1997; Joner et al., 2000; Meharg and Cairney, 2000). Detoxification mechanisms enable the plant and fungus to avoid toxic effects. Addition of inorganic or organic amendments to culture media or soils and resultant competitive or binding interactions for adsorption sites also may alter the effective level of metal toxicity (Timmer and Leyden, 1980; Gildon and Tinker, 1983a, b; Dosskey and Adriano, 1992; Shetty et al., 1995; Smith and Read, 1997; Joner et al., 2000). Biological factors proposed to affect bioavailability and potential toxicity of metals to arbuscular mycorrhizal fungi and plants include plant and fungal genera, species, genotype and ecotype as well as interactions between plants and mycorrhizal fungi and other rhizosphere or bulk soil microbes (Gildon and Tinker, 1983; Baker and Walker, 1989; Kothari et al., 1991; Shetty et al., 1994; Dıaz et al., 1996).

There are few reports on direct effects of heavy metals on AMF. Gildon and Tinker (1983) reported on differences in sensitivity of *Glomus* sp. to metals. They found that germination and growth of isolates obtained from metal-contaminated soils tended to be more tolerant to higher concentrations of metals in agar media than isolates obtained from sites with low metal concentrations. In greenhouse and field studies, Leyval et al. (1995) generally found lower spore numbers, mycorrhizal infectivity potential and germination of spores isolated from more contaminated soils compared with fungi obtained from less contaminated sites. They also reported a delay in colonization in soils with increased metal content. Most reports note a positive effect of mycorrhizal inoculation on growth of plants in metal-contaminated soils. This protective benefit may be related to the adsorptive or binding capability for metals of the relatively large fungal biomass associated with host plant roots, which may physically minimize or exclude the entry of metals into host plants (Gadd, 1993; Leyval et al., 1997; Hildebrandt et al., 1999; Joner et al., 2000; Meharg and Cairney, 2000; Cairney and Meharg, 2000).

Several biological and physical mechanisms have been proposed to explain the generally lower metal toxicity to plants colonized by arbuscular mycorrhizal fungi. These include adsorption onto plant or fungal cell walls present on and in plant tissues or on or into extraradical mycelium in soil (Gildon and Tinker, 1983; Dueck et al., 1986; Dehn and Schuepp, 1989; Galli et al., 1994; Hildebrandt et al., 1999; Kaldorf et al., 1999; Joner et al., 2000; Meharg and Cairney, 2000), chelation by such compounds as siderophores and metallothionens released by fungi or other rhizosphere microbes, and sequestration by plant-derived compounds like phytochelatins or phytates (Joner and Leyval, 1997; Van Steveninck et al., 1987). Other possible metal tolerance mechanisms include dilution by increased root or shoot growth, exclusion by precipitation onto polyphosphate granules, and compartmentalization into plastids or other membrane-rich organelles (Van Duin et al., 1991; Turnau et al., 1993; Galli et al., 1994; Kaldorf et al., 1999). Both fungal isolates and plants may vary in their individual or combined tolerances to metals. Optimizing the use of arbuscular mycorrhizal fungi to permit growth of plants in soils contaminated with metals may require careful selection of specific fungal and host plant combinations for a given set of soil conditions. It will also likely require skillful use of inorganic and organic amendments to maximize plant growth and to capitalize on interactions or competitions between metals and elements such as P and sulfur, whose uptake is generally enhanced in mycorrhizal plants. For example, increased P may increase plant biomass and thus perhaps detoxify the potential effects of heavy metals by dilution, precipitation or adsorption of metals onto polyphosphate granules. The non-target ecological effects of plants or fungi which have adsorbed, translocated and sequestered metals also need to be considered in parallel with efforts to revegetate soils contaminated with high levels of metals. Efforts to phytoremediate, reclaim or restore vegetation to soils contaminated with metals by use of mycorrhizal plant species and inocula is gaining acceptance. Since metal uptake and tolerance depend on both plant and soil factors, including soil microbes, interactions between plant root and their symbionts such as AM fungi can play an important role in successful survival and growth of plants in contaminated soils. Mycorrhizal associations increase the absorptive surface area of the plant due to extrametrical fungal hyphae exploring rhizospheres beyond the root-hair zone, which in turn enhances water and mineral uptake. AM fungi can further serve as a filtration barrier against transfer of heavy metals to plant shoots. The protection and enhanced capability of uptake of minerals result in greater biomass production, a prerequisite for successful remediation.

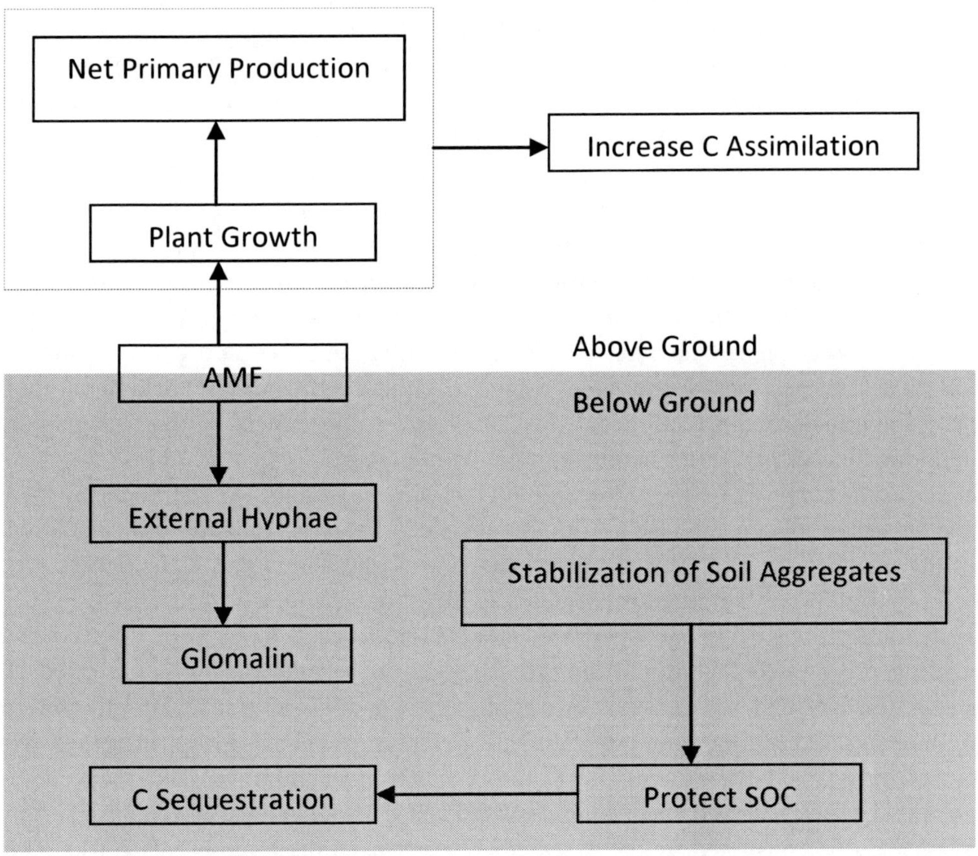

Figure 1. Contributions of AMF hyphae to soil carbon sequestration carbon sequestration (Zhu and Miller, 2003).

Critical Role of AMF in Carbon Sequestration

During the past 100 years we have seen continuous increase in the carbon dioxide concentration in the atmosphere, as well as an increase in mean temperature. Mycorrhizal associations, ubiquitous symbioses in terrestrial ecosystems, are potentially important in various ecosystem services provided by soils, contributing to sustainable plant production, ecosystem restoration, and soil carbon sequestration and hence mitigation of global climate changes. In the current context of climate change, largely due to atmospheric CO_2 increase, agroforestry practices seem to be a promising mitigation strategy. Increasing the soil organic carbon has a double benefit - first, it will improve soil quality (better fertility, reduction in erosion and nutrient leaching). Second, storing more carbon in the soil will reduce the atmospheric CO_2 concentration, so soils can buffer climate change effects. In terms of the global environmental change, mycorrhizal fungi prove to play a critical role in carbon sequestration in soils. Arbuscular mycorrhizal fungi (AMF) are widespread and agronomically important plant symbiont and often stimulate plant uptake of nutrients like: P, Zn, Cu, and Fe in deficient soils (Kothari et al. 1991; Liu et al., 2000) and increases resistance of plants to metals and salts (Dupré de Boulois et al., 2005; Colla et al., 2008; Subramanian et al. 2008). Mycorrhizal hyphae can significantly improve NPK uptake, and translocated less Na to the shoots in soil overlying areas of coal fly ash (Bi et al., 2003; Cheng and Baumgartner 2006). The inoculation of soybeans with AM fungi increased P uptake and decreased Pb concentrations by 30% in shoots (Andrade et al., 2004). Therefore, today, mycologists are typically well versed in the function of arbuscular mycorrhizas and their consequences for nutrient cycling and plant productivity. Arbuscular mycorrhizal fungi are obligate biotrophic symbionts associated with roots of most plants. An external hyphal network proliferates from the root into the soil and enhances plant growth by transporting phosphorus from the soil to the host plant (Smith and Read, 1997). The filamentous hyphae of these fungi extend beyond the nutrient-depleted zone in the soil, absorbing and supplying nutrients to the plants in exchange for photosynthetically made carbon compounds manufactured by the plant. The hyphae also attach to litter and decompose the organic matter, releasing the mineral ions sequestered in the structural polymers. The mineral ions are absorbed and translocated to the roots. Most plants are so dependent on their fungal partners for supplying nutrients that they languish or die without them.

Increases in atmospheric CO_2 (Keeling et al., 1995) highlight the need to explore ways to trap and sequester this greenhouse gas in terrestrial ecosystems. Plants fix CO_2 and allocate a part of the photosynthate to roots (Rogers et al., 1994) and soil (Jones et al., 1998). Organic carbon in soil plays an important role in soil aggregation (Kemper and Rosenau, 1986). Soil aggregates are groups of primary particles that adhere to each other more strongly than to surrounding soil particles (Martin et al., 1955). Relatively labile carbon is protected in soil aggregates (Jastrow and Miller, 1997; Six et al., 1998) and has a turnover of 140 - 412 years in a pasture soil, depending upon the aggregate size (Jastrow et al., 1996). Increased fixation of CO_2 by plants may have a direct effect on root symbionts that utilize plant-fixed carbon for growth -the arbuscular mycorrhizal fungi (AMF). 20–30% of the carbon assimilated by the host plant may be consumed by the mycorrhizal partner (Bidartondo et al., 2001). Zhu and Miller in 2003 gave a model showing the contribution of AMF hyphae to soil carbon sequestration (Figure 1). They are ubiquitous symbionts of the majority of land plants and are also important in soil aggregate stabilization (Jastrow and Miller, 1997). The contribution of

AMF to stabilization of aggregates was thought to be through entrapment of soil particles by fungal hyphae, the filamentous structures making up the body of the fungus. Hyphae extend several centimeters from the root into soil. Rillig et al., (1999) provided evidence for a change in aggregate stability under elevated CO_2 in three natural ecosystems. Glomalin is abundant in soils and is closely correlated with aggregate water-stability. It contains carbon and hence constitutes a non-trivial portion of the terrestrial carbon pool. Possibly far more importantly, however, stabilization of aggregates amplifies the role of glomalin in soils because carbonaceous compounds are protected from degradation inside of aggregates. Increased atmospheric CO_2 can lead to increased production of glomalin because of the symbiotic association that exists between plants and producers of glomalin, AMF. We have also shown that its concentrations in soils are influenced by management practices, for example in agro-ecosystems, further highlighting the role of this protein in carbon storage. Glomalin is an unusual molecule that has proven difficult to analyze biochemically due to its recalcitrance and possible complexity. Future research will be directed towards the elucidation of its structure and the controls on its production.

Microbial Interactions and Environment

The symbiotic association between mycorrhizal fungi and the roots of plants is widespread in the natural environment. There are a number of different types of fungi that form these associations, but for agriculture, it is the arbuscular mycorrhizal fungi (AMF) (Schussler et al., 2001) that are most important. Low input systems such as organic farming are generally more favorable to AMF and AMF have the potential to substitute for the fertilizers and biocides which not permitted in organic systems. Plant roots secrete "food" for bacteria and fungi, which attracts nematodes (worms) to the roots, because nematodes eat bacteria and fungi, and excrete nitrogen, sulphur and phosphorus in a form that the plants can use (URS, 2001). The nematodes only keep 1/6 of the nitrogen that they process 5/6 is excreted to the plant. Once the nematodes have excreted the nutrients, the hyphae of the mycorrhizal fungi pick them up and transfer them into the plant. Because of this symbiotic relationship, the least-leachable form of nitrogen you can apply is bacteria and fungi, and bacteria are the most nitrogen-rich organisms on earth (URS, 2001). High-quality food production rich in protein is highly dependent on the availability of sufficient N. Nitrogen deficiency is the major constraint to productivity in small holder cropping systems. AM fungi are known to assimilate and transport both NH_4^+ ions and some organic-nitrogen compounds to their host plants, particularly under conditions of low N availability and low pH.

Mycorrhiza also benefits plants indirectly by enhancing the structure of the soil. The high soil porosity (large spaces between soil particles) caused by microbes is important, because it aids water infiltration. If pore spaces are too small, they cannot break the surface tension of a water droplet, and water will run off, instead of saturating the soil, where it can be taken up by plant roots. Mycorrhizal fungi form a bridge between the roots and the soil, gathering nutrients from the soil and giving them to the roots. One of the most beneficial properties of mycorrhiza is its ability to "mine" the soil great distances from the roots for nutrients, especially those, such as Phosphorus, that are poorly mobile in the soil. Mycorrhiza also assist in picking up water further away from the roots, and block pest access to roots (Peters,

2002).AMF hyphae excrete gluey, sugar-based compounds called Glomalin, which helps to bind soil particles, and make stable soil aggregates. This gives the soil structure, and improves air and water infiltration, as well as enhancing carbon and nutrient storage (Peters, 2002). Most natural, undisturbed soils have an adequate supply of mycorrhiza for plant benefits.

Sustainable production of food crops in the tropics is often severely constrained by the fragility of soils, being prone to several forms of degradation. Making better use of the biological resources in these soils can contribute to enhanced sustainability. Mycorrhizal fungi constitute an important biological resource in this respect. Soil structure creates a unique three dimension framework solid and voids (pores). The configuration of particles within peds forms numerous crevices and cavities (pores) that trap and store soil water. The abundance and size of the soil pore are probably the most important aspect of soil structure. Preservation of soil structure is difficult. Rather, weak forces bind particles into peds, making them very susceptible to natural and man-caused physical/chemical disruption. Raindrop impact, machinery compaction, dispersion, oxidation losses of organic binding agents through excessive tillage are probably the most prevalent aggregate destroying factors. Aggregation of soil particles can occur in different patterns, resulting in different soil structures. The circulation of soil water varies greatly according to soil structure (Auge, 2001), thus, making it an important element for plant growth and development. Interestingly, AMF associations have a direct effect on soil structure, which is especially important in mixed culture systems, where cultivations and low levels of soil organic matter tend to result in damaged soil structure. AMF association with crop components in mixed culture systems have been reported to have a great impact in soil structure (Borie et al., 2000; Franzluebbers et al., 2000; Wright and Anderson, 2000; Rillig et al., 2003; Rillig, 2004). For example, through the enmeshing effects of hyphae, AMF bind soil microaggregates into macroaggregates (Tisdall et al., 1997), produces glomalin which upon accumulation in soils, sticks hyphae to soil directly (Rillig et al., 2001), release exudates into the soil, and promote aggregate stability as a result of rapid hyphal turnover which provide C to other soil micro organisms (Johnson et al., 2002; Staddon et al., 2003). Although the overall effect of hyphal enmeshment and C inputs in mixed culture systems reflects a significant increase in soil structural stability (Bethlenfalvay et al., 1999; Piotrowski et al., 2004), the observed wide range of results suggests more understanding of different combinations in the association of host-fungal in mixed culture systems to further improve the soil structure for the benefits of plant growth and development. As well as interacting with disease causing soil organisms AMF also interact with a whole range of other microorganisms in soils. Bacterial communities and specific bacterial strains promote germination of AM fungal spores and can increase the rate and extent of root colonisation by AM (Johansson et al., 2004). Once the arbuscular symbiosis has developed, AM hyphae influence the surrounding soil, which has been termed the mycorrhizosphere (Linderman, 1988), resulting in the development of distinct microbial communities relative to the rhizosphere and bulk soil (Andrade et al., 1997). Within the mycorrhizosphere AMF interact with beneficial rhizosphere microorganisms including free living N fixing bacteria and general plant growth promoting rhizobacteria (PGPR) (Galleguillos et al., 2000; Tsimilli- Michael et al., 2000; Biro et al., 2000). The *Rhizobium* symbiosis is dependant on high concentrations of P and so the enhanced P nutrition arising from the AM colonization can result in an increase in nodulation and N_2 fixation (Va´zquez et al., 2002). However, the AMF may differ in their ability to elicit specific host plant or host soil responses, suggesting that a combination of fungi is needed to function as an adequate plant-soil interface.

RESPONSIVENESS OF AMF IN AGRICULTURE

The productivity of agricultural systems is influenced by environmental stresses and stresses resulting from the cultural practices employed by man. The current requirement in agriculture for high yields as quickly as possible may be an ongoing necessity for the future of food production. The increasing consumer demands for organic or sustainably produced food, however, will require changes to incorporate cultural practices which increase AMF diversity. In agricultural and horticultural crop production, application of large amounts of fertilizers and pesticides is accepted, and is normal practice (Smith and Read, 1997). The possible benefits of mycorrhiza in plant production are: (1) increased crop yield; (2) reduced fertilizer and pesticide inputs; and (3) maintenance of a healthier soil system and resulting benefits such as improved water relations and reduced severity of some plant diseases. The AMF inoculation could significantly increase plant growth (including plant height, leaf area, and fresh and dry mass), enhance relative leaf water content, photosynthetic rates, transpiration rates and stomatal conductance, and improve plant drought tolerance. The symbiosis, they form, is potentially valuable not only for developmental programmes based on low-input agriculture, but also as a complex experimental model, where both fungal and host plant growth are regulated (Bethenfalvay and Linderman, 1992).

Mycorrhizal associations between a fungus and a plant root are ubiquitous in the natural environment. Several types of mycorrhizal associations occur in horticultural crops including AMF, ectomycorrhiza, ericoid mycorrhiza, and arbutoid mycorrhiza (Smith and Read, 1997). The benefits from root colonization by mycorrhizal fungi are thought to be highest when colonization occurs as early as possible during plant growth (Sahay and Varma, 2000). In horticultural production systems, this means that inoculum should be present during radicle emergence in seed germination, prior to the weaning or acclimation phase of tissue culture production, or during adventitious root formation in cutting propagation.

An important consequence of colonization by mycorrhizal fungi is alteration of root system architecture, with increased initiation and branching reported for many species (Hooker et al., 1992). During vegetative propagation, the number of roots initiated influences the length of the production cycle and the quality of the rooted cutting produced. Verkade and Hamilton (1987) found that the presence of AMF in the rooting medium increased root development and growth of *Viburnum dentatum* L. and *Ligustrum obtusifolium* var. *regelianum* but not root initiation. Mycorrhizal inoculation and colonization has been found to increase crop uniformity, reduce transplant mortality, and increase productivity.

Plant growth is generally depressed by severe conditions that include excessive H^+ ions, the toxicity of aluminium (Al) and/or manganese (Mn) and the deficiency of some essential mineral nutrients, primarily phosphorus (P), calcium (Ca), magnesium (Mg) and molybdenum (Mo) (Marschner, 1991). Growth depression in acid soils might also result from lack of activity of micro-organisms that form mutualistic associations with the plants, such as bacteria that form root nodules, e.g. *Rhizobium* spp. (Robert, 1995), and soil fungi that form arbuscular mycorrhizas (Abbott and Robson, 1985). They are distributed widely in native and man-made ecosystems and colonise roots of most plant species (Smith and Read, 1997). AM fungi may become important in increasing crop productivity in acid soils because they may improve P uptake in highly P-fixing conditions. Enhanced uptake of P is generally regarded as the most important benefit that AMF provide to their host plant, and plant P status is often the main controlling factor in the plant–fungal relationship (Graham, 2000). AMF can play a significant role in crop P nutrition, increasing total uptake and also found associated with increased growth and yield (Koide et al., 2000).

Moisture and Drought Stress

Either high or low levels of water can be stresses to plants. Plant response to colonization by arbuscular mycorrhizal fungi depends on the severity and periodicity of drought and other edaphic conditions. Arbuscular mycorrhizal fungi may affect host plant function and productivity under both high and low moisture conditions (Auge, 2000). AM fungi can affect the water balance of both amply watered and droughted host plants. Mycorrhizal tree seedlings can often resist drought better than non-mycorrhizal seedlings. Several studies indicate that *Rhizopogon* can help plants to tolerate and recover from water deficits and this can aid seedling establishment. Plants colonised by AM fungi may have increased tolerance to drought. There is considerable evidence to suggest that AMF are able to increase the host plant's tolerance to water stress (Davies et al., 2002; Auge´, 2004), including that caused by high salinity (Al-Karaki et al., 2001; Feng et al., 2002; Mohammad et al., 2003). Several mechanisms have been proposed to explain the effect, including increased root hydraulic conductivity, improved stomatal regulation, osmotic adjustment of the host and improved contact with soil particles through the binding effect of hyphae, enabling water to be extracted from smaller pores (Auge, 2001; 2004). Often both water and nutrient uptake are higher in drought stressed mycorrhizal plants than in non-mycorrhizal plants (Al-Karaki and Clark, 1999). However, AMF can only alleviate moderate drought stress, and in more severe drought conditions they become ineffective (Bryla and Duniway, 1997). Mycorrhizal inoculation has proven to be the better alternative to the chemical fertilizers especially phosphorus to increase and maintain the productivity of crops in the past endeavors and so do in agricultural research. Besides yield, mycorrhizal plants made their survival in drought conditions during dry months after planting (Gautam et al., 2009).

Plant growth in mixed culture systems is closely associated with moisture supply and the amount of storage that is readily available in the root zone for plant consumptive use. To maximise plant growth, soil moisture content should be maintained at field capacity either naturally through rainfall or artificially by irrigation throughout the critical stages of plant growth. In stressed conditions, there is substantial evidence suggesting that AM may increase the host plant's tolerance to water stress (Davies et al., 2002; Auge, 2004), including that caused by high salinity (Al-Karaki et al., 2001; Feng et al., 2002; Mohammad et al., 2003). The mechanisms used by AMF include increased root hydraulic conductivity, improved stomatal regulation, osmotic adjustment in the host, enabling extraction of water from smaller pores through improved contact with soil particles as a result of the hyphae binding effect (Auge, 2001, 2004), increased evaporative leaf surface area (Nelsen, 1987), and increased finely divided roots for greater root surface area to increase water absorption (Okon et al., 1996). These mechanisms suggest that AMF association with component crops growing in moisture stressed mixed culture systems, may benefit from improved moisture supply, thus, improving grain yield. An association between mycorrhiza fungus and plant roots is beneficial to the plant when it is grown under low phosphorus or dry land (i.e. low rainfall, non-irrigated) conditions (Fageria 2009) and grows well under relatively harsh mineral stress conditions (Clark and Zeto 2000) prevailed in subsistence agriculture.

pH

Arbuscular mycorrhizal fungi responded to pH in a highly variable fashion. The response of arbuscular mycorrhizal plants to pH has been studied for some very practical reasons including potential negative effects of H on plant productivity via direct effects on the endophyte and host plant physiology, and indirect effects via changes in soil processes, e.g. metal and base cation availability (Habte and Soedarjo, 1996; Clark et al., 1999a, b). Some AM fungi did poorly in low-pH soils, while other fungi did poorly after acid-soils were limed (Mosse, 1972a, b). Yet, in other studies, plants forming associations with arbuscular mycorrhizal fungi had improved plant growth in acid soils that were limed (Clark et al., 1999a, b; Davis et al., 1983), while in other acid soils, a positive AM fungal effect was found without the need to increase *pH* (Guzman-Plazola et al., 1988). Arbuscular mycorrhizal fungi play an important role in improving plant productivity by enhancing the nutrient uptake, particularly phosphorus (Clark et al., 1999a, b; Smith and Read, 1997; Bago et al., 1996). The influence of the symbiosis in nutrient absorption depends on the uptake capabilities of the host and the endophyte, extent of colonization of the root and the surrounding soil, and factors affecting the formation and reproduction of the endophyte symbiosis (Wang et al., 1985; Habte, 1995). The H activity affects most, if not all, of these characteristics. For example, availability of P in soil is low at all pH values because P reacts with soil constituents forming insoluble compounds. In general, mechanisms of cation absorption (e.g. NH^{+}_{4} Johansen et al., 1992; Frey and Schuepp, 1993; Johansen et al., 1993a, b; Clark et al., 1999a;)and anions (e.g. NO^{-}_{3} Tobar et al., 1994; Bago et al., 1996; Clark et al., 1999b) by arbuscular mycorrhizal fungi appear similar to those found in other organisms. Some studies have also suggested that mycorrhizal fungi tolerate adverse external *pH* conditions by modifying the pH of the mycorhizosphere during the uptake process (Tinker, 1975; Pacovsky, 1986). Klironomos in 1995 speculated that AM may protect acid-sensitive sugar maple (*Acer sac- charum*) in conditions which otherwise would be detrimental. He examined propagule levels and colonization of *A. saccharum* in forests located in southern Ontario on three soil types i.e. brunisols, luvisols, and podzols. The more acidic, organically-enriched and moist podzolic soils with humus are considered less favorable for arbuscular mycorrhizal fungi, and generally support ectomycorrhizal associations. Brunisols and luvisols are considered more favorable for arbuscular mycorrhizal fungi. In luvisolic soils, various colonization levels were similar, and spore densities were lower compared with values found for brunisols. Patterns were nearly opposite for roots in podzolic soils; where low occurrence of arbuscules, high levels of hyphal coils and vesicles, and much higher spore densities prevailed. Other studies indicate that arbuscular mycorrhizal fungi characteristics found in some podzolic soils by Klironomos (1995) may be indicative of stress (Duckmanton and Widden, 1994). These results indicate that many arbuscular mycorrhizal fungi have the capability to tolerate low *pH* conditions. The response of arbuscular mycorrhizas to soil pH seems to be dependent primarily on the fungal species. Some fungal species readily form arbuscular mycorrhizas in low *pH* soils, while other species form mycorrhizas in higher *pH* soils.

Tillage

Conservation tillage frequently results in increased use of herbicides to control weeds. This has discouraged the incorporation of conservation tillage into low-input agriculture. A field experiment initiated to examine the utility of different tillage regimes in conventional and low input agriculture was used to examine effects tillage regimes had upon AM fungi. Disturbance effects vary with soils and vegetation, and affect the density of infective propogules needed to reestablish AMF colonization differently (Jasper et al., 1991).

Tillage affects AM fungi, obligate symbionts potentially useful in sustainable agriculture. Many studies have shown that efficacy of AMF mycelia in the soil to act as propogules and organs for P uptake decreases with increasing soil disturbance (Fairchild and Miller, 1990). Abbott and Robson in 1991 define that, no tilled soil had more spores in the top 8 cm whereas, tilled soil had more spores in the 8 to 15 cm depth. Soil tillage in agricultural production may reduce the subsequent rate of colonisation of plants by AMF by breaking up the living extraradical mycelium in the soil. The result of this disturbance will be a reduction in propagules of susceptible AMF (*Acaulosporaceae* and *Gigasporaceae*) but may increase those of more resistant species of *Glomus*. The survival of the extraradical mycelium and its non-disturbance seems to be a vital agronomic practice for the subsequent colonisation of spring crops and optimum functioning of the mycorrhizas. Contact with the common mycorrhizal network (CMN) is the main method by which seedlings are colonised by AMF in natural ecosystems (Read et al., 1976; Read and Perez, 2003). Soil tillage causes severe disruption to the CMN (Evans and Miller, 1988), resulting in delayed or reduced root colonisation and a reduction in the volume of soil that is exploited by the AMF (Jasper et al., 1989a, b, Evans and Miller, 1990; Jasper et al., 1991). This can in turn translate into reduced plant nutrient uptake (Evans and Miller, 1988), in some cases crop growth (Kabir et al., 1999) and sometimes though not always crop yield (Kabir et al., 1998). However, the effects on growth and nutrient uptake are in some cases only transitory (McGonigle and Miller, 1993) and the exact effect of tillage on AMF may be dependant on soil type (Kabir et al., 1998). Deep inversion tillage is also likely to bury propagules below the depth of early seedling root growth, also delaying colonisation (Kabir, 2005). Reducing tillage has been repeatedly shown to increase AM colonisation and nutrient uptake. Galvez et al., (2001) compared mould board ploughed soils with chisel disked and no-till soil.

The presence of mycorrhizal fungi in agriculture offers potential benefits to the crops. AMF have a direct effect on soil structure, which is especially important in an agricultural context, where cultivations, trafficking and low levels of soil organic matter all tend to result in damaged soil structure. Soil typically contains several species of AM fungi, all of which may colonize the roots of most crop plants. The host plant transfers as much as 20% of all fixed C to the fungal partner (Jakobsen and Rosendahl, 1990) and in agricultural soils it can produce significant biomass (Rillig et al., 1999). Plant growth increases by AMF at low pH have been reported for several agricultural crops, including cowpea (Clark and Zeto, 1996; Mendoza and Borie, 1998). The beneficial effect of the symbiosis, however, varies extensively and is closely related to the ability of the fungi to deal with soil stresses (Clark et al., 1999).

Fertilizers and Pesticides

There is increasing interest worldwide in the maintenance of soil quality and remediation strategies for management of soil contaminated with fertilizers. The protective effects of AMF on host plants under conditions of fertilizer contamination have raised the prospect of utilizing the mutalistic association in soil revegetation programmes. Agricultural fertilizers are successful in controlling pathogens, but their application often also results in the indiscriminate killing of beneficial microorganisms, such as AMF. It is well-documented that high inputs of chemical fertilizers (especially phosphates and high nitrates) along with certain fungicides (e.g. benomyl) and soil sterilants have negative effects on AMF. This is primarily observed as reductions in colonisation levels in plant roots over time but also in P-uptake (Olsen et al., 1999). Fertilizers may therefore, affect AM fungi directly or indirectly through their effects on the host plant or on the host soil biota (Garcia-Romera, 1988). These complex interactions are of particular interest in sustainable agriculture, since they induced retardation of AMF hyphal growth, the inhibition of root colonization, and a shift in species combinations in the soil can be important for soil stability and plant production (Dehn and Schuepp, 1989). The uptake of trace elements by mycorrhizal plants depends on several factors such as physicochemical properties of the soil, particularly its fertility level, pH, the host plants and the fungi involved, and above all, the concentration of the trace elements in soil. Under deficiency conditions, most studies point to an increase in trace element uptake by mycorrhizal plants. For example, increased Zn uptake by maize inoculated with mycorrhizal fungus, *Glomus mosseae,* has been reported compared to the non-mycorrhizal plants at low Zn levels (Kothari *et al.*, 1991). However, when the soil contained potentially toxic amounts of trace elements, mycorrhizal formation usually induces lower concentrations of these elements in the aerial part of the host plant, thus consequently lead to a beneficial effect on plant growth. This has been reported for Zn when growing white clover in Zn contaminated soils (Zhu *et al.*, 2001).

Research involving the interaction of toxic organic pollutants, including pesticides, with arbuscular mycorrhizas has been largely limited to assessing effects of fungicides on mycorrhizal formation and P uptake. Interaction of multiple fungicides on the function of

mycorrhizas is important in the context of minimizing effects of plant pathogens while maximizing beneficial effects of mycorrhizas to plant nutrition (Sukarno et al., 1996; Abd-Alla et al., 2000;). Several studies have found that benomyl inhibits arbuscular mycorrhizal infection and P uptake in crop plants. Sukarno et al., (1996) reported that 31 mg benomyl kg^{-1} soil (manufacturer's recommended rate) reduced *Glomus* sp. formation on onion (*Allium cepa* L) and P acquisition. However, Larsen et al. (1996) reported that 10 mg benomyl kg^{-1} soil reduced formation of mycorrhizas by *G. caledonium* on *Cucumis sati_us* L. Merryweather and Fitter (1996) found that 63 mg benomyl kg^{-1} inhibited arbuscular mycorrhizal colonization on *Hyacinthoides non-scripta* L. Chouard ex. Roth. (bluebell). When mycorrhizal formation was reduced, these studies found that P uptake was also reduced. Several other studies also showed that benomyl reduced mycorrhizal formation in annual grasses leading to reduced plant biomass, but, P acquisition was not affected (Carey et al.,

1992; West et al., 1993; Newsham et al., 1994). Fumigation of soil with methyl bromide inhibits arbuscular mycorrhizal formation and P uptake by all plants that have been studied. Jawson et al., (1993) reported that soil fumigation with methyl bromide substantially reduced mycorrhizal formation on corn roots to a depth of 15 cm; below 15 cm roots became mycorrhizal. Afek et al., (1991) found that fumigation of soil with methyl bromide inhibited mycorrhizal formation in cotton, onion and pepper (*Capsicum annuum* L.). However, if soil was inoculated with *G. intraradices* spores after fumigation, these plants had higher rates of colonization by mycorrhizal fungi and greater plant biomass. Buttery et al., (1988) also found that methyl bromide fumigation of soil reduced mycorrhizal formation on, and P uptake by, peas *(Phaseolus ulgaris* L.) and soybean (*Glycine max* L.*)* Hass et al., (1987) reported that methyl bromide fumigation inhibited mycorrhizal formation on pepper. Menge, (1982) and Brown et al., (1974) reported that methyl bromide fumigation inhibited mycorrhizal formation on, and P uptake by, wheat (*Triticus aesiium* L.) and potatoes (*Solanum tuberosum* L.). Sukarno et al., (1993, 1996), Despatie et al., (1989), Jabaji-Hare and Kendrick, (1985, 1987) and Clark, (1978) all reported that application of fosetyl-Al (aliette) reduced root growth, but stimulated mycorrhizal formation on, and P uptake by, several crop plants. However, application of approximately 13 mg ridomil kg^{-1} soil did not reduce colonization by *Glomus* sp. on onion roots (Sukarno et al., 1996). The sensitive response of AMF to changes of above mentioned aspects presented variation at root colonization, fungal biomass and sporulation, which, in turn, determined the magnitude of AMF propagule reserve.

Protective Role of Mycorrhizal Fungi

Pests and diseases in many cropping systems has been a hindrance factor to most of the farmer's economy in the entire world. Many techniques including chemical measures to arrest the situation have not until recently been 100% successful. However, researchers are using different and/or a combination of techniques to further reduce this problem. It has been reported that AMF association plays a role in the suppression of crop pests and diseases, in particular soil-borne fungal diseases (Harrier and Watson, 2004; Whipps, 2004). A variety of pathogenic fungi, bacteria, viruses and nematodes drive their energy from plant roots. Yet in recent years, several reports have indicated that a host plant previously inoculated with a AM fungal symbiont exhibits increased resistance to several root diseases (Borowicz, 2001). The reductions in disease severity is reported to occur and caused a significant increase in grain yield compared to plants not inoculated with AMF (Karagiannidis et al., 2002). In some cases, the apparent resistance of plants to pests or diseases may be simply the result of improved mineral elements (Cordier et al., 1996; Karagiannidis et al., 2002), and/or multiple mechanisms of resistance, probably operating simultaneously (Whipps, 2004). AMF colonisation does not in itself cause a significant defensive response by the plant but induces the plant to respond faster to infection by pathogenic fungi (Whipps, 2004), by evolving several mechanisms. For example, most effective control is achieved when colonisation by AMF takes place before attack by the pathogen (Matsubara et al., 2001; Sylvia and Chellemi, 2001), and when there are changes in root exudates (Norman and Hooker, 2000). Released root exudates may lead to changes in the rhizosphere microbial community and populations (Dar et al., 1997), host root architecture (Vigo et al., 2000) or root biochemistry connected with plant defense mechanisms (Gianinazzi-Pearson et al., 1996). It is an important part of the microbial flora, when it begins its dependent relationship with plants, it joins in the metabolism of many psychology and biochemistry, obviously helps the plants absorb more water and mineral element especially the P element in the soil and improves the plant's resistance such as drought resistance and salt resistance to enable plants grow and develop well. Mycorrhizal plants are often more resistant to diseases, such as those caused by microbial soil-borne pathogens, and are also more resistant to the effects of drought. These effects are perhaps due to the improved water and mineral uptake in mycorrhizal plants. Though nutrient uptake has been the focus of much research on the AM association there is evidence that AMF also play a role in the suppression of crop pests and diseases, particularly soil-borne fungal diseases (Linderman, 1994; Borowicz, 2001). Once colonisation of a root cell by AMF occurs the pathogen is excluded from that cell. As a result, the most effective control is achieved when colonisation by AMF takes place before attack by the pathogen (Matsubara et al., 2001; Sylvia and Chellemi, 2001). Other types of pest and disease causing organism which may be suppressed by AMF include pathogenic nematodes (Habte et al., 1999; Nagesh et al., 1999; Talavera et al., 2001), above ground fungal disease (Feldmann and Boyle, 1998) and herbivores (Gange et al., 2002). Often the degree of control achieved with AMF varies between AMF species (Matsubara et al., 2000; Gange et al., 2003), which may be the result of host or disease/pest specificity. Ryan et al., (2000) showed that AM inoculation of potato plants stimulated production of the potato cyst nematode-selective hatching chemical. It has the potential to biologically suppress root pathogens and benefit the host plant in conditions that are deleterious to host root growth and development (e.g. compensation for disease and counteracting toxicities).

Role of Mycorrhizal in Afforestation and Agroforestry

Deforestation has become an extremely important global environmental issue over the past five years. The causes of deforestation are varied, for example population pressure, shifting cultivation, agricultural development, transmigration, forest fires and unsupervised, poor logging practices. The negative impact of deforestation leads to increased soil erosion, loss of biological diversity, damage of wildlife habitats, degradation of watershed areas and deterioration in the quality of life (UNCED, 1992). Actions that can help to reduce the rate of deforestation of the forest and help to support preservation and restoration of existing different forest ecosystems are urgently needed. Concern about the negative effects of deforestation has lead to a reforestation programe. India is facing to perform intensive aforestation. New forests will be planted on dry, poorly fertile soils non-profitable for agricultural use (Barna, 2002). Applying artificially mycorrhized seedlings may considerably increase the effectivity of aforestation and decrease costs. Agricultural practices aimed to reduce soil erosion and improve crop yield have been suggested to influence the activity of arbuscular mycorrhizal (AM). It plays an essential role in the composition of plant community and construction and balance of the diversity and stability of species in ecosystem. They are crucial components of ecosystems as they transport, store, release and cycle nutrients. A good example of the potential of mycorrhizal fungi to capture and deliver nutrients to their host comes from studies using inoculated eucalypts in field trials in sub-tropical China.

The ecological principles that define the competitive and complementary interactions among trees, crops, and fauna in agroforestry systems have received considerable research attention during the recent past. The search for highly productive, yet sustainable and environmentally responsible agricultural systems has led to a renewed interest in agroforestry practices in all over the world (Matson et al., 1997). Most of the land available for plantation forestry have been degraded over recent centuries with extensive loss of the A horizon caused by population pressure, inadequate management and over-harvesting. Top soil crusting is common, contributing to enhanced erosion, reduced soil water storage, compaction and poor root development (Xu et al., 2000). To overcome this problem, we need urgently to consider how to recover microbial biodiversity as there is no doubt that this will be important for improving long-term soil fertility. The capacity of some mycorrhizal fungi to promote both early growth and survival of crops is very important for commercial plantations on disturbed and difficult sites. Agroforestry is an age old agricultural system practiced in various parts of Asia. It is a traditional pacific island practice of integrating tress with crops and/or animals. Starting in the colonial period, trees were systematically removed from agricultural systems in the desire to maximize outputs from cash crops. However, the short-term profits were often offset by long-term environmental problems, escalating dependence on chemical fertilizers and inputs, and high levels of economic risk. The concept is aimed at the overall management of land by combining trees with food crops to control soil erosion, improve soil conditions and conserve soil and soil water to meet the needs of the crops as well as people. The belowground interactions that occur due to the high density of tree roots within the same region as the crop roots may have an effect on the community composition of AMF, which play an essential role in the functioning and productivity of agroecosystems. Arbuscular mycorrhizal fungi (AMF) are important components of agroecosystems since they form symbiotic associations with the majority of annual crops in temperate regions. Agronomic management can strongly affect AMF abundance in agro-ecosystems, its populations will develop in soils where certain conditions are met. These conditions include (i) avoidance of bare-soil fallow, (ii) low inputs of tillage, synthetic fertilizers, and certain high-phosphorus animal manures, and (iii) minimal rotation to crops that are poor or non-hosts to AMF (Boswell et al., 1998; Douds and Millner, 1999). In forests, litter is an important nutrient reservoir, mycorrhizal fungi can mobilise P, N and other nutrients from litter to tree roots (Perez-Moreno and Read, 2000). Fogel, (1980) estimated that AMF account for 43% of the annual turnover of N in a *Pseudotsuga menziesii* forest in Oregon. It is commonly a mutualistic interaction involving the transfer of mineral nutrients from the fungus to the host plant in exchange for carbon. In all ecosystems, rhizosphere organisms act in support of plant growth and productivity in several ways. Arbuscular mycorrhizal fungi (AMF) are one such soil organisms which play a crucial role in linking plants and soil, transporting mineral elements to plants and carbon compound to the soils and its biota (Reid, 1990; Fagbola et al., 1998) and are therefore important in tropical food crop based sustainable agriculture (Bethlenfalvey and Linderman, 1992; Oyetunji, 2003). They are a main component of the soil edaphon in most agroecosystems.

Furthermore, AMF play an important role in the formation of stable soil aggregates that allow water and air infiltration and prevent soil erosion (Miller et al., 2000). Because of these beneficial effects it is a challenge. Considerable evidence suggests that AMF can affect the nature of weed communities in agro-ecosystems in a variety of ways, including changing the relative abundance of mycotrophic weed species (hosts of AMF), and non-mycotrophic species (non-hosts). These affects may merely change the composition of weed communities without affecting the damage that these communities cause. Weed and AMF interactions may reduce crop yield losses to weeds, limit weed species shifts, and increase positive affects of weeds on soil quality and beneficial organisms. The combination of the AM fungi and the forest trees enhance root production of crops growing over their, more than AMF inoculation alone (Oyetunji, 2003). It could also be due to numerous reasons. The trees may improve the soil physical conditions for mycorrhiza to perform better. The release of nitrogen from the nodules of the multipurpose trees may enhance the efficiency of mycorrhiza. Agroforestry with mycorrhizal integration could go a long way in enhancing sustainable production of plants with little or no chemical fertilizer application. The integration of multipurpose trees and AMF if appropriately applied can solve some ecological problems (such as damaging of the natural resource base upon which agricultural production is based) confronting food production. The trees replenish the soil nutrients and improve physico-chemical properties of the soil while mycorrhiza transports the nutrients (particularly phosphorus) to the associated crops. Today, integrating trees with agricultural production is known to be a beneficial practice, one that can combine continuous agricultural production with conservation and land improvement.

CONCLUSION

The importance of mycorrhizal symbioses in nature cannot be overestimated. It is now very well established that AMF improve plant growth mainly through phosphorous nutrition; other beneficial effect are in the biological control of root pathogen, biological nitrogen fixation, hormone production and greater ability to with stand water stress. Because of their unique ability to increase the uptake of phosphorous by plant, mycorrhizal fungi can be utilized as practical substitute for phosphotic fertilizer. The significance of these interactions in the nutrition and well-being of the individual plant partner is well established and recent evidence also indicates that they may also have major effects on the structure of the plant community. In sustainable and organic agricultural systems, the role of AMF in maintaining soil fertility and bio-control of plant pathogens may be more important than in conventional agriculture where their significance has been marginalized by high inputs of agrochemicals. The fungi of the mycorrhizal symbiosis are crucial components of both plant and soil development and health. AM fungi interact with a wide variety of organisms during all stages of their life. They are an integral part of ecosystems and provide physical links between the primary producers, the consumers and the decomposers. For example, arbuscular mycorrhizal (AM) fungi, root symbionts of most higher plant species that can often have positive effects on plants, can alter reproductive traits of flowering plants (Koide and Dickie 2002). Given that changes in reproductive traits can influence plant interactions with pollinators (Elle and Carney 2003; Mitchell et al. 2004), AM fungal effects on host plants could change the dynamics of the interactions between the plant and its pollinators. Furthermore, nutrients can also be passes from dying roots directly to living roots via hyphal connections. However, numerous investigations showed grazing of soil microarthropod on AMF hyphae. These hyphae are not only involved in nutrient translocation but are also a significant plant carbon sink and contribute to the fungal energy channel of the soil food web.

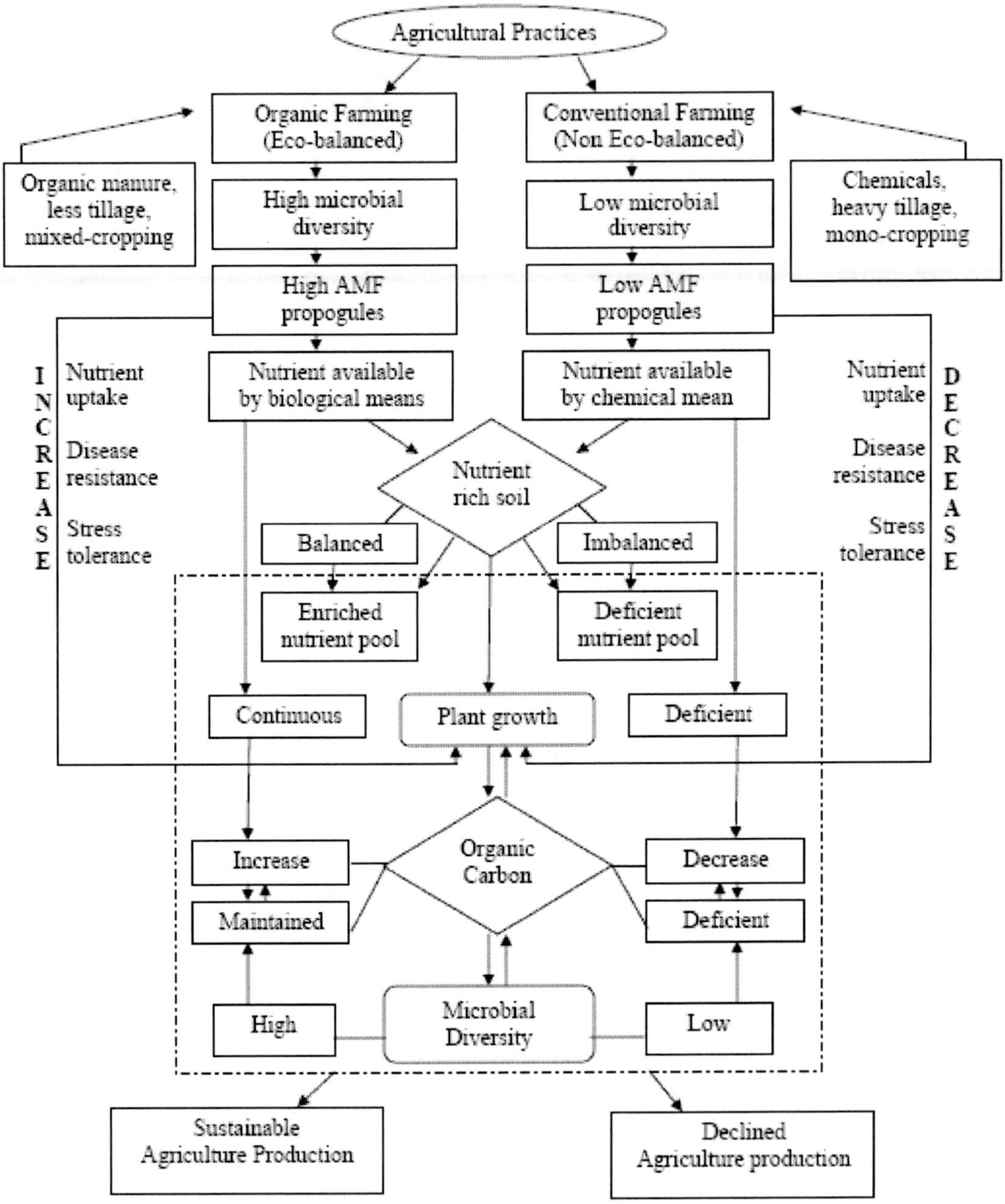

Figure 2. Multifarious properties of Arbuscular mycorrhizal fungi.

AMF forms a link between the biotic and geochemical components (Figure 2). The contribution of these fungal hyphae to biochemical cycling is significant and could be summarized as: (1) external hyphae constitute a rather large carbon sink and hence their turnover may represent a significance control point for carbon and nutrient cycling with in the rhizosphere, (2) External hyphae can chemically alter the soil matrix around their hyphal network, thereby increase the availability of mineral ions to plants and microbial biomass, (3) The physical placement of external hyphae with in the soil matrix allows asses to nutrients that are beyond the root and root hair zone, (4) Hypal linked between two or more plant allow for the potential transfer of nutrients directly from one root system to another, (5) Mycorrhizal fungal contributes to the information of stable soil aggregates, a necessary process for the accrual of organic matter and nutrients in soil. Plant breeders may need to find newer more appropriate plant cultivars which can maximize the role of AMF and other beneficial micro-organisms in agriculture in the 21st century. A recent discussion of conventional versus "organic" production systems highlighted how the Green Revolution helped to meet the needs of an ever increasing world population but at the price of environmental pollution (Tillman, 1999). The aim is therefore for practices that can give high enough yields with fewer environmental costs. The success noted in the enhancement of soil fertility and stability by adding manure, reducing tillage etc. in organic systems may, indirectly, be the result of optimisation of the soil microbiota and specifically AMF.

As agents of plant productivity and soil conservation, AMF have a useful role to play in agriculture. Its symbiosis is a key component in helping plants to establish in degraded soils. AMF inoculum is easily produced for application in forest nurseries, but the necessity of its inoculum production via a host plant is still an obstacle to ample utilization of AM fungi in agricultural crops. Nevertheless, progress is being made in this area and some commercial inoculum is currently marketed in the world. With increasing concerns about excessive nutrient application to the environment, the use of mycorrhizal symbioses to promote plant growth while reducing the inputs of fertilizer and pesticides may have great potential for crops, which respond very well to inoculation. Improved drought tolerance and related rapid recovery from wilting appear to be the most significant, but increased growth and establishment rates, greater chlorophyll content, and a lowered phosphorus requirement are also worthy of note. Biological systems such as those involving mycorrhizal may be used to supplement the expensive chemical fertilizers and in the alleviation of various biotic and abiotic constraints. There is an urgent need to study the below-ground microbiology of soils in agro- and natural ecosystems as AMF are pivotal in closing nutrient cycles and have a proven multi-functional role in soil-plant interactions.

References

Abbott, LK; Robson, AD. The effect of soil pH on the formation of vesicular arbuscular mycorrhiza by two species of *Glomus*. *Aust. J. Agric*. 1985. 23, 235–261.

Abott, LK; Robson, AD. Field management of VA mycorrhizal fungi. In: D. L. Keister and P. B. Cregan (Ed.), *The rhizosphere and plant growth.* Kluwer, Norwell, MA, 1991. 355-362.

Abd-Alla, MH; Omar, SH; Karanxha, S. The impact of pesticides on arbuscular mycorrhizal and nitrogen-fixing symbioses in legumes. *Appl. Soil. Ecol.* 2000. 14, 191-200.

Afek, U; Menge, JA; Johnson, ELV; 1991. Interaction among mycorrhizae, soil solarization and metalaxyl, and plants in the field. *Phytopathology.* 1991. 75, 655-671.

Al-Karaki, GN; Clark, RB. Varied rates of mycorrhizal inoculum on growth and nutrient acquisition by barley grown with drought stress. *J. Plant Nutr.* 1999. 22, 1775–1784.

Al-Karaki, GN; Hammad, R; Rusan, M. Response of two tomato cultivars differing in salt tolerance to inoculation with mycorrhizal fungi under salt stress. *Mycorrhiza*. 2001. 11, 43–47.

American Society of Agronomy, Madison, WI, 57-79.

Andrade, SAL; Abreu, CA; de Abreu, MF; Silveira, APD. Influence of lead additions on arbuscular mycorrhiza and Rhizobium symbioses under soybean plants. *Appl Soil Ecol.* 2004. 26,123–131.

Andrade, G; Mihara, KL; Linderman, RG; Bethlenfalvay, GJ. Bacteria from rhizosphere and hydrosphere soils of different arbuscular mycorrhizal fungi. *Plant Soil.* 1997. 92, 71–79.

Auge, RM. Stomatal behavior of arbuscular mycorrhizal plants. In: Kapulnik, Y., Douds, D.D. (Ed.), *Arbuscular Mycorrhizas: Physiology and Function.* Kluwer Academic Publishers, Dordrecht, The Netherlands. 2000. 384.

Auge, RM. Water relations, drought and vesicular-arbuscular mycorrhizal symbiosis. *Mycorrhiza*. 2001.11, 3–42.

Auge, RM. Arbuscular mycorrhizae and soil/plant water relations. *Can. J. Soil Sci.* 2004. 84, 373–381.

Bago, B; Vierheilig, H; Piche, Y; Azcon-Aguilar, C. Nitrate depletion and pH changes induced by the extraradical mycelium of the arbuscular mycorrhizal fungus *Glomus intraradicies* grown in monoxenic culture. *New Phytol.* 1996. 133, 273-280.

Baker, AJM; Walker, PL. Physiological responses of plants to heavy metals and the quantification of tolerance and toxicity. *Chem. Speciat.* 1989. 1, 7-17.

Barna, T. The Role of Mycorrhizae in Afforestation (A Review). *Acta Microbiologica et Immunologica Hungarica,* 2002. 49, 215–226.

Bethlenfalvey, GJ; Linderman, RG. Mycorrhizae in Sustainable Agriculture. ASA Spec. Publ. Maidson, WI, USA., 1992. 124.

Bethlenfalvay, GJ; Cantrell, IC; Mihara, KL; Schreiner, RP. Relationships between soil aggregation and mycorrhizae as influenced by soil biota and nitrogen nutrition. *Biol. Fertil. Soil.* 1999. 28, 356–363.

Bi YL; Li XL; Christie P; Hu ZQ; Wong MH. Growth and nutrient uptake of arbuscular mycorrhizal maize in different depths of soil overlying coal fly ash. *Chemosphere.* 2003. 50: 863–869.

Bidartondo, MI; Ek, H; Wallander, H; Soderstrom, B. Carbon sink strength in relation to ammonium and apatite additions of *Suillus pungens, Rhizopogon* section *Amylopogon* spp., and *Paxillus involutus* in ectomycorrhizal association with *Pinus muricata. New Phytol.* 2001.151, 543–550.

Biro, B; Koves-Pechy, K; Voros, I; Takacs, T; Eggenberg, P; Strasser, RJ. Interrelations between *Azospirillum* and *Rhizobium nitrogenfixers* and arbuscular mycorrhizal fungi in the rhizosphere of alfalfa at sterile, AMF-free or normal soil conditions. *Appl. Soil Ecol.* 2000. 15, 159–168.

Borie, FR; Rubio, R; Morales, A; Castillo, C. Relationships between arbuscular mycorrhizal hyphal density and glomalin production with physical and chemical characteristics of soils under no-tillage. *Rev. Chil. Hist. Nat.* 2000. 73, 749–756.

Borowicz, VA. Do arbuscular mycorrhizal fungi alter plant–pathogen relations? *Ecology.* 2001. 82, 3057–3068.

Boswell, EP; Koide, RT; Shumway, DL; Addy, HD. Winter wheat cover cropping,VA mycorrhizal fungi, and maize growth and yield. *Agri. Eco. Environ.* 1998. 67, 55- 65.

Brown, G; Corbett, DCM; Hide, GA; Webb, RM. Bromine residues in wheat crops grown in soil fumigated with methyl bromide. *Pesticide Sci.* 1974. 5, 25-29.

Bryla, DR; Duniway, JM. Effects of mycorrhizal infection on drought tolerance and recovery in safflower and wheat. *Plant Soil.* 1997. 197, 95–108.

Buttery, BR; Park, SJ; Findlay, WI; Dhanvantari, BN. Effects of fertilization and fertilizer on growth, yield, chemical composition and mycorrhizae in white bean and soybean. *Plant Soil.* 1988. 68, 677-686.

Cairney, JWG; Meharg, AA. Influences of anthropogenic pollution on mycorrhizal fungal communities. *Environ. Pollut.* 2000. 106, 169-182.

Carey, PD; Fitter, AH; Watkinson, AR. A field study using the fungicide benylate to investigate the effect of mycorrhizal fungi on plant fitness. *Oecologia.* 1992. 90, 550-555.

Cheng XM, Baumgartner K. Effects of mycorrhizal roots and extraradical hyphae on 15N uptake from vineyard cover crop litter and the soil microbial community. *Soil Biol Biochem.* 2006. 38, 2665– 2675.

Clark, CA. Effects of pesticides on vesicular arbuscular mycorrhizae. In: Clark, CA. (Ed.), 1978 Report of the Rothamsted Experiment Station. Part I, 1978. 236-237.

Clark, RB; Zeto, SK. Growth and root colonization of mycorrhizal maize grown on acid soil and alkaline soil. *Soil Biol. Biochem.* 1996. 28, 1505–1511.

Clark, CA; Zeto, SK; Zobel, RW. Arbuscular mycorrhizal fungal isolate effectiveness on growth and root colonization of *Panicum irgatum* in acidic soil. *Soil. Biol. Biochem.* 1999a. 31, 1757_1763.

Clark, CA; Zobel, RW; Zeto, SK. Effects of mycorrhizal fungus isolates on mineral acquisition by of *Panicum irgatum* in acidic soil. *Biol. Fertil. Soil.* 1999b.9, 167_196.

Clark, RB; Zeto, SK. Mineral acquisition by arbuscular mycorrhizal plants. *Journal of Plant Nutrition* 2000. 23: 867 – 902.

Colla, G; Rouphael, Y; Cardarelli, M; Tullio, M; Rivera, CM; Rea, E. Alleviation of salt stress by arbuscular mycorrhizal in zucchini plants grown at low and high phosphorus concentration. *Biol Fertil Soils.* 2008. 44:417–424.

Cordier, C; Trouvelot, A; Gianinazzi, S; Gianinazzi-Pearson, V. Arbuscular mycorrhiza technology applied to micropropagated *Prunus avium* and to protection against *Phytopthora cinnamomi. Agronomie.* 1996.16, 679–688.

Dar, GH; Zargar, MY; Beigh, GM. Biocontrol of *Fusarium* root rot in the common bean (Phaseolus vulgaris L.) by using symbiotic *Glomus mosseae* and *Rhizobium leguminosarum. Microb. Ecol.* 1997. 34, 74–80.

Davies, FT; Olalde-Portugal, V; Aguilera-Gomez, L; Alvarado, MJ; Ferrera-Cerrato, RC; Boutton, TW. Alleviation of drought stress of Chile ancho pepper (*Capsicum annuum* L. cv. San Luis) with arbuscular mycorrhiza indigenous to Mexico. *Sci. Hortic.* 2002. 92, 347– 359.

Davis, EA; Young, JL; Linderman, RG. Soil lime level pH and VA-mycorrhiza effects on growth responses of sweetgum seedlings. *Soil Sci. Soc. Am. J.* 1983. 47, 251-256.

Dehn, B; Schuepp, H. Influence of VA mycorrhizae on the uptake and distribution of heavy metals in plants. *Agric. Ecosyst. Environ.* 1989. 29, 79-83.

Del Val, C; Barea, JM; Azcòn-Aguilar, C. Assessing the tolerance to heavy metals of arbuscular mycorrhizal fungi isolated from sewage sludge-contaminated soils. *Appl. Soil Ecol.* 1999. 11, 261–269.

Despatie, S; Furlan, V; Fortin, JA. Effects of successive applications of fosetyl-Al on growth of *Allium cepa* L. associated with endomycorrhizal fungi. *Plant Soil.* 1989. 113, 175-180.

Dıaz, G; Azcon-Aguilar, C; Honrubia, M. Influence of arbuscular mycorhizae on heavy metalŽZn and Pb.uptake and growth of *Lygeum spartum* and *Anthyllis cytisoides*. *Plant Soil.* 1996. 180, 241-249.

Dosskey, MG; Adriano, DC. Trace metal impact on plants: Mediation by soil and mycorrhizae. In: Verry, E.S., Vermette, S.J. (Ed.), The Deposition and Fate of Trace Metals in our Environment. MN General Technical Report NC 150. USDA North Central Forest Experiment Station, Saint Paul, MN, 1992.105-115.

Douds, DD; Millner, PD. Biodiversity of arbuscular mycorrhizal fungi in agro ecosystems. *Agri. Eco. Environ*. 1999. 74, 77-93.

Duckmanton, L; Widden, P. Effect of ozone on the development of vesicular-arbuscular mycorrhizae in sugar maple saplings. *Mycologia.* 1994. 86, 181-186.

Dueck, TA.; Visser, P; Ernst, WHO; Schat, H. Vesicular-arbuscular mycorrhizae decrease zinc toxicity to grasses growing in zinc polluted soil. *Soil Biol. Biochem.* 1986. 18, 331-333.

Dupre de Boulois, H; Delvaux, B; Declerck, S. Effects of arbuscular mycorrhizal fungi on the root uptake and translocation of radiocaesium. *Environ. Pollut.* 2005. 134, 515–524.

Elle, E; Carney, R. Reproductive assurance varies with flower size in *Collinsia parviflora* (Scrophulariaceae). *Am. J. Bot.* 2003. 90, 888–896.

El-Tohamy, W; Schnitzler, WH; El-Behairy, U; El-Beltagy, MS. Effect of VA mycorrhiza on improving drought and chilling tolerance of bean plants (*Phaseolus vulgaris* L.). *J. Appl. Bot.* 1999. 73: 178–183.

Evans, DG; Miller, MH.Vesicular-arbuscular mycorrhizas and the soil-disturbance-induced reduction of nutrients absorption in maize. Causal relations. *New Phytol.* 1988. 110, 67–74.

Evans, DG; Miller, MH. The role of the external mycelial network in the effect of soil disturbance upon vesicular-arbuscular mycorrhizal colonisation of maize. *New Phytol.* 1990. 114, 65–71.

Fagbola, O; Osonubi, O; Mulongoy, K. Growth of cassava cultivar TMS 30572 as affected by alley-cropping and mycorrhizal inoculation. *Bio. Fert. Soils.* 1998. 27: 9-14.

Fageria, NK. The use of nutrients in crop plants. CRC Press, Taylor and Francis Group. 2009. 448.

Fairchild, GL; Miller, MH. Vesiular arbuscular mycorrhizas and the soil disturbance- induced reduction of nutrient absorption in maize. III. Influence of phosphorous amendments of soil. *New Phytol.* 1990.114, 641-650.

Feldmann, F; Boyle, C. Concurrent development of arbuscular mycorrhizal colonization and powdery mildew infection on three Begonia hiemalis cultivars. Z. Pflanzenk. Pflanzen. 1998.105,121–129.

Feng, G; Zhang, FS; Li, XL; Tian, CY; Tang, C; Rengel, Z. Improved tolerance of maize plants to salt stress by arbuscular mycorrhizal is related to higher accumulation of soluble sugars in roots. *Mycorrhiza.* 2002.12, 185–190.

Fogel, R. Mycorrhizae and nutrient cycling in natural forest ecosystems. *New Phytol.* 1980. 86, 199-212.

Foy, CD. Physiological effects of hydrogen, aluminum and manganese toxicities in acid soil. In: Adams, F.(Ed.), Soil Acidity and Liming. Agronomy Monograph No. 1983.12.

Franzluebbers, AJ; Wright, SF; Stuedemann, JA. Soil aggregation and glomalin under pastures in the southern Piedmont, USA. *Soil Sci. Soc. Am. J.* 2000. 64, 1018–1026.

Frey, B; Schuepp, H. Acquisition of nitrogen by external hyphae of arbuscular mycorrhizal fungi associated with *Zea mays* L. *New Phytol.* 1993. 124, 221-230.

Gadd, GM. Interactions of fungi with toxic metals. Tansley Review No. 47. *New Phytol.* 1993.124, 25-60.

Galleguillos, C; Aguirre, C; Barea, JM; Azcon, R. Growth promoting effect of two Sinorhizobium meliloti strains (a wild type and its genetically modified derivative) on a non-legume plant species in specific interaction with two arbuscular mycorrhizal fungi. *Plant Sci.* 2000.159, 57–63.

Galli, U., Schuepp, H., Brunold, C., 1994. Heavy metal binding by mycorrhizal fungi. Physiol. Plant 92, 364-368.

Galvez, L; Douds Jr, DD; Drinkwater, LE; Wagoner, P. Effect of tillage and farming system upon VAM fungus populations and mycorrhizas and nutrient uptake of maize. *Plant Soil.* 2001.118, 299–308.

Gange, AC; Brown, VK. All mycorrhizas are not equal. *Trends Ecol. Evol.* 2001.16, 671–672.

Gange, AC; Brown, VK; Aplin, DM. Multitrophic links between arbuscular mycorrhizal fungi and insect parasitoids. *Ecol. Lett.* 2003. 6, 1051– 1055.

Garcia-Romera, I; Garcia-Garrido, JM; Martin, J; Fracchia, S; Mujica, MT; Godeas, A; Ocampo, JA. Interactions between saprotrophic Fusarium strains and arbuscualr mycorrhiza of soy bean plants. *Symbiosis.* 1998. 24, 235–245.

Gautam, S; Shrestha, G; Vaidya; Rajbhandari, B. P. Effects of Arbuscular Mycorrhiza in the Productivity of Maize and Fingermillet Relay Cropping System. *Nepal J. Sci. Tech.* 2009.10, 51-55.

Gianinazzi-Pearson, V; Dumas-Gaudot, E; Gollotte, A; Tahiri-Alaouia, A; Gianinazzi, S. Cellular and molecular defense related root responses to invasion by arbuscular mycorrhizal fungi. *New Phytol.* 1996. 133, 45–57.

Gildon, A; Tinker, PB. Interactions of vesiculararbuscular mycorrhizal infections and heavy metals in plants. I. The effects of heavy metals on the development of vesicular-arbuscular mycorrhizas. *New Phytol.* 1983a. 95, 247_261.

Gildon, A; Tinker, PB. Interactions of vesicular arbuscular mycorrhizal infections and heavy metals in plants. II. The effects of infection on uptake of copper. *New Phytol.* 1983b. 95, 263_268.

Graham, JH. Assessing cost of arbuscular mycorrhizal symbiosis in agroecosystems. In: Podila, G.K., Douds, Jr., D.D. (Ed.), Current Advances in Mycorrhizal Research. APS Press, St. Paul, NM, 2000.127–140.

Guzman-Plazola, RA; Ferrera-Cerrato, R; Etchevers, JD. *Leucaena leucocephala*, a plant of high mycorrhizal dependence in acid soils. Leucaena *Res. Rep.* 1988. 9, 69-73.

Habte, M; Soedarjo, M. Response of *Acacia mangium* to vesicular-arbuscular mycorrhizal inoculation, soil pH, and soil P concentration in an oxisol. *Can. J. Bot.* 1996. 74, 155-161.

Habte, M; Zhang, YC; Schmitt, DP. Effectiveness of Glomus species in protecting white clover against nematode damage. *Can. J. Bot.* 1999. 77, 135–139.

Harrier, LA; Watson, CA. The potential role of arbuscular mycorrhizal (AM) fungi in the bioprotection of plants against soil-borne pathogens in organic and/or other sustainable farming systems. *Pest Manage. Sci.* 2004. 60, 149–157.

Hass, JH; Bar-Yosef, B; Krikun, J; Barak, R; Markovits, T; Kramer, J. Vesicular-arbuscular mycorrhizal fungus infestation and phosphorus fertigation to overcome pepper stunting after methyl bromide fumigation. *Agron. J.* 1987. 79, 905_910.

Hildebrandt, U; Kaldorf, M; Bothe, H. The zinc violet and its colonization by arbuscular mycorrhizal fungi. *J.Plant Physiol.* 1999. 154, 709_717.

Ho, I. Vesicular-arbuscular mycorrhizae of halophytic grasses (Harley and Smith, 1983). A balance between both plant in the Alvord Desert of Oregon (USA). *Northwest Sci.* 1987. 61,148–159.

Hodge, A; Campbell, CD; Fitter, AH. An arbuscular mycorrhizal fungus accelerates decomposition and acquires nitrogen directly from organic material. *Nature.* 2001. 413, 297-299.

Hooker, JE; Munro, M; Atkinson, D. Vesicular-arbuscular mycorrhizal fungi induced alteration in poplar root system morphology. *Plant Soil.* 1992. 145, 207-214.

Huang, JK; Hu, FR; Pray, C, Qiao, FB; Rozelle, S. Biotechnology as an alternative to chemical pesticides: a case study of Bt cotton in China. *Agric Econ.* 2003. 29, 55–67.

Jabaji-Hare, SH; Kendrick, WB. Effects of fosetyl-Al on root exudation and on composition of extracts of mycorrhizal and nonmycorrhizal leek roots. *Can. J. Plant athol.* 1985. 7, 118-126.

Jabaji-Hare, SH; Kendrick, WB. Response of an endomycorrhizal fungus in *Allium porrum* L. to different concentrations of systematic fungicides, metalaxyl, ŽRidomil.and fosetyl-AlŽAliette. *Soil Biol. Biochem.* 1987. 19, 95-99.

Jakobsen, I; Rosendahl, L. Carbon flow into soil and external hyphae from roots of mycorrhizal cucumber plants. *New Phytol.* 1990. 115, 77–83.

Jamal, A; Ayub, N; Usman, M; Khan, AG. Arbuscular mycorrhizal fungi enhance zinc and nickel uptake from contaminated soil by soyabean and lentil. *Int. J. Phytoremed.*, 2002. 4, 205–221.

Jasper, DA; Abbott, LK; Rabson, AD. The effect of soil disturbance on vesicular-arbuscular mycorrhizal fungi in soils from different vegetation types. *New Phytol.* 1991. 118, 471-476.

Jastrow, JD; Boutton, TW; Miller, RM. Carbon dynamics of aggregate-associated organic matter estimated by carbon-13 natural abundance. *Soil Sci. Soc.A m. J.* 1996. 60, 801-807.

Jastrow, JD; Miller, RM. Soil Processes and the Carbon Cycle, CRC Press, Boca Raton, FL. 1997. 207-223.

Jawson, MD; Franzlubbers, AJ; Galusha, DK; Aiken, RM. Soil fumigation within monoculture and rotations: response of corn and mycorrhizae. *Agron. J.* 1993. 85, 1174-1180.

Johansen, A; Jakobsen, I; Jensen, ES. Hyphal transport of 15N-labelled nitrogen by a vesicular-arbuscular mycorrhizal fungus and its effect on depletion of inorganic soil N. New Phytol. 1992. 122, 281-288.

Johansen, A; Jakobsen, I; Jensen, ES. External hyphae of vesicular-arbuscular mycorrhizal fungi associated with *Trifolium subterraneum* L.3. Hyphal transport of 32 P and 15N. *New Phytol.* 1993a. 124, 61_68.

Johansen, A; Jakobsen, I; Jensen, ES. Hyphal transport by vesicular-arbuscular mycorrhizal fungus on N applied to the soil as ammonium or nitrate. *Biol. Fertil. Soils.* 1993b. 16, 66-70.

Johnson, D; Leake, JR; Ostle, N; Ineson, P; Read, DJ. In situ 13CO2 pulse-labelling of upland grassland demonstrates a rapid pathway of carbon flux from arbuscular mycorrhizal mycelia to the soil. *New Phytol.* 2002. 153, 327–334.

Johansson, J; Paul, LR; Finlay, RD. Microbial interactions in the mycorrhizosphere and their significance for sustainable agriculture. *Microbiol. Ecol.* 2004.48, 1–13.

Joner, EJ; Leyval, C. Uptake of 109Cd by roots and hyphae of a *Glomus mosseae_Trifolium subterraneum* mycorrhizal from soil amended with high and low concentrations of cadmium. *New Phytol.* 1997.135, 353_360.

Joner, EJ; Briones, R; Leyval, C. Metal binding capacity of arbuscular mycorrhiza mycelium. *Plant Soil.* 2000. 226, 227-234.

Joner, EJ; Ravnskov, S; Jakobsen, I. Arbuscular mycorrhizal phosphate transport under monoxenic conditions using radio-labelled inorganic and organic phosphate. *Biotech. Lett.* 2000. 22, 1705-1708.

Jones, TH; Thompson, LJ; Lawton, HH; Bezemer, TM; Bardgett, RD; Blackbum, T M; Bruce, KD; Cannon, PF; Hall, GS; Hartley, SE; Howson, G; Jones, CG; Kampichler, *C*; Kandeler, E; Ritchie, DA. *Science.*1998. 280,441- 443.

Kabir, Z; O'Halloran, IP; Fyles, JW; Hamel, C. Dynamics of the mycorrhizal symbiosis of corn (Zea mays L.), effects of host physiology, tillage practice and fertilization on spatial distribution of extra-radical mycorrhizal hyphae in the field. *Agri. Ecosyt. Environ.* 1998. 68, 151– 163.

Kabir, Z; O'Halloran, IP; Hamel, C. Combined effects of soil disturbance and fallowing on plant and fungal components of mycorrhizal corn (Zea mays L.*) Soil Biol. Biochem.* 1999. 31, 307–314.

Kabir, Z. Tillage or no-tillage: impact on mycorrhizae. *Can. J. Plant Sci.* 2005. 85, 23–29.

Kaldorf, M; Kuhn, AJ; Schroder, WH; Hildebrandt, U; Bothe, H. Selective element deposits in maize colonized by a heavy metal tolerance conferring arbuscular mycorrhizal fungus. J. Plant Phys. 1999. 154, 718-728.

Karagiannidis, N; Bletsos, F; Stavropoulos, N. Effect of Verticillium wilt (Verticillium dahliae Kleb.) and mycorrhiza (Glomus mosseae) on root colonization, growth and nutrient uptake in tomato and egg plant seedlings. *Sci. Hortic.* 2002. 94, 145–156.

Keeling, CD; Whorf, TPI; Whalen, M; van der Plicht, J. Inter annual extremes in the rate of rise of atmospheric carbon dioxide since 1980. *Nature.* 1995. 375, 666-670.

Kemper, WD; Rosenau, RC. Methods of Soil Analysis (Part I), American Society of Agronomy, Madison, WI. 1986. 425-442,

Khan, AG; Kuek, C; Chaudhry, TM; Khoo, CS; and Hayes, WJ. Role of plants, mycorrhizae and phytochelators in heavy metal contaminated land remediation. *Chemosphere.* 2000. 41, 197–207.

Klironomos, JN. Arbuscular mycorrhizae of *Acer saccharum* in different soil types. *Can. J. Bot.* 1995. 73, 1824_1830.

Koide, RT; Goff, MD; Dickie, IA. Component growth efficiencies of mycorrhizal and nonmycorrhizal plants. *New Phytol.* 2000. 148, 163–168.

Koide, RT; Dickie, IA. Effects of mycorrhizal fungi on plant populations. *Plant Soil,* 2002. 244, 307–317.

Kothari, SK; Marschner, H; Romheld, V. Contribution of VA mycorrhizal hyphae in acquisition of phosphorus and zinc by maize grown in a calcareous soil. *Plant Soil.* 1991a. 131, 177-185.

Kothari, SK; Marschner, H; Romheld, V. Effect of a vesicular-arbuscular mycorrhizal fungus and rhizosphere micro-organisms on manganese reduction in the rhizosphere and manganese concentrations in maize *Zea mays* L. *New Phytol.* 1991b. 177, 649_655.

Krupa, P. Znaczenie ektomikoryzy w ochronie roślin przed skażeniem metalami ciężkimi. *Chemia i Inżynieria Ekologiczna.* 1999. 6, 607.

Larsen, JL; Thingstrup, I; Jakobsen, I; Rosendahl, S. Benomyl inhibits phosphorus transport but not fungal alkaline phosphatase activity in a Glomus-cucumber symbiosis. *New Phytol.* 1996. 132, 127-133.

Leake, JR. Mycorrhizas and the terrestrial carbon cycle: roles in global carbon sequestration and plant community composition. Chapter 8 In: Fungi in the Environment. Gadd, G.M., S.C. Watkinson, and P.S. Dyer (Ed.). 2007. 161-184. Cambridge University Press, Cambridge, UK.

Leyval, C; Singh, BR; Joner, EJ. Occurrence and infectivity of arbuscular mycorrhizal fungi in some Norwegian soils influenced by heavy metals and soil properties. *Water Air Soil Pollut.* 1995. 84, 203-216.

Leyval, C; Turnau, K; Haselwander, K. Effect of heavy metal pollution on mycorhizal colonization and function: physiological, ecological and applied aspects. *Mycorrhiza.* 1997. 7, 139-153.

Li, XL; Geoege, E; Marschner, H. Extension of the phosphorus depletion zone in VAM white clover in a calcareous soil. *Plant Soil.* 1991. 136:41–48.

Linderman, R. Mycorrhizal interactions with the rhizosphere microflora: the mycorrhizosphere effect. *Phytopathology.* 1988. 78, 366–371.

Linderman, RG. Role of VAM in biocontrol. In: Pfleger, F.L., Linderman, R.G. (Ed.), Mycorrhizae and Plant Health. APS Press, St. Paul, NM, USA, 1994. 1–26.

Liu, A; Hamel, C; Hamilton, RI; Ma, BL. Acquisition of Cu, Zn, Mn and Fe by mycorrhizal maize (Zea mays L.) grown in soil at different P and micronutrient levels. *Mycorrhiza.* 2000. 9, 331–336.

Liu, B; Zeng, Q; Yan, FM; Xu HG; Xu CR. Effects of transgenic plants on soil microorganisms. *Plant Soil.* 2005. 271:1–13.

Liu, WK. Effects of Bt transgenic crops on soil ecosystems: a review of a 10 years research in China. *Front Agric China.* 2009. 3, 190–198

Liu, WK; Du L. Interactions between Bt transgenic crops and arbuscular mycorrhizal fungi: a new urgent soil ecology issue in agro-ecosystems. *Acta Agric Sc B Soil Plant Sci.* 2008. 58, 187–192.

Malcova, R; Vosátka, M; Gryndler, M. Effects of inoculation with *Glomus intraradices* on lead uptake by *Zea mays* L. and *Agrostis capillaris* L. *Appl. Soil Ecol.* 2003. 23, 55–67.

Marschner, H. Mechanisms of adaptation of plants to acid soils. *Plant Soil.* 1991. 134, 1–20.

Marschner, H; Dell, B. Nutrient uptake in mycorrhizal symbiosis. *Plant Soil.* 1994. 159, 89–102.

Martin, JP; Martin, WP; Page, JB; Ranley, WA; DeMent, JD. Soil aggregation. *Adv. Agro.* 1955. 7, 1-37.

Matson, PA; Parton, WJ; Power, AG; Swift, MJ. Agricultural intensification and ecosystem properties. *Science.* 1997. 277, 504–509.

Matsubara, Y; Kayukawa, Y; Yano, M; Fukui, H. Tolerance of asparagus seedlings infected with arbuscular mycorrhizal fungus to violet root rot caused by *Helicobasidium mompa. J. Jpn. Soc. Hortic. Sci.* 2000. 69, 552–556.

Matsubara, Y; Ohba, N; Fukui, H. Effect of arbuscular mycorrhizal fungus infection on the incidence of *fusarium* root rot in *asparagus* seedlings. *J. Jpn. Soc. Hortic. Sci.* 2001. 70, 202–206.

McGonigle, TP; Miller, MH. Mycorrhizal development and phosphorus absorption in maize under conventional and reduced tillage. *Soil Sci. Soc. Am. J.* 1993. 57, 1002–1006.

Meharg, AA; Cairney, JWG. Co-evoloution of mycorrhizal symbionts and their hosts to metal-contaminated environments. *Adv. Ecol. Res.* 2000. 30, 69-112.

Mendoza, J.; Borie, F. Effect of *Glomus etunicatum* inoculation on aluminium, phosphorus, calcium and magnesium uptake of two barley genotypes with different aluminium tolerance. Commun. *Soil Sci. Plant Anal.* 1998. 29, 681–695.

Menge, JA. Effect of soil fumigants on vesicular arbuscular fungi. *Phytopathology* 1982. 72, 1125_1132.

Merryweather, J; Fitter, AH. Phosphorus nutrition of an obligately mycorrhizal plant treated with benomyl in the field. *New Phytol.* 1996. 132, 307-311.

Miller, RM. Nontarget and ecological effects of transgenically altered disease resistance in crops-possible effects on the mycorrhizal symbiosis. *Mol Ecol.* 1993. 2, 327–335.

Miller, RM; Jastrow, JD. Mycorrhizal fungi influence soil structure. In: Kapulnik, Y., Douds, D. D. (Ed.), *Arbuscular Mycorrhizae: Molecular Biology and Physiology*, Kluwer Academic Publishers, The Netherlands, 2000. 3-18.

Mitchell, R.J; Karron, JD; Holmquist, KG; Bell, JM. The influence of Mimulus ringens floral display size on pollinator visitation patterns. *Funct. Ecol.* 2004. 18, 116–124.

Mohammad, MJ; Malkawi, HI; Shibli, R. Effects of mycorrhizal fungi and phosphorus fertilzation on growth and nutrient uptake of barley grown on soils with different levels of salts. *J. Plant Nutr.* 2003. 26, 125–137.

Mosse, B. The influence of soil type and Endogone strain on the growth of mycorrhizal plants in phosphate deficient soils. *Rev. Ecol. Biol. Sol.* 1972a. 9, 529-537.

Mosse, B. Effects of different Endogone strains on the growth of *Paspalum notatum. Nature.* 1972b. 229, 221_223.

Nagesh, M; Reddy, PP; Kumar, MVV; Nagaraju, BM. Studies on correlation between *Glomus fasciculatum* spore density, root colonization and Meloidogyne incognita infection on *Lycopersicon esculentum. Z. Pflanzenk. Pflanzen.* 1999. 106, 82–87.

Nelsen, CE. The water relations of vesicular-arbuscular mycorrhizal systems. In: Safir GR (Ed.), Ecophysiology of VA mycorrhizal plants. CRC, Boca Raton, Fla, 1987. 71–91.

Newsham, KK; Fitter, AH; Watkinson, AR. Root pathogenic and arbuscular mycorrhizal fungi determine fecundity of fitness in asymptomatic plants in the field. *J. Ecol.* 1994. 82, 805_814.

Norman, JR; Hooker, JE. Sporulation of *Phytopthora fragariae* shows greater stimulation by exudates of nonmycorrhizal than by mycorrhizal strawberry roots. *Mycol. Res.* 2000. 104, 1069–1073.

Okon, IE; Osonubi, O; Sanginga, N. Vesicular-arbuscular mycorrhiza effects on *Fliricidia sepium* and *Senna siamea* in a fallowed alley cropping system. *Agrofor. Syst.* 1996. 33: 165–175.

Olsen, J K; Schaefer, JT; Edwards, DG; Hunter, MN; Galea, VJ; Muller, LM. Effects of mycorrhizae, established from an existing intact hyphal network, on the growth response of capsicum (*Capsicuum annuum* L.) and tomato (*Lycopersicon esculentum* Mill.) to five rates of applied phosphorus, *Aus J Agri Res*. 1999. 50, 223-237.

Olson, PE; Reardon, KF; Pilon-Smits, EAH. Ecology of rhizosphere bioremediation. Chapter 10 *In* Phytoremediation: Transformation and control of Contaminants. S.C. McCutcheon and J.L. Schnoor (Ed.). Wiley-Interscience, John Wiley and Sons, Inc. Hoboken, New Jersey. 2003. 317-353.

Oyetunji, OJ; Osonubi, O; Ekanayake, I.J. Contribution of an alley cropping and arbuscular mycorrhizal fungi to maize productivity under cassava intercrop in the derived savannah zone. J Agri Sci. 2003.140, 311-316.

Pacovsky, RS. Micronutrient uptake and distribution in mycorrhizal or phosphorus-fertilized soybeans. *Plant Soil.* 1986. 95, 379-388.

Perez-Moreno, J; Read, DJ. Mobilization and transfer of nutrients from litter to tree seedlings via the vegetative mycelium of ectomycorrhizal plants. *New Phytologist.* 2000. 145, 301-309.

Peters, S. Reforestation Technologies International, Salinas, CA. Mycorrhiza. 2002. 101.

Piotrowski, JS; Denich, T; Klironomos, N; Graham, JM; Rillig, MC. The effects of arbuscular mycorrhizas on soil aggregation depend on the interaction between plant and fungal species. *New Phytol.* 2004.164, 365–373.

Powell, JR; Gulden, RH; Hart MM; Campbell, RG; Levy-Booth, DJ; Dunfield, KE; Pauls, KP; Swanton, CJ; Trevors, JT; Klironomos, JN. Mycorrhizal and rhizobial colonization of genetically modified and conventional soybeans. *Appl Environ Microbiol.* 2007. 73:4365–4367.

Read, DJ; Koucheki, HK; Hodgson, J. Vesicular-arbuscular mycorrhizas in natural vegetation systems. I. The occurrence of infection. *New Phytol.* 1976. 77, 641–653.

Read, DJ. The mycorrhizal mycelium. In: *Mycorrhizal Functioning.* MF Allen, (Ed.), Chapman and Hall, London, UK, 1992. 102-133.

Read, DJ; Perez-Moreno, J. Mycorrhizas and nutrient cycling in ecosystems—a journey towards relevance? *New Phytol.* 2003. 157, 475–492.

Reforestation Technologies International. MycoPak Product Information. 2003.

Reid, CPP. Mycorrhiza. In: The rhizosphere Lynch, J.M. (Ed.), John Wiley and Sons, New York. 1990.

Rillig, MC; Wright, SF; Allen, MF; Field, CB. Rise in carbon dioxide changes soil structure. *Nature*. 1999a. 400, 628.

Rillig, MC; Field, CB; Allen, MF. Soil biota responses to long term atmospheric CO_2 enrichment in two California annual grasslands. *Oecologia.* 1999b. 119, 572–577.

Rillig, MC; Wright, SF; Nichols, KA; Schmidt, WF; Torn, MS. Large contribution of arbuscular mycorrhizal fungi to soil carbon pools in tropical forest soils. *Plant Soil.* 2001. 233, 167–177.

Rillig, MC; Maestre, FT; Lamit, LJ. Microsite differences in fungal hyphal length, glomalin, and soil aggregate stability in semiarid Mediterranean steppes. *Soil Biol. Biochem.* 2003. 35, 1257–1260.

Rillig, MC. Arbuscular mycorrhizae, glomalin, and soil aggregation. *Can. J. Soil Sci.* 2004. 84, 355–363.

Robert, M. Aluminum toxicity: A major stress for microbes in the environment. *In* Environmental impact of soil component interactionsMetal, Other Inorganics, and Microbial Activities. (Ed.), PMHuang, J Berthelin, JMBollag, WBMcGill and A L Page. CRC Press, Boca Raton.1995. 2, 227–242.

Rogers, HH; Runion, G B; Krupa, S V. Plant responses to atmospheric CO2 enrichment with emphasis on roots and the rhizosphere. *Environ. Pollution.* 1994. 83, 155-189.

Ryan, MH; Small, DR; Ash, JE. Phosphorus controls the level of colonisation by arbuscular mycorrhizal fungi in conventional and biodynamic irrigated dairy pastures. *Aust. J. Exp. Agric.* 2000. 40, 663– 670.

Sahay, NS; Varma, A. A biological approach towards increasing the rates of survival of micropropagated plants. *Cur Sc.* 2000. 78, 126-129.

Schussler, A; Schwarzott, D; Walker, C. A new fungal phylum, the Glomeromycota: phylogeny and evolution. *Mycol. Res.* 2001. 105, 1413–1421.

Sharma, AK; Srivastava, R. Effect of transgenic crops in rhizosphere microorganisms: An overview. In: Biosafety Issues related to practicing agriculture biotechnology. Department of Molecular Biology and Genetic Engineering, GBPUAT, Pantnagar. Vikrant Offset, Haldwani. 2008. 75 – 84.

Shetty, KG; Hetrick, BAD; Figge, DAH; Schwab, AP. Effects of mycorrhizae and other soil microbes on revegetation of heavy metal contaminated mine spoil. *Environ. Pollut.* 1994. 86, 181-188.

Shetty, KG; Hetrick, BAD; Schwab, AP. Effects of mycorrhizae and fertilizer amendments on zinc tolerance of plants. *Environ. Pollut.* 1995. 88, 307-314.

Six, J; Elliott, ET; Paustian, K; Domn, JW. Aggregation and soil organic matter accumulation in cultivated and native grassland soils. *Soil Sci.* Soc. *Am.* 1998. 62, 1367-1377.

Smith, SE; Read, DJ. Vesicular-arbuscular mycorrhizas in agriculture and horticulture. In: Mycorrhizal Symbiosis. Second edition. Smith, S.E. and D.J. Read (Ed.), Academic Press, London, UK.1997. 453-69.

Staddon, PL; Ramsey, CB; Ostle, N; Ineson, P; Fitter, AH. Rapid turnover of hyphae of mycorrhizal fungi determined by AMS microanalysis of C-14. *Science.* 2003. 300, 1138–1140.

Subramanian, KS; Charest, C. Influence of arbuscular mycorrhizae on the metabolism of maize under drought stress. *Mycorrhiza.* 1995. 5, 273–278.

Subramanian KS; Bharathi, C; Jegan, A. Response of maize to mycorrhizal colonization at varying levels of zinc and phosphorus. *Biol Fertil Soils.* 2008. 45,133–144.

Sukarno, N; Smith, SE; Scott, ES. The effect of fungicides on vesicular-arbuscular mycorrhizal symbiosis. I. The effects on vesicular-arbuscular mycorrhizae and plant growth. *New Phytol.* 1993. 125, 139-147.

Sukarno, N; Smith, FA; Smith, SE; Scott, ES.The effect of fungicides on vesicular-arbuscular mycorrhizal symbiosis. II. The effects on area of interface and efficiency of P uptake and transfer to plant. *New. Phytol.* 1996. 132, 583-592.

Sylvia, DM., Chellemi, DO. Interactions among root-inhabiting fungi and their implications for biological control of root pathogens. *Adv. Agron.* 2001. 73, 1–33.

Talavera, M; Itou, K; Mizukubo, T. Reduction of nematode damage by root colonization with arbuscular mycorrhiza (Glomus spp.) in tomato-Meloidogyne incognita (Tylenchida, Meloidognidae) and carrot- Pratylenchus penetrans (Tylenchida, Pratylenchidae) pathosystems. *Appl. Entomol. Zool.* 2001. 36, 387–392.

Tillman, D. The greening of the green revolution, *Nature*. 396, 1999. 211-212.

Timmer, LW; Leyden, RF. The relationship of mycorrhizal infection to reduced phosphorus induced copper deficiency in sour orange seedlings. *New Phytol.* 1980. 85, 15-23.

Tinker, PB. Soil chemistry of phosphorus and mycorrhizal effects on plant growth. In: Sanders, F.E., Mosse, B., Tinker, P.B.(Ed.), Endomycorrhizas. Academic Press, New York, 1975. 353-371

Tisdall, JM; Smith, SE; Rengasamy, P. Aggregation of soil by fungal hyphae. *Aust J. Soil Res.* 1997. 35, 55–60.

Tobar, R; Azcon, R; Barea, JM. Improved nitrogen uptake and transport from 15N-labelled nitrate by external hyphae of arbuscular mycorrhiza under water-stressed conditions. *New Phytol.* 1994. 126, 119-122.

Tonin, C; Vandenkoornhuyse, P; Joner, E J; Straczek, J; Leyval, C. Assessment of arbuscular mycorrhizal fungi diversity in the rhizosphere of *Viola calaminaria* and effect of these fungi on heavy metal uptake by clover. *Mycorrhiza*. 2001. 10, 161–168.

Tsimilli-Michael, M; Eggenberg, P; Biro, B; Koves-Pechy, K; Voros, I; Strasser, RJS. Synergistic and antagonistic effects of arbuscular mycorrhizal fungi and Azospirillum and Rhizobium nitrogen-fixers on the photosynthetic activity of alfalfa, probed by the polyphasic chlorophyll a fluorescence transient O-J-I-P. *Appl. Soil Ecol.* 2000. 15, 169–182.

Turnau, K; Kottke, I; Oberwinkler, F. Element localization in mycorrhizal roots of *Pteridium aquilinum* ŽL. Kuhn collected from experimental plots treated with cadmium dust. *New Phytol.* 1993. 123, 313-324.

Turnau, K; Mesjasz-Przybylowicz, J. Arbuscular mycorrhizal of *Berkheya coddii* and other Ni-hyperaccumulating members of Asteraceae from ultramafic soils in South Africa. *Mycorrhiza*. 2003. 13, 185–190.

Turrini, A; Sbrana, C; Nuti, MP. Development of a model system to assess the impact of genetically modified corn and aubergine plants on arbuscular mycorrhizal fungi. *Plant Soil.* 2004. 266, 69–75.

UNCED. Combating deforestation. *The Forest Chronicle*. 1992. 68, 1-7.

URS. Caltrans Erosion Control Training for Landscape Architects, Section 8 – Seed and Fertilizer. California Department of Transportation. 2001.

Vazquez, MM; Barera, JM; Azcon, R. Influence of arbuscular mycorrhizae and a genetically modified strain of Sinorhizobium on growth, nitrate reductase activity and protein content in shoots and roots of Medicago sativa as affected by nitrogen concentrations. *Soil Biol. Biochem.* 2002. 34, 899–905.

Van Duin, WE; Groeffen, WA; Ietswaart, JH. Occurrence and function of mycorrhizae in environmentally stressed soils. In: Rosema, J., Verkliej, J.A.C. (Ed.), Ecological Response to the Environment. Kluwer Academic Publishers, Norwell, MA USA, 1991. 223_254.

Van Steveninck, RFM; Van Steveninck, ME; Fernand, DR; Horst, WJ; Marshner, H. Deposition of zinc phytate in globular bodies in roots of *Deschampsia caepitose* ecotypes: a detoxification mechanism? *J. Plant Physiol.* 1987. 131, 247-257.

Vangronsveld, JJ; Colpaert, V; Van Tichelen, KK. Reclamation of a bare industrial area contaminated by non-ferrous metals: physico-chemical and biological evaluation of the durability of soil treatment and revegetation. *Environ. Pollut.* 1996. 94: 131-140.

Verkade, SD; Hamilton, DF. Effect of endomycorrhizal inoculum on root initiation and development of Viburnum dentatum L. cuttings. *J. Environ. Hort.* 1987. 5, 80-81.

Vigo, C; Norman, JR; Hooker, JE. Biocontrol of Phytopthora parasitica by arbuscular mycorrhizal fungi is a consequence of effects on infection loci. *Plant Pathol.* 2000. 49, 509–514.

Wang, GM; Stribley, DP; Tinker, PB. Soil pH and vesicular-arbuscular mycorrhizas. In: Fitter, A.H. (Ed.), Ecological Interaction in Soil, Plants, Microbes and Animals. Blackwell Scientific Publications Ltd, Oxford, UK, 1985. 219-224.

Weissenhorn, I; Glashoff, A; Leyval, C; Berthelia, J. Differential tolerance to Cd and Zn of arbuscular mycorrhizalŽAM. fungal spores isolated from heavy metal polluted and unpolluted soils. *Plant Soil.* 1994. 167, 189-196.

Weissenhorn, I; Leyval, C. Spore germination of arbuscular mycorrhizal fungi in soils differing in heavy metal content and other parameters. *Eur. J. Soil Biol.* 1996. 32, 165_172.

West, HM; Fitter, AH; Watkinson, AR. Response of *Vulpia ciliata* ssp. ambigua to removal of mycorrhizal infection and to phosphate application under natural conditions. *J. Ecol.* 1993. 81, 351-358.

Whipps, JM. Prospects and limitations for mycorrhizas in biocontrol of root pathogens. *Can. J. Bot.* 2004. 82, 1198–1227.

Wolfenbarger, LL; Phifer, PR. The ecological risks and beneficial engineered plants. *Science.* 2000. 290, 2088–2093.

Wright, SF; Anderson, RL. Aggregate stability and glomalin in alternative crop rotations for the central Great Plains. *Biol. Fertil. Soil.* 2000. 31, 249–253.

Xu, D; Bai, J; Dell, B. Overcoming the constraints to productivity of plantation eucalypts in southern China. p. 93-100. In Mycorrhizal Fungal Biodiversity and Applications of Inoculation Technology. Proceedings of Guangzhou ACIAR International Workshop, 1998. China Forestry Publishing House, Beijing. 2000.

Zhu, YG; Christie, P; Laidlaw, AS. Uptake of Zn by arbuscular mycorrhizal white clover from Zn-contaminated soil. *Chemosphere.* 2001. 42: 193-199.

Zhu, YG; Miller, RM. Carbon cycling by arbuscular mycorrhizal fungi in soil–plant systems. Trends. *Plant Sci.* 2003. 8, 407–409.

Reviwers:

1. Prof. B. N. Johri. Department of Biotechnology. Barkatullah University, Bhopal – 462026; Email: bhavdishnjohri@rediffmail.com
2. Dr. Alok Adholeya. Bioresources and Biotechnology Division, TERI, India Habitat Centre, Lodhi Road, New Delhi – 110003; E-mail: aloka@teri.res.in

In: Mycorrhizal Fungi
Editor: Susanne M. Fulton

ISBN 978-1-61122-659-1
© 2011 Nova Science Publishers, Inc.

Chapter 2

ARBUSCULAR MYCORRHIZAS IN AGROECOSYSTEMS

***Marcela Claudia Pagano*[1] *and Fernanda Covacevich*[2]**
[1]Biology Department, Federal University of Ceará, Fortaleza, Ceará, Brazil
[2]CONICET-Unidad Integrada EEA INTA- Facultad de Ciencias Agrárias UNMP, Balcarce, Argentina

Abstract

This review was done to explore the current information on the benefit of arbuscular mycorrhizal symbioses in agricultural systems, drawing on results of research on agroecosystems in Argentina and Brazil. The increasing recognition that agricultural intensification and use of agrochemicals can adversely affect soil quality, modifying the number, diversity and activity of the soil microbiota, including symbiotic fungal populations, have improved the research aimed at crop yield improvement and sustainability. Thus, microbial-based approaches have been proposed and quantitative relationships between soil or plant phosphorus (P) levels and arbuscular mycorrhizas (AM) are being investigated in order to adjust the fertilizer dose to obtain acceptable yield production without depressing the symbiosis. Biodiversity of AM is high in agricultural soils from Argentina both under tillage or non-tillage management. Soils cultivated with wheat, one of the most important crops in Argentina, contain indigenous AM that colonize this crop; however the selection of effective fungus, as well as the study of the mycorrhizal populations in soils supplied with P may allow the attainment of acceptable yield levels with low fertilizer inputs, while also reducing costs and environmental pollution risk. Fertilization with synthetic P sources as superphosphate decreases both mycorrhizal potential and indigenous arbuscular mycorrhizal colonization (AMC) of grain and grasses crops. However, fertilization with rock phosphate doesn't negatively affect both mycorrhizal potential and indigenous AMC. Additionally, the positive effects of mycorrhizal inoculation are shown with the rock phosphate fertilization, while when the fertilizer was superphosphate mycorrhizal benefit was only evident at moderate/low doses of P. Alternatively, fertilizer placement also affects indigenous AMC. Applications with banded superphosphate depress the AM formation of wheat roots as compared with broadcasted applications both under no-tillage and under conventional tillage. In Brazil, reports for maize, peanut, soybean and sorghum as well as for green manure species have pointed their mycotrophy and AM dependency. However, more research is needed in relation to germplasm banks of AM for increase the production of agricultural plants as well as to optimize the use of renewable P sources, which is essential to the Brazilian agribusiness. This chapter discusses the AM parameters, such as root colonization and spore number in agricultural systems and the relationship with some soil characteristics, such as organic matter and glomalin contents. Relevant findings are emphasized, such as inoculation and management benefits for farmers by increasing soil fertility without loss of yield or added costs. Research directions that are needed to increase understanding of mycorrhizal associations in agriculture are indicated.

Keywords: Arbuscular mycorrhizae; Crops; Microbial-based approaches; Sustainable agriculture; Latin America

Introduction

Agricultural technology is nowadays changing their management due to the need to link conservationist viewpoint with cost reductions. Interest in more ecologically sustainable agricultural systems is rising with increasing recognition that agricultural intensification can adversely affect soil quality (Borie et al. 2006). The agricultural plants are largely arbuscular mycorrhizal (AM) in their associations, although a few species do not present this symbiosis, such as radish (Brassicaceae) (Miranda 2008). Since the earliest observations on arbuscular mycorrhizal fungi (AMF), arbuscules (Figure 1) have been the defining morphological characteristic of the symbiosis (Gallaud 1905, in Burleigh 2000). Main function of arbuscules is the transfer of inorganic nutrients, including P, Zn, Cu and N from the fungus to the host plant, and the carbon movement from the plant to the fungus. Traditionally, Glomeraceae is the main family of AM involved in agricultural symbiotic associations (Oehl et al. 2003, Oehl et al. 2005, Bedini et al. 2007).

The potential for improved establishment of agricultural seedlings with the inoculation of mycorrhiza has been demonstrated for several species, thus, attention in AM propagation for sustainable agriculture is increasing due to its role in the promotion of plant health, and improvements in soil fertility, minimizing use of pesticides and inorganic fertilizers (Siddiqui and Pichtel 2008). Previous findings by Helgason et al. (1998) have showed that AM fungal diversity is low in arable crops, Daniell et al. (2001) further pointed that the causes and consequences of the low AM fungal diversity remain to be established.

The knowledge of the taxonomy-biology-physiology of these fungi and their symbiosis is still scarce and we do not know all the mechanisms by which the symbiosis works. For this reason it is often considered as a “black box” and this affects the management of the symbiosis. So that research must focus to increase understanding of mycorrhizal associations in agriculture.

This chapter reports the latest advances of mycorrhizal symbiosis, particularly in agricultural systems of Argentina and Brazil (Table 1). We report the effects of agricultural management practices on the occurrence and the benefit of mycorrhizal symbiosis in crops of interest. Parameters that may be using for quickly assess the presence and benefits from the symbiosis and the relationship with some soil characteristics are discussed. Relevant findings are emphasized, such as inoculation and management benefits for farmers by increasing soil fertility without loss of yield or added costs.

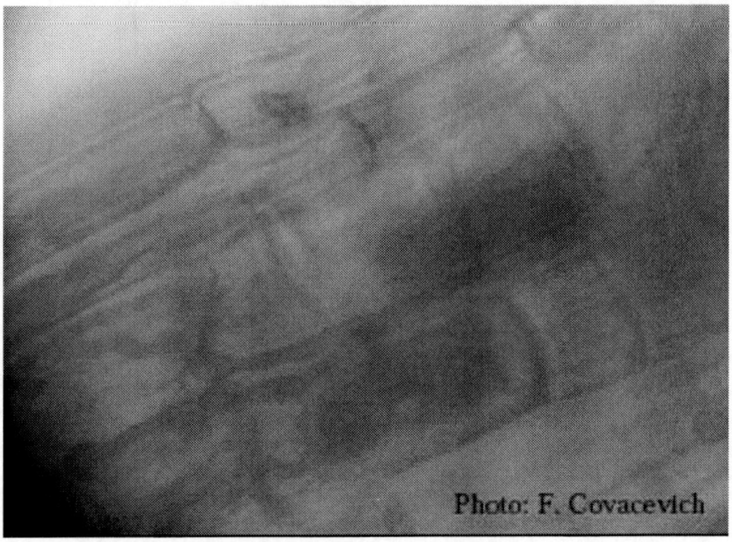

Figure 1. Intercellular arbuscules in *Brachiaria decumbens* roots.

Mycorrhizal Symbiosis in Agroecosystems

Effect of Tillage, Crop Rotation and Fertilization

Studies have confirmed that AM colonize most agricultural plants and that they can have a substantial impact on crop productivity (Sieverding 1991, Miranda 2008, Tchabi et al. 2008), although many questions remain how to optimally manage these beneficial fungi (Cardoso and Kuyper 2006). Many crops, expected to be facultatively mycorrhizal, are highly dependent on AM (eg. many tropical crops and forage species) (see Brundrett 1991). For example wild oat is less dependent on mycorrhizas than cultivated oat (Koide et al. 1988).

In agroecosystems, AM density varies according to the host plant (culture), edaphic and climatic factors, age of the culture and agricultural management. Agricultural practices like tillage (T), crop sequence, plant breeding, fertilizer and pesticide applications alter the physiology and growth of crops, and may also alter AM fungal populations, species composition and root colonization (Schwab and Reeves 1981, Kurle and Pfleger 1994, Miranda 2008). Consequently, suitable management is desirable to obtain benefits from AM communities. It is required the knowledge of culture systems which may affect the survival and multiplication of AM fungi (AMF) (Beauregard et al. 2008). Due to the potential of AMF in agroecosystems (almost all tropical crops are mycorrhizal, most of them are strongly responsive and dependent on AM), the study of both the manipulation of indigenous AMF in the field through appropriate agricultural practices and the development of a successful inoculation, if necessary, is relevant.

Table 1. Summary of actual evidence on AM in agricultural soils worldwide

Location	Source	Agroecosystem type	Crops	AM spores	S*	AMC#	GC	Others
Argentina/ Buenos Aires	Schalamuk and Cabello (2010a‡,b)	Agroecosystem (agricultural experimental station)	Wheat	IN	21	NI	NI	Tillage, Propagule Bank
Argentina/ southeast Buenos Aires	Covacevich and Echeverría (2000)	Agroecosystem (agricultural experimental station)	Orchardgrass, Ryegrass, *Poa anua*, clover, sunflower	ND	ND	15-60	ND	Conventional agriculture, intensive livestock and organic livestock
Argentina/ southeast Buenos Aires	Covacevich et al.(2006)	Agroecosystem (agricultural experimental station)	Tall fescue, wheatgrass	IN	IN	7-59	ND	Sources and rates of P fertilization
Argentina/ southeast Buenos Aires	Covacevich et al. (2007)	Agroecosystem (agricultural experimental station)	Wheat	IN	IN	10-50	ND	Rates of P fertilization
Argentina/ Buenos Aires	Covacevich et al. (2005, 2008)	Field experiment (low and high P soil, NT and CT)	Wheat	ND	ND	3-35	ND	P placement
Brazil	Pagano (in press) ‡	Agroecosystems, agroforestry	Various	IN	IN	IN	IN	Cover crop, tillage, soil compaction
India	Muthukumar and Prakash (2009)	Agroecosystems (crops and weeds)	Maize, sugarcane, associated weeds	ND	ND	62.64 (crops) 52.92 (weeds)	ND	
Brazil / Minas Gerais State	Oliveira et al. (2009) †	Field experiment (low and high P soil)	Maize	40 to 120	14	>60	ND	DGGE
Mainly Asia study cases...	Siddiqui et al. (2008) ‡	Agroecosystems, agroforestry	Various	IN	IN	IN	IN	AM, EM
North America	Boomsma and Vyn (2008) ‡	Field and greenhouse	Maize	IN	IN	IN	NI	Inoculation, Leaf relative water content, etc.
Iran / Karaj	Miransari et al. (2009a,b)	Experimental research field soil (pots)	Corn, wheat	ND	ND	ND	ND	Inoculation, soil compaction, nutrient uptake
West Africa	Tchabi et al.	Natural and cultivated	Yam, maize,	200-	11 to	70-95	ND	

	(2008, 2009)	sites	peanut	1700	60			
Brazil	Miranda (2008) ‡	Agroecosystems	Soybean, cassava	IN	IN	IN	ND	
Southeastern Argentine Pampas	Covacevich et al. (2007)	Field crop	Wheat	NI	NI	>22.5	ND	P fertilization
Canada Southern Ontario	Manns et al. (2007)	Test plots (farm)	Oat	ND	ND	NI	ND	soil aggregation, moisture, OC, fungal hyphal
Mid-Atlantic area of the U.S.	Wright et al. (2007)	Faming systems	Corn/soybean wheat/soybean	ND	ND	ND	0.53-0.66	
Italy / Siena, Tuscany	Bedini et al. (2007)	Monoculture	Maize	464-58.5	7	ND	2	
Chile	Borie et al. (2006)	No-tillage; reduced and conventional tillage	Wheat	372-755	NI	53	5.7-7.2	
Brazil	Stürmer and Siqueira (2006) ‡	Agroecosystems (crops and green manure)	Various	IN	IN	IN	NI	AMF frequently recovered / used for inoculation
Tropics, Brazil	Cardoso and Kuyper (2006)‡	Agroecosystems, agroforestry	Maize, soybean, coffee, banana,	IN	IN	IN	IN	Inoculation, soil compaction nutrient uptake, AM, EM
Canada / Saskatchewan	Ma et al. (2005)†	Organic farm fields	NI	ND	0–4	ND	ND	
Brazil	Siqueira et al. (2002)	Agroecosystems	Various	NI	40	NI	NI	Inoculation
Argentina/ Buenos Aires	Menéndez et al. (2001)	Field site	Wheat, barley, red clover	<20 to 50	17	ND	ND	
UK/ North Yorkshire	Daniell et al. (2001) †	Field sites	Winter wheat, barley, maize, peas	ND	ND	4 root sequences types	ND	

AM spores = spore number 100 g soil-1; #Maximal AM colonization (%) reported; S = AM species richness; *Spores; ND = not determined in the study; IN= Informed in review; NI = Not informed. †Molecular identification; ‡Review or book; GC = total glomalin content mg g soil-1; OC = organic carbon; AM = Arbuscular mycorhizae, EM = Ectomycorhizas; DGGE = denaturing gradient gel electrophoresis; MR = mycorrhizal responsiveness; NT = non- till; CT = conventional tillage.

The rate of colonization is influenced by the ability of AMF to spread rapidly and extensively in plant roots, and is affected by factors linked to spore germination, presymbiotic mycelial growth and appressorium (Figure 2) formation (Giovannetti 2000). Colonization of roots (cortical tissue) by AMF can initiate from three sources of inoculum called 'propagule bank': spores, colonized root fragments and free hyphae of bulk soil. This 'propagule bank' is 'waiting' for suitable soil conditions to germinate, grow and eventually colonize plant roots (Schalamuk 2010b). Moreover, a single root can be colonized simultaneously by various, commonly 5 to 6, AM species in the rhizospheric soil (Miranda 2008). It is known that crops, such as sweet potato, soybean, maize, sorghum, barley, sugarcane, tobacco, cotton, and cacao, frequently exhibit high colonization rates under natural conditions. On the other hand, wheat, beans, coffee and tomato can present moderate colonization rates, small intraspecific differences being observed in the colonization percentages between different ecotypes, cultivars or clones of the same crop (Velázquez and Cabello 2010). Useful information on methods for clearing different crop roots is available in the book by Miranda (2008).

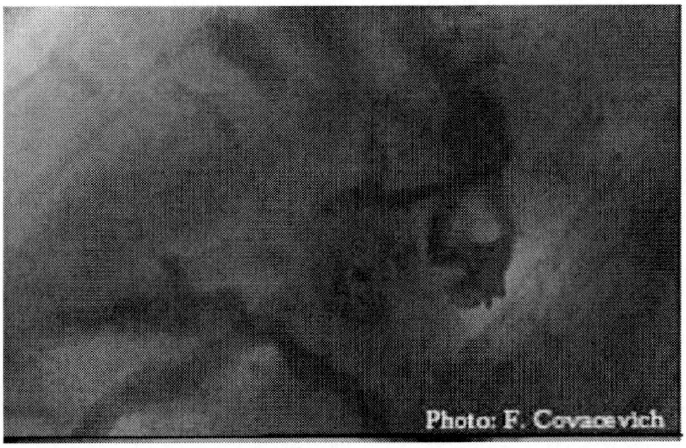

Figure 2. Formation of appressoria by arbuscular mycorrhizal fungi on roots of tall fescue.

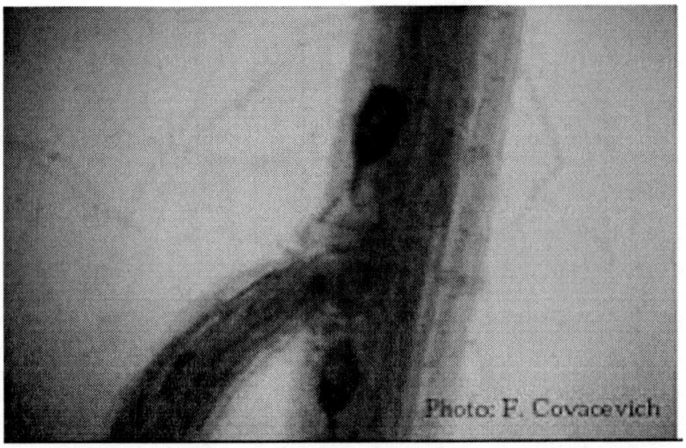

Figure 3. Mycorrhizal external hyphae and intracellular vesicles of *Brachiaria decumbens* roots.

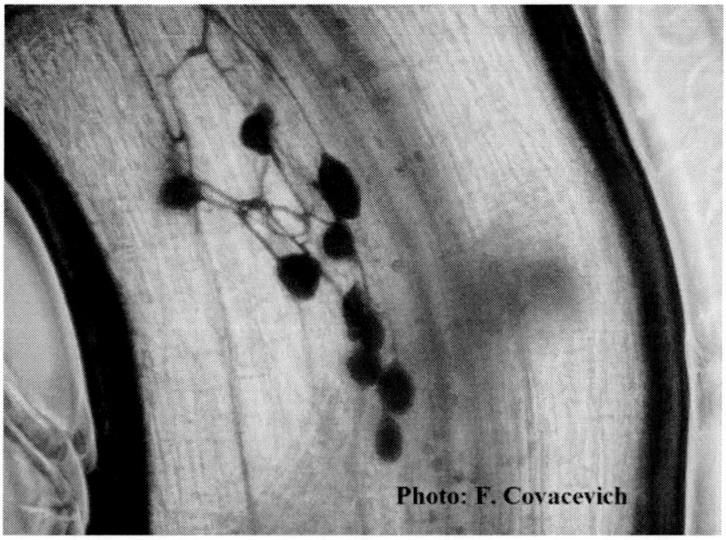

Figure 4. Spread of mycorrhizal colonization through extracellular AMF hyphae and vesicles of ryegrass roots.

It is therefore necessary to focus on increasing awareness of native fungi of agronomic interest, and this could be achieved with propagule multiplication. Networks of soil hyphae are an important source of inoculum in ecosystems that are highly susceptible to disturbance as agroecosystems where fast sporulating *Glomus* would be selected, being able to propagate by means of mycelium fragments and vesicles (Figures 3 and 4) (de Souza et al. 2005). Although the spore count method have some limitations of, it was used to evaluate the effect of cultures and agriculture practices on the AM communities (Siqueira et al. 1987, Siqueira et al. 1989, Johnson et al. 1992, Cuenca and Meneses 1996, de Souza et al. 1999, Boddington and Dodd 2000, Carrenho et al. 2001, Franke-Snyder et al. 2001, Miranda et al. 2005, Purin et al. 2006, Schalamuk and Cabello 2010b).

Tillage affects the soil physical and chemical properties, modifying the number, diversity and activity of the soil microbiota, including both free and symbiotic fungal populations. Sustainable agriculture is more favorable to the prevalence of AMF than conventional agriculture (Bethlenfalvay and Schuepp 1994, Smith and Read 2008), because disturbs will affect their distribution. It is well established that mycorrhizal hyphae extend from infected roots, allowing plants to explore greater soil volumes. Damage to the hyphal network by T can reduce AMF growth and root colonization due to death or reduced infectivity of the hyphal fragments compared with intact networks (Evans and Miller 1990; Goss and de Varennes 2002). Thus, T directly affects all types of AMF propagules through different mechanisms acting together: (i) disruption of the hyphal network; (ii) dilution of the propagule-rich topsoil; and (iii) accelerated root decomposition (Schalamuk and Cabello 2010b).

In Switzerland Jansa et al. (2002) found in differently tilled soils from a long-term field experiment rotated with different crops that the effect of T on mycorrhizal symbiosis was different in relation to crops plants and AMF species. They reported that in reduced tillage (RT), the presence of certain AMF species, especially those which do not belong to *Glomus*, had a tendency to increase. The largest contributor to spore counts of its experiment (*Glomus intraradices*) did not show significant differences with respect to soil T treatments, but T decreased the abundance of *Glomus invermaium* and *Glomus microaggregatum*. Although more AMF spores were observed in soil from no-tilled wheat field than from the tilled field, the total spore count was not significantly affected by T following the maize season. Similar to Jansa et al. (2002), Crovo et al. (2001) found no differences in indigenous mycorrhizal colonization of maize crop under T or RT from long field experiment in Argentina.

In Argentina, no-tillage (NT) farming has currently become widespread for most of crop production (Figure 5a,b). NT produces physical, chemical and biological changes in the soils compared to soil under conventional tillage (CT) (Figure 6a-c), i.e. increases of P available and organic matter (OM) in soil surface layers (Calviño et al. 2000), higher soil phosphatase activity (Doran 1980) and greater root density in the soil surface layers (Qin et al. 2004). In cultivated sites (wheat, barley or clover) of the Pampas (Buenos Aires Province) the most widely recognized influence of T on AM fungal species diversity and spore number were the decrease of some undentified species of *Glomus*, plus *Glomus aggregatum*, *G. microaggregatum*, *Acaulospora denticulata*, *Entrophospora* spp. and *Glomus coremioides*. In contrast, other unidentified species (*Glomus mosseae* and *Scutellospora pellucida*) with larger size persisted after several decades of T (Menéndez et al. 2001). They also showed that the composition of the AM fungal population decreased compared to a grassland soil.

a

b

Figure 5. a. Wheat crop under non-tillage in Buenos Aires Province, Argentina; b. Details of the residues present in wheat crop in tillers formed stage (Feekes 3.0) under non-tillage in Buenos Aires Province, Argentina.

Figure 6. a. Soil preparation for sowing of wheat crop under conventional tillage and for mycorrhizal inoculation assays in Buenos Aires Province, Argentina; b. Mycorrhizal inoculation of wheat crop under conventional tillage in Buenos Aires Province, Argentina; c. Wheat crop in beginning

of erect growth stage (Feekes 4.0) under conventional tillage in the Buenos Aires Province, Argentina for mycorrhizal inoculation assays.

Figure 7. Harvest of soybean grown in rotation with wheat under direct seeding, Argentina.

Also in Buenos Aires province Schalamuk et al. (2006) identified, in agricultural field crops, 24 species of Glomeromycota belonging to six genera of AMF: *Acaulospora*, *Archaeospora*, *Entrophospora*, *Gigaspora*, *Glomus* and *Scutellospora*. Contrarily to others, they showed that AMF biodiversity was higher in tilled plots than in untilled ones and that the contribution of species belonging to Glomeraceae increased in NT plots to the detriment of Acaulosporaceae and Gigasporaceae, contradicting reports by Jansa et al. (2002). Those authors suggested that the higher contribution of Glomeraceae previously found by Jansa et al. (2002) in NT systems could be related to the lack of disruption of the hyphal network and the composition of the soil propagules in that system.

For maximum effect, NT requires crop rotation (CR) and the avoidance of plowing. As the growing of the same crop year after year leads to the accumulation of plant's diseases and pests, CR has been the time-honored way to reduce crop damage due to pests and diseases, especially under NT management.

The major change in the agricultural production of Argentina in the last 20 years was the dominance of soybean as the principal crop of the Pampa (Solbrig 2005). Cropping schemes have developed in Argentina for rainfed soybean (*Glycine max* Merr.) production that invariably employ NT management and CR. Often these schemes include growing soybean in a sequence of crops including wheat (Figure 7) and maize.

Moreover, past research in multiple plant species has shown that the common use of monocultures in crop production leads both a spore population decrease and changes in AMF community (Johnson et al. 1992, Rao et al. 1995, Oehl et al. 2003). In CR, preceding crops affect growth and yield of subsequent crops (Karlen et al. 1994) as well as mycorrhizal development, for example a non-mycorrhizal crop can decrease both AMF colonization and yield of subsequent crops (Douds et al. 1997, Arihawa and Karasawa 2000).

In Brazil, NT systems are increasingly used, especially in the cerrado savannas (Prado 2001). With regard to AMF, research interest in cerrado savannas and dry forest (caatinga biome) is nowadays more favored, with increasing selection of the crop, cultivars and the rotation to be used. As some annual crops (soybean, bean, maize) and green manure (mucuna, *Crotalaria*, *Canavalia ensiformes*, *Cajanus*, sunflower, millet, *Ricinus communis*) as well as some fodder (estilosantes, andropogon) present high AMF dependency, they can increase the native AMF community, which benefit subsequent crops (Miranda 2010). On the other hand, rice and *Raphanus sativus* (green manure), show low or no AMF dependency, respectively, and the introduction of them in rotation decrease the fungal community in soil and their benefits to the subsequent crop. Therefore, to avoid the use of low or no AMF dependent crops will be more benefic (Miranda 2010). Nevertheless, effects of hyphae on soil structure can persist in the absence of AM plants, thus non-mycorrhizal (NM) plants can benefit from the improved soil when they are cultivated after mycorrhizal crops (Smith et al. 2010).

In a six years field experiment in Brazil, Miranda et al. (2005) reported changes in the indigenous AMF populations after crop rotation (andropogon grass / soybean) and seasonal variations in cerrado soils. They observed that spore numbers and AMC varied according to cropping time, soil moisture, crops and rotation system and were higher in the cultivated soil than in undisturbed cerrado soil. Moreover, they showed that the number of AMF genus and species increased according to cropping time and management, being higher under annual crops in rotation. In the same study, they estimated that AMF contributed 53% and 95% to the growth of soybean and andropogon grass, respectively, in a greenhouse experiment.

In the semiarid of Brazil, studies showed the effect of selective logging (thinning) on propagules in cultivated dry forest areas (Lima et al. 2007). Further, in the South of Brazil, Benedetti et al. (2005) showed that covering plant legumes as *Canavalia ensiformes* favored AMF during and after maize cultivation. For that, Carrenho et al. (2002) remarked the use of sorghum and maize as the best choice and they also defined procedures for AMF multiplication in pots.

In fertile agricultural soils of the Pampas region, Argentina, soil use and yield average of main crops were remarkably increased in the last 20 years (Echeverria and García 1998). This has generated a progressive diminution of nutrient availability such as phosphorus (P), and therefore, a generalized response to P application. The impact of soluble fertilizers on AMF colonization and function is contradictory (Beauregard et al. 2008). The environmental conditions also affect the formation of AMF and/or the AMC, crops with high supplies of fertilizers rich in nitrogen, phosphorus and potassium, being inhibited in their mycorrhizal development (Velázquez and Cabello 2010).

The application of soluble P generally decrease root colonization with occasional reports of increases. In Argentina, one of the most important crops highly colonized by indigenous AM is wheat; therefore, the selection of effective fungus, as well as the study of the mycorrhizal populations in soils supplied with P may allow the attainment of acceptable yield levels with low fertilizer inputs, while also reducing costs and environmental pollution risk. Covacevich et al. (2007) showed that P fertilization with an inorganic soluble P source (superphosphate -SP-) increases available soil P but depress indigenous AMC of wheat monoculture, suggesting that the primary effect of P on mycorrhizal symbiosis seems to be a reduction in AMC caused by high available soil P concentration. Alternatively, fertilizer placement also affects indigenous AMC of agricultural crops. Covacevich et al. (2005, 2008) found that fertilization of wheat crops with banded SP depressed indigenous mycorrhizal formation as compared with broadcasted applications both under NT and under CT.

Effect of Fertilization on Pastures

At the Pampas region, the main area of livestock production in Argentina, pastures represent approximately 70% of the grazed surface. Tall fescue and wheatgrass are the most seedling perennial crops pastures. With the current tendency for a reduced use of agrochemicals, a renewed interest in fertilization with natural minerals as rock phosphate (RP) has arisen.

Furthermore, some farmers that produced forage grass under organic management choose to use RP because is considered a suitable source of P for acidic soils (like Argentina Pampas soils, pH 5.7). Work done by Covacevich et al. (2006) and Covacevich and Echeverría (2009) both under field and greenhouse conditions showed that fertilization with SP reduced (more than 10%) the mycorrhizal potential of agricultural tall fescue and wheatgrass crops and fertilization with RP had a high (12–14 AM propagules g soil^{-1}) mycorrhizal potential. They also showed that although fertilization with SP or RP (at equivalent rate of 30 kg P ha^{-1}) produced similar available soil P content and grasses production, mycorrhiza root colonization of both crops decreased after fertilization with SP, but fertilization with RP did not decrease colonization. Additionally, Covacevich and Echeverría (2000) found that agricultural practices, mainly fertilization with SP, decreased ryegrass and orchardgrass indigenous mycorrhizal colonization under intensive livestock in comparison with organic livestock in Buenos Aires province of Argentina.

Greenhouse studies for pastures in Brazil (Carneiro et al. 2009), showed that the rates of AMC decreased with the increase of P fertilization (KH_2PO_4, NaH_2PO_4) in alfalfa (*Medicago sativa* L. cv. Crioula), which was highly dependent on the mycothrofic condition to maximize its yield, indicating the higher benefit of AMF inoculation of this plant species under low natural soil fertility. Earlier studies looking at the importance of nutrient in pastures such those by Santos et al. (2002) showed that in mixed plantations of *Brachiaria brizantha* and *Arachis pintoi* the mycorrhization resulted in an increase of plant nutrients for the last species.

Effect of N Fertilization in Crops

Reports regarding to nitrogen (N) fertilization on mycorrhizal symbiosis in agroecosystems (Gryndler et al. 1990, Liu et al. 2000) also showed contradictory results. Egerton-Waburton and Allen (2000) and Johnson et al. (2003) have found differences in species composition with N fertilization. Conversely, Schalamuk et al. (2006) did not found differences in AMF species composition of wheat field from Argentina among different N fertilization conditions. Also Covacevich et al. (2007) reported that N fertilization did not affect growth or AMC of wheat plants grown in pots.

For instance, studies on the effect of N fertilization on AM symbioses in crops have not been reported for Brazil (Cardoso and Kuyper 2006), contrarily to reports for Brazilian tree species (see Urquiaga et al. 2005). Nevertheless, the symbiotic nitrogen fixation (SNF) was extensively studied in soybean and bean (Hungria et al. 2002), legumes with recognized contribution of *Bradyrhizobium* or *Rhizobium* to host-plant. The SNF is not discussed here, but major aspects of their importance are briefly described in this chapter. The presence of legumes in mixed crops has a benefic effect such as to increase the AM inoculum potential in soil, as well as the continuity of the physiological activity of both plant and AMF (Miranda 2008).

Other Crops

The status of the Brazilian inventory of glomalean fungi in agroecosystems has been reviewed (Stürmer and Siqueira 2006), coffee systems showing up as the best studied crop; however a high proportion of undescribed AMF species was recorded in the studied systems.

Existing preliminary reports (Quadros et al. 2004) on mycorrhizal associations in potato crop, soybean, bean and maize, in organic systems in Brazil, showed a higher plant production. Furthermore, under different crop systems, differences in AMC were as follows: bean > soybean > maize > potato. Moreover, winter cover plants presented the following decrease AMC: vetch > black oat > ryegrass > turnip. It is notably that soybean showed higher nodules in the integrated > conventional > organic crop systems; however more detailed reports are needed. Also, for cassava plantations (Silva et al. 2006) a mean of 666 spores/50 cm^3 soil was reported. Recently reports in sugar cane (Azevedo 2008), also showed a predominance of *Glomus* in the rhizospheric soils, and a higher AMC without pre-harvesting burning, as compared to pre-harvesting burning; however these results still need more divulgation. Lastly, reports for cucumber inoculated with *Glomus* isolates resulted in a large intraspecific diversity for mycelium growth and improvement of P uptake, suggesting that AMF communities of low species diversity may still contain considerable functional heterogeneity (Munkvold et al. 2004).

Taxonomical and Biodiversity Studies of Mycorrhizas

Little is known about the extent of genetic variation in natural populations of AMF and the environmental influence on biodiversity (Stukenbrock and Rosendahl 2005). Arbuscular mycorrhizal fungi are obligate biotrophes that complete their life cycle only when they colonize roots of the host plants (Siddiqui and Pichtel, 2008). Therefore, for the handling and characterization of AMF we must work with the fungus and the plant host together. This is also the main cause of the lack of inoculum production and their application at large-scale. A successful colonization reflects a successful spore formation, which are indicator of the ecological processes and anthropic influences on the AM communities. Identification of species in Glomeromycota has relied extensively on the morphology of spores collected from soil. However, spores cannot be directly linked to one vegetal species, unless in monoespecific plant communities (de Souza et al. 2008). Moreover, spores are not present along the life history of the AMF. De Souza et al. (2008) remarked that the successfully use of the spore collection method depends on the taxonomic background of the researcher as well as the previous knowledge of the site diversity and use of trap-cultures. Morphotyping requires considerable experience of both culturing and spore morphology of a wide range of

AMF. The lack of specialist in taxonomy able to determine species of AM presents a considerable constraint for their management.

While identification of some of the most common species is less problematic, identification of rare species, and especially native ones collected for either ex-situ conservation, can be a problem. Given the importance of site- or host-specific species for the success for use in agricultural programs, correct determination of AM to the species level is relevant. However, even when spores can be taxonomically identified, the AM fungal community knowledge based on spores in the soil is restricted to the soil seed bank available. Furthermore, spore counts may not reflect the *in planta* composition of an AMF community at the time of sampling (Clapp et al. 1995). These limitations often lead to underestimate fungal diversity, population number and species richness.

To characterize AM communities, there are traditional methods (mainly based in spore morphology) and molecular techniques, which represent one tool to objectively characterize complex fungal communities in agroecosystems (Ma et al. 2005). More than 150 species of AMF have been described (Walker and Trappe 1993) by traditional-taxonomical studies. However there is accumulating evidence that the morphological diversity of AMF spores may not reflect their physiological and genetic plasticity. This indicates that identification of morphospecies may not be sufficient or appropriate for ecological studies (Clapp et al. 2001). Over the last two decades biochemical and molecular techniques based on DNA analysis (e.g. isozyme profiles or sequencing of ribosomal genes) have been extensively exploited in order to identify (Simon et al. 1993, Geue and Hock 2004, Corradi et al. 2007) and determine phylogenetic relationships (e.g. Redecker et al. 2000) and genetic heterogeneity of AMF.

In AMF as well as in all eukaryotic organisms there are multiple copies of the ribosomal deoxyribonucleic acid (rDNA) per cell; the ribosomal genes possess sections highly conserved which facilitate the design of primers to hybridize successfully in the region desired to be amplified by the polymerase chain reaction (PCR). In addition, there are variable regions that allow the differentiation of taxa at different levels. Ribosomal genes are multicopy genes tandemly organized in the genome, separated from each other by an Inter Genic Spacer (IGS). Ribosomal genes are comprised of three subunits of coding regions (18S[small subunit: SSU], 5.8S and 25S[large subunit: LSU]), separated from each other by an Inter Non Transcribed region (ITS). The 18S rDNA is the most useful region for phylogenetic and biodiversity studies. However, it has some limitations because it is the most variable region of the nuclear ribosomal genes, and show high intraspecific variation. The first molecular study on AMF was made by Simon et al. (1992), and the polymorphic nature of the ribosomal DNA in AMF was first described by Sanders et al. (1995). Comparative studies of the 18S rDNA allowed the identification of AMF (Bruns et al. 1991, Simon et al. 1993). These comparisons are made with the nucleotide sequence of rDNA after PCR amplification (Saiki 1988) and sequencing. This approach is well suited to the study of AMF in their natural environment (Simon et al. 1992, Helgason et al. 1998). Schüßler et al. (2001) reported that the within-isolate variation of the 18S rDNA is relatively small; thus, that AM fungal phylogeny could be based on this gene. However, the 5' end of the 25S ribosomal subunit also harbors two informative polymorphic domains. The polymorphism observed in these domains between and within taxa allowed the identification of some AMF species (Clapp et al. 2001, Geue and Hock 2004, Corradi et al. 2007).

Direct evidence shows that individual spores of AMF are multinucleated; that is, one AMF species is heterokaryotic containing thousands of nuclei genetically different (Kuhn et al. 2001). Thus, molecular studies for taxonomy purposes are preferably conducted on a single spore. Several of these studies (Hijri et al. 1999, Hosny et al. 1999, Pringle et al. 2000) have indicated that the level of genetic diversity in AMF rDNA genes is very high. Corradi et al. (2007) found by using a quantitative PCR approach in spores, that *Glomus intraradices* have high gene copy number polymorphism. Given that a segment of root can be colonized by several species (5 or 6) of mycorrhizal fungi simultaneously, for biodiversity studies, DNA extraction and PCR amplification from AM colonized roots are the most common approach (Simon et al. 1992, Geue and Hock 2004). The analyses of rDNA gene cluster aimed to identify AMF colonizing plant roots, and it made possible to study the diversity of AMF *in planta* with a high degree of precision and reproducibility. Some authors (Clapp et al. 1995, Merryweather and Fitter 1998) consider unhelpful to correlate spore information to specific colonization in plant roots as spores are produced on external hyphae that are easily detached from the root, and prefer to analyze sequences from the root samples (Daniell et al. 2001). Both strategies showed to offer a wider range of advantages and are important tools to increase the actual knowledge on AMF in agriculture.

Recent advances in DNA identification and designed primers (reviewed by Covacevich 2010) are likely to revolutionize AMF species delimitation in the future. Simon et al. (1993) designed primers to identify AMF in colonized roots by PCR fragment amplification of the 18S rDNA. Simon and Lalonde (1995) designed and patented a primer, which amplified part of the 18 S rDNA from AMF (*Glomus intraradices* and *Gigaspora margarita*) directly from colonized roots. Helgason et al. (1998) designed the general fungal primer AM1 which detect AMF colonizing plant roots because the primer targets the 18S rDNA of AMF, and exclude plant DNA sequences. The AM1 primer has been shown to amplify 3 families of the AMF (Glomeraceae, Gigasporaceae, and Acaulosporaceae). Geue and Hock (2004) designed two pairs of specific primers which targeted against the D2 domain of the 25S rDNA and allowed identification of *Acaulospora* and *Glomus* colonizing roots of grasses/herbs (e.g. *Trifolium, Plantago*) in pastures and Meadow, and roots of crops (wheat, maize) in arable land of Germany. Gamper and Leuchtmann (2007) designed new PCR Taxon-specific primers for the 25S rDNA which detected two inconspicuous AMF (*Acaulospora paulinae* and an undescribed member of the Diversisporaceae) from temperate agricultural grasslands of Swiss. Molecular evidence indicates that *Glomus* is commonly associated with most agricultural plants (Kjøller and Rosendahl et al. 2001, Daniell et al. 2001), and that several genotypes of *Glomus* can coexist in agricultural fields (Stukenbrock and Rosendahl 2005). Moreover, Daniell et al. (2001) observed that two dominant sequences (corresponding to *Glomus geosporum* and *Glomus mosseae*) colonized arable crop roots, highlighting the ability of these species to survive under conditions that may be unfavourable for other AMF. Jansa et al. (2002) pointed out the needs of other methods, such as specific primer quantitative PCR or specific primer PCR coupled with multiple sampling, in order to attain a reliable quantification of the AMF in the rhizosphere by targeting soil mycelium network or intraradical colonization structures.

With regard to the occurrence of natural root-colonizing AMF, Öpik et al. (2006) reviewed about 26 publications that report on the AMF identified using 18S rDNA region sequences from communities across 36 host plant species and 25 locations in different ecosystems around the globe, concluding that the number of AM fungal taxa per host plant species differed between habitat types: a significantly higher richness was found in tropical forests (18.2 fungal taxa per plant species), followed by grasslands (8.3), temperate forests

(5.6) and habitats under anthropogenic influence (arable fields and polluted sites, 5.2). The authors focused on AM fungal communities inside plant roots, which are considered to comprise the symbiotically active fungi.

Information of AM in impacted areas such as semiarid gypsite mining areas in Araripina, Pernambuco State, Brazil, where *Paraglomus occultum*, *Glomus mosseae*, *Glomus intraradices* and *Glomus etunicatum* were identified by using the restriction fragment length polymorphism analysis (PCR/RFLP) approach (Mergulhão et al. 2008), has recently received more attention. These studies focusing on the role of AMF in the restoration of highly disturbed areas, where AMF propagules are reduced or eliminated are important to the understanding of the functionality of AMF, also to be applied in agriculture.

It is well established that AMF extract nutrients from the soil and deliver those nutrients to plant roots and, in return, the fungi obtain carbohydrates produced by photosynthesis (Parniske 2005). However, Bidartondo et al. (2002) found that the non-photosynthetic plant *Arachnitis uniflora* from subantarctic forest in Nahuel Huapi National Park (Río Negro province, Argentina) forms AM with a narrow lineage within a clade (group A) of *Glomus*. They amplified the fungal nrITS region by using the universal fungal primers ITS1F and ITS4 directly from *Arachnitis* roots, and suggested that *Glomus* detected in *Arachnitis* also formed mycorrhizae in adjacent photosynthetic plants from three different families, mediating inter-plant carbon transfer in natural environments (constituting the only report for molecular analysis in AMF from Argentina).

Despite the global importance of AMF and their potential in agriculture, research into their molecular genetics is still in its infancy (Parniske 2005). In particular, Argentina and Brazil still have not implemented programs to investigate the biodiversity and the role of mycorrhizal associations in agroecosystems by molecular techniques. These techniques that allow discriminate genetic sequences of glomalean fungi excluding plants and other soil microorganisms proved to be difficult. Furthermore, many groups have not yet been identified and some signaling and control mechanisms of the symbiosis have been not yet understood, which are subject of further research.

Application of AMF in Agriculture

The process of technological incorporation happened in the agriculture reflect different interests and opinions about the models of the countries development and their own sustainability (Hespanhol 2008). Both in Argentina and Brazil, interest in alternative agriculture and in maintainable production systems in environmental and social terms in the agriculture of small scale, such as organic production is increasing. Moreover, the effect of cultural practices over AMF communities in crop fields is being studied (Siqueira et al. 1987, Siqueira et al. 1989, de Souza et al. 1999, Covacevich and Echeverria 2000, Carrenho et al. 2001, Miranda et al. 2005, Purin et al. 2006, Covacevich et al. 2006, 2007, Schalamuk et al. 2006).

Numbers of AMF surviving propagules in soils decline with time in the absence of host plants (Brundrett 2002). As well we need to include AMF inoculation to guarantee plant performance in revegetation programmes for degraded areas where the AM inoculum potential is low or when the natural colonization have low efficiency (Alguacil et al. 2005).

However, agriculture practices decrease AMF diversity (Schalamuk et al. 2006, Schalamuk and Cabello 2010a), root colonization (Covacevich et al. 2006, 2007, 2009) as well the inoculum potential in the soil (Covacevich et al. 2006).

Mycorrhizal efficiency is correlated with the ability of different AMF isolates to promote plant growth by improving mineral nutrition and increasing tolerance to biotic and abiotic stresses (Giovannetti and Avio 2002). In agricultural ecosystems the inoculation with greater number of AM fungal taxa (higher AMF biodiversity), can also result in higher productivity and biomass yield. Therefore, we note the need to test agricultural land restoration in soils affected by intensive agriculture practices, which could be achieved by native inoculum propagation and field inoculation. Furthermore, the P uptake and translocation by hyphae of AM associated with perennial plants is expected to differ from that of annuals such as crops (Smith et al. 1994); but the great functional diversity of the roots of perennial hosts remains virtually unexplored (Brundrett 2002). Owing to some AM fungi increase the uptake of P and plant growth, in glasshouse experiments using P-deficient soils, it is generally assumed that the same happens on a larger scale in agriculture, revegetation sites and in natural environments.

In Argentina, although soils of the Pampas region showed high native AMF that colonize crop plants under different management systems (Covacevich and Echeverria 2000, Covacevich et al. 2006, 2007, Schalamuk et al. 2006) they are not yet manipulated in order to increase plant productivity. The high diversity of AMF (Schalamuk and Cabello 2010a) and high (about 12 to 14 AMF propagules g^{-1} soil) inoculum potential (Covacevich et al. 2006) in the humid Pampa, Argentina, suggest that it is not necessary to inoculate these fungi. Covacevich and Echeverría (2009) reported that mycorrhizal growth responses (MR) of indigenous AMF from tall fescue and wheatgrass grassland of Argentinean Pampas were lower than the achieved by the inoculation with *Acaulospora longula* or *Glomus claroideum*. So, this could be an evidence that indigenous AMF were not efficient for improving plant growth and the need for inoculation in these soils. They also pointed that AMF inoculation plus fertilization with RP, significantly stimulated plant growth, mycorrhiza and MR over the single treatments. However, the growth benefit was shown only when plants were growing without indigenous microorganisms from Argentinean Pampas soil. In contrast, AMF inoculation with indigenous microflora did not significantly increase plant growth and P uptake. So even with the need for inoculation, the practice under field conditions may not always result in increased pasture yield. In this regard, it was established that the ground biota also can exert antagonistic, synergistic or neutral interactions on AMF (Pfleger and Linderman 1996, Jeffries et al. 2003, Godeas et al. 1999)

Martinez et al. (2004) reported that *Trichoderma pseudokoningii* inhibited (both at in vitro and in greenhouse experiments) the germination of *G. mosseae* and *Gigaspora rosea* and the colonization of soybean plants grown in open pots filled with an Argiudol type soil collected from the Province of Buenos Aires, Argentina. Little is known of the functional diversity of soil microbial communities (Pankhurst et al. 1996) resulting from limited access to suitable techniques. The functional diversity of microbial communities includes the range and relative expression of activities involved in decomposition, nutrient transformation, and plant growth promotion (see Siddiqui et al. 2008). It is also interesting the synergism between AMF and diazotrophic bacteria found in coffee soils of Brazil (Siqueira et al. 2002). Babana and Antoun (2006) also showed that by inoculating wheat seeds with phosphate rock-solubilizing microorganisms and *Glomus intraradices* under field conditions, the grain yields were comparable to those produced by using diammonium phosphate fertilizer (see Siddiqui et al. 2008). Another possibility for the lack of MR after inoculation with AMF in the presence of indigenous soil microflora in fertile Argentinean Pampas (Covacevich and Echeverría 2009) is the existence of different levels of compatibility between host plants and AMF (Roldán et al. 1992, Smith and Read 2008), hence, the effectiveness of AMF will depend on the plant species inoculated (Caravaca et al. 2003). Thus, the selection of efficient AMF for a particular crop and soil condition are key prerequisite in inoculation programs. In general the use of native mycorrhizas may be considered a preferential inoculation strategy (Caravaca et al. 2003), since native AMF are presumably physiologically and genetically adapted to the whole ecosystems.

The use of AMF in agriculture implies an increase in the colonization rates of the plants through selected inoculation or management practices of native AMF populations. However, field crop inoculation is not widely practiced due to operational difficulties associated with both the production of inoculum (where it should be considered the obligate nature of the symbiosis) and the placement of large-scale inoculant. In some situations, mycorrhizal management reducing soil disturbance or CR may be a better option than inoculation due to the costs of inoculum production (Cardoso and Kuyper 2006). Due to the fact that inoculation implies benefits for crops it can be carried out in the soil, during plant sowing (in annuals and pastures) or in seedlings (oleraceous). Interestingly, Benedetti et al. (2005) observed the influence of various covering plants on AMF during maize cultivation, pointing *Canavalia ensiformis* as a good species to be cultivated before maize. Still, a recent book by Miranda (2008) extensively examined AM advances on crop plants in Brazil.

Lastly, interactions between transgenic crops and AMF, a new urgent issue of soil ecology in agroecosystems in six countries (USA, Argentina, Brazil, Canada, India and China) that account for 86% of the genetic transformed crops species area over the world, including soybean, maize, cotton, canola (Wenke and Lianfeng 2008), need more research. Only one report (Galdino et al. 2009) on soybean suggested differences in AMC of two varieties of genetic modified soybean in relation with non modified plants from Brazil. The impact of soybean expansion on biodiversity, which is related to crops reviewed in this chapter, is also a topic of special interest (see Fearnside 2001). However, the ecological risks of crops were critically highlighted for potential adverse effects on agroecosystems, in particular, non-target effects on soil microorganisms as AMF. So that, interactions between AMF and crops at individual and community level are a new urgent soil ecological issue.

Soil Conditions in the Agroecosystems

Several plant species of great economic importance such as coffee, citrus, cassava, sweet potato and soybean, do not survive in unfertile soils without the AMF symbiosis (de Souza et al. 2008). AMF are diverse, both systematically and functionally, exhibiting ecological differentiation and specialization to both their biotic and abiotic environments (Fitter et al. 2004).

In relation to N, AM colonized plants absorb more available soil N and AMF are able to mineralize organic N, facilitating plant nutrition (Siqueira et al. 2002).

The efficient use of AM symbiosis may allow the attainment of acceptable yield levels with the minimum fertilizer dose, while also reducing costs and environmental pollution risk. This is a promising way for obtaining high crop yields with low fertilizer inputs in order to support a sustainable agriculture (Covacevich et al. 2007).

Agriculture can be considered as an anthropogenic disturbance which impact on AMF (Alexander et al. 1992). Agricultural practices such as tillage, long fallow periods, soil compaction, or growth of non-mycorrhizal crops may also be detrimental to AMF associations (see Brundrett 2002). Recent work by Miransari et al. (2007) showed that AMF inoculation of corn reduce the stressful effects of soil compaction. This is due to the unique properties of AMF (very thin and high volume of hyphae), which are much finer even than root hairs.

AMF contributes for soil C stock (Rillig et al. 2001) and aggregate stability, through physic and chemical actions, by means of the mycelium network formed and an arbuscular-mycorrhizal fungal soil protein (glomalin) secreted by them (see Rillig 2004). Additionally, organic compounds produced by decomposition of vegetal residues accumulated in the soil surface can increase the soil base saturation and crop production (Prado 2001). Glomalin responds to land-use change (Rillig et al. 2003) and positive correlation between glomalin-related soil protein (GRSP) and AMF biomass parameters (spore density and biovolume) can be found (Bedini et al. 2007) (Table 1). More recent previous reports on higher contents of glomalin in soils from organic than from conventional farms in Korea (Kil et al. 2009) enhance the use of GRSP as a useful biochemical parameter for the assessment of biological soil health in sustainable agriculture.

In maize and other species, the main contribution of AMF to host-plant nutrition concerns their ability to extract P from outside the P depletion zone around plant roots (reviewed by Boomsma and Vyn 2008). For example, recent work by Oliveira et al. (2009) highlighted that effective AM colonization of maize may influence the crop yield in the savannas of Brazil. They found different AM species of *Acaulospora* and *Glomus* only in the rhizosphere of P efficient maize genotypes cultivated in low P soil, using denaturing gradient gel electrophoresis (DGGE) technique. Boomsma and Vyn (2008) reviewed the benefits of AM symbiosis in maize production systems, highlighting that the management of native AMF seems achievable even in soils previously exposed to intensive crop production. In another recent review, Cardoso and Kuyper (2006) compiled the biological potential of tropical soils and enumerate previous works on AM symbiosis from an agronomist's viewpoint.

Fertilizer may select for less beneficial AM species, and mildly nutrient-stressed vegetal species can release more soluble carbohydrate in root exudates being better symbionts (Johnson 1993). According to Gosling et al. (2006) and Miranda and Miranda (2006) AMF can be used in agriculture to increase crop yields while minimizing the requirements on chemical fertilizers. However, AMF technology is still far from being routinely applied in agricultural practices due to problems in identifying and tracking fungal species in the field, the poor understanding of the basic biology of AMF and the inability to grow AMF in pure cultures.

AMF can also survive soil freezing (Addy et al. 1997), or wetting and drying cycles (Braunberger et al. 1996), their spores being more resistant to topsoil disturbance and storage than root fragments and much more resistant than hyphae (Daft et al. (1987). However, it was highlighted that mycorrhizal propagules can be severely influenced by damage to vegetation and soils resulting from human intervention, as well as intense fires, topsoil removal and flooding (Brundrett 2002). Moreover, AMF affect plant water relations in both water-limiting and well-watered conditions (reviewed by Boomsma and Vyn 2008). For further discussion of plant water relations in numerous host species colonized by AMF refer to Augé (2001) and articles cited within.

Challenges to Melhoration of Agricultural Systems

The understanding of the impacts of agriculture on belowground AM fungi is just beginning and an effective collaboration between mycorrhizobiologist and other specialist groups are required.

It is important to note that propagules of AMF species promoting crop growth can be increased managing agrosystems before cultivating crops, which will be more feasible to small farmers (Stürmer and Siqueira 2006). Miranda and Miranda (2006) pointed that each production system needs to be programmed including selected cultures and cultivars as well as an adequate rotation. Annual cultures (soybean, bean, maize) and cover plant (*Mucuna, Crotalaria*, *Canavalia ensiformis*, *Cajanus cajan*, sunflower, millet *Pennisetum glaucum*, *Ricinus communis*) as well as some fodder plants (*Stylosanthes, Andropogon*) are highly AM dependent (Table 2), increasing the populations of native AMF, and benefit subsequent cultures.

Selection of appropriate AMF, production of inoculum in quality and quantity, and the ecology of AM inoculation are critical issues for the application of AMF technology in agriculture production (Simon et al. 1992, Gianinazzi et al. 2002, Barea et al. 2005). However, more tools are required for the translation of the AMF diversity information into functional properties of the communities. Progress in this respect may only be expected from the ability to link functionally uniform AMF groups to taxonomic groups of AMF (Munkvold et al. 2004). Additionally, in Latin America the culture collections of AMF are maintained by researchers; in spite of the fact that more research is needed in relation to germplasm banks of AM for increase the production of agricultural plants (see Stürmer and Siqueira 2006). This is an urgent issue to fulfill requirements of ecosystem services created by a modern agriculture (Mercy et al. 2009).

Outlook and Conclusion

The interactions among crops, microbes and abiotic factors are still not well understood but there are great expectations regarding the fact that biofertilizers and mycorrhizas benefit sustainable crop systems. This chapter mentioned that previous reports revealed that AMF are a common and important microbial component in agricultural systems both in Argentina and Brazil, and should be included in future agricultural programs. In addition, efforts have been made in recent years to accrue benefits from AMF for agriculture and horticulture, with some difficulties due to the fact that AM interactions are complex. However, in Brazil, there are few reports on AM fungi for crop soils in spite of the fact that under tropical climate AM constitute an important biological resource and the high prices for fertilizer and technical equipment constrain their use. AM symbiosis can improve crop growth and final grain yield even in conventional production systems and with synthetic fertilizer application. Although Argentina have conducted several studies on the impact of agricultural practices on mycorrhizal symbiosis, even the use of this symbiosis has not been implemented by farmers and remains primarily in the scientific field (Figure 6a-c).

Table 2. Summary of actual evidence on AM in major agricultural plants and green manures

Source	Crops / green manure	RC†	Spores‡	Glomeraceae dominant	Total AM species number	MPN of AMF	MD
Schalamuk and Cabello (2010b)	Wheat	NI	IN	+	21		NI
Sturmer and Siqueira (2006)	Coffee	NI	NI	+	22	NI	NI
Miranda (2008)	Rice	NI	4-17 spores 50g soil-1	NI	NI	NI	L
	Maize	88	53-62 spores 50g soil-1	-	4 to 11	9-119	H
	Soybean	26-60	32-97 spores 50g soil-1	NI	NI	119	H
	Soybean/ Maize#	50/ 58	96/ 187 spores 50g soil-1	NI	NI	NI	NI
	Bean	76	67-97 spores 50g soil-1	NI	NI	NI	H
	Sunflower	51	20 spores 50g soil-1	NI	NI	NI	NI
	Cassava	NI	NI	NI	NI	NI	H
	Millet	49	NI	NI	NI	NI	NI
	Pigeonpea*	52	NI	NI	NI	NI	NI
	Mucuna* (*Mucuna* sp.)	49	NI	NI	NI	99	NI
	Crotalária	49	NI	NI	NI	NI	H
	Sorghum	92	1028 spores 50g soil-1	NI	NI	NI	H
	Wheat	NI	NI	NI	NI	36.8	M
Cardoso and Kuyper (2006)	Mungbean (Vigna radiata)	IN	Inoculated *Glomus* sp.	+	1	NI	AM
Tchabi et al. (2009)	Yam (*Dioscorea* spp.)	ND	2-17 spores 50g soil-1	+	14	NI	NI
Tchabi et al. (2008)	Yam, peanut, maize	ND	2-8 spores 50g soil-1	+	~15	NI	NI

† % maximal AM root colonization; *green manure; IN= Informed; NI = Not informed. #rotation; MPN of AMF = the most probable number of AMF propagules; MD = mycorrhizal dependency; H = high; M = medium; L = low (Adapted from Miranda 2008); AM = arbuscular mycorrhizal present.

Recent attention has been done to the change in composition of the AMF propagule banks in agricultural soils, soil compaction and glomalin content over the world. The studies of AMF in natural ecosystems are also important to be compared with those in different managed ecosystems (same biome) as reference sites.

In the present chapter it was also tried to highlight the challenges and the perspectives that come in terms of production systems in the agriculture, to maintain ecosystem function and a consistent food to a growing global population, effective management of soil microorganisms, especially AM fungi, which needs to be developed.

Finally, only when we know about AMF genotypes, AMC, and their tempo-spatial dynamics in crop fields can we attempt to make predictions about the comportment of the agroecosystem. The lack of congruence in studies on AMF (fertilization vs. AMC vs. sporulation, for example) imposes uncertainties about true relationships and can lead to the generation of unexpected decisions, notwithstanding our comprehension of AMF must change even more.

Acknowledgments

Dr Marcela Pagano is grateful to Council for the Development of Higher Education at Graduate Level, Brazil (CAPES), for Post-doctoral scholarship granted. Dr Fernanda Covacevich is Investigador Adjunto of CONICET, Argentina.

References

Addy, H; Miller, MH; Peterson, RL. Infectivity of the propagules associated with extraradical mycelia of two AM fungi following winter freezing. *New Phytologist,* 1997, 135,745–753.

Arihawa, J; Karasawa, T. Effect of previous crops on arbuscular mycorrhizal formation and growth of succeeding maize. *Soil Science and Plant Nutrition*, 2000, 46, 43-51.

Augé, RM. Water relations, drought and vesicular-arbuscular mycorrhizal symbiosis. *Mycorrhiza*, 2001, 11, 3–42.

Azevedo, LCB. *Arbuscular mycorrhizal fungi communities in soil and sugarcane roots*. Doctoral thesis, Escola Superior de Agricultura Luiz de Queiroz (ESALQ), 2008, (in Portuguese).

Barea, JM; Pozo, MJ; Azcón, R; Azcón-Aguilar, C. Microbial co-operation in the rhizosphere. *Journal of Experimental Botany*, 2005, 56, 1761–1778.

Beauregard, MS; Hamel, C; St.-Arnaud M. Arbuscular Mycorrhizal Fungi Communities in Major Intensive North American Grain. In: Siddiqui ZA, Akhtar MS, Futai K, editors. *Mycorrhizae: Sustainable Agriculture and Forestry*, Springer, New Delhi; 2008; 135-158.

Bedini, S; Avio, L; Argese, E; Giovannetti M. Effects of long-term land use on arbuscular mycorrhizal fungi and glomalin-related soil protein. *Agriculture, Ecosystems & Environment*, 2007, 120, 463–466.

Benedetti, T; Antoniolli Z I; Giracca EMN; Steffen RB. Diversity of arbuscular mycorrhizal fungi in maize following different kind of winter covering plant species. *Revista de Ciências Agroveterinárias*; 2005, 4, 1, 44–51 (in Portuguese).

Bethlenfalvay, GJ; Schüepp, H. Arbuscular mycorrhizas and agrosystem stability. In: *Impact of arbuscular mycorrhizas on sustainable agriculture and natural ecosystems.* Gianinazzi, S; Schüepp, H. editors. Birkhäuser Verlag, Basel, 1994, 117–131.

Bidartondo, MI; Redecker, D; Hijri, I; Wiemken, A; Bruns, TD; Domínguez, L; Sérsic, A; Leakek, JR; Readk, DJ. Epiparasitic plants specialized on arbuscular mycorrhizal fungi. *Nature*, 2002, 419, 389–392.

Boomsma, CR; Vyn, TJ. Maize drought tolerance: Potential improvements through arbuscular mycorrhizal symbiosis? *Field Crops Research*, 2008, 108, 14–31.

Borie, F; Rubio, R; Rouanet, JL; Morales, A; Borie, G; Rojas, C. Effects of tillage systems on soil characteristics, glomalin and mycorrhizal propagules in a Chilean Ultisol. *Soil & Tillage Research*, 2006, 88, 253–261.

Brundrett, MC. Mycorrhizas in natural ecosystems. In: *Advances in Ecological Research*, Begon M, Fitter AH, MacFadyen A, editors, 1991, 171-313.

Brundrett, MC. Arbuscular mycorrhizas in plant communities. In: Sivasithamparam K Dixon KW, Barrett RL, editors. *Microorganisms in Plant Conservation and Biodiversity*, Kluwer Academic Publishers; 2002; 151–193.

Bruns, TD; White, TJ; Taylor, JW. Fungal molecular systematics. *Annual Review of Ecology and Systematics*, 1991, 22,525–564.

Burleigh, S. 2000. Cloning arbuscule-related genes from mycorrhizas. *Plant and Soil*, 226, 287–292.

Calviño, P; Echeverría, HE; Redolatti M. Estratificación de fósforo en el suelo y diagnóstico de la fertilización fosfatada en trigo en siembra directa. In: Actas CD XVII Congreso Argentino de la Ciencia del Suelo. AACS. Mar del Plata, Buenos Aires, Argentina, 2000, (in Spanish).

Carneiro, RFV; Evangelista, AR; Araújo ASF. Vegetative growth and acquisition of nutrients for the alfalfa in response to mycorrhiza and levels of phosphorus. *Revista Brasileira de Ciências Agrárias*, 2009, 4, 3, 267–273 (in Portuguese).

Caravaca, F; Alguacil, MM; Azcón, R; Díaz, G; Roldán A. Comparing the effectiveness of mycorrhizal inoculation and amendment with sugar beet, rock phosphate and *Aspergillus niger* to enhance field performance of the leguminous shrub *Dorycnium pentaphyllum* L. *Applied Soil Ecology*, 2004, 25 169–180.

Cardoso, IM; Kuyper, TW. Mycorrhizas and tropical soil fertility. *Agriculture, Ecosystems & Environment*, 2006, 116, 72–84.

Carrenho, R; Silva, ES; Trufem, SFB, Bononi, VLR. Successive cultivation of maize and agricultural practices on root colonization, number of spores & species of arbuscular mycorrhizal fungi. *Brazilian Journal of Microbiology*, 2001, 32, 262–270.

Carrenho, R; Trufem, SFB; Bononi, VLR. Effects of using different host plants on the detected biodiversity of arbuscular mycorrhizal fungi from an agroecosystem. *Brazilian Journal of Botany*, 2002, 25,93–101.

Clapp, JP; Young, JPW; Merryweather, JW, Fitter, AH. Diversity of fungal symbionts in arbuscular mycorrhizas from a natural community. *New Phytologist*, 1995, 130, 259–265.

Clapp, JP; Rodriguez, A; Dodd, JC. Inter- and intra-isolate rRNA large subunit variation in *Glomus coronatum* spores. *New Phytologist*, 2001, 149, 539–554.

Corradi, N; Croll, D; Colard, A; Kuhn, G; Ehinger, M; Sanders, IR. Gene Copy Number Polymorphisms in an Arbuscular Mycorrhizal Fungal Population. *Applied and Environmental Microbiology*, 2007, 73, 366–369.

Covacevich, F; Echeverría, HE. 2000. Caracterización química y biológica de un Hapludol bajo diferentes manejos. *Ciencia del Suelo*, 2000, 18, 2, 144–148 (in Spanish).

Covacevich, F; Sainz Rozas, H; Barbieri, P; Echeverría, HE. Formas de colocación de fósforo sobre el crecimiento y la micorrización espontánea del cultivo de trigo. *Ciencia del Suelo*, 2005, 23, 1, 39–45 (in Spanish).

Covacevich, F; Marino, MA; Echeverría, HE. The phosphorus source determines the arbuscular mycorrhizal potential and the native mycorrhizal colonization of tall fescue and wheatgrass in a moderately acidic Argentinean soil. *European Journal of Soil Biology*, 2006, 42, 127–138.

Covacevich, F; Echeverría, HE; Aguirrezabal, LAN. Soil available phosphorus status determines indigenous mycorrhizal colonization into field and glasshouse-grown spring wheat in Argentina. *Applied Soil Ecology*, 2007, 35, 1–9.

Covacevich, F; Echeverría, HE. Receptivity of an Argentinean pampas soil to arbuscular mycorrhizal *Glomus* and *Acaulospora* strains. *World Journal of Agricultural Sciences*, 2008, 4, 6, 688–698.

Covacevich, F; Sainz Rozas, H; Barbieri, P; Echeverría, H. Crecimiento y micorrización arbuscular nativa de trigo en siembra directa bajo distintas formas de colocación de fósforo. *Ciencia del Suelo*, 2008, 26, 2, 169–175 (in Spanish).

Covacevich, F; Echeverría, HE. Mycorrhizal occurrence and responsiveness in tall fescue and wheatgrass are affected by the source of phosphorus fertilizer and fungal inoculation. *Journal of Plant Interactions*, 2009, 4, 2, 101–112.

Covacevich, F. Molecular tools for biodiversity and phylogenetic studies in mycorrhizas: The use of primers to detect arbuscular mycorrhizal fungi. In: Thangadurai E, Hijri M, Busso CA, editors. *Perspectives in Mycorrhizal Research*. Bioscience Publications; 2010; 186–202.

Crovo, V; Covacevich, F; Echeverría, HE; Studdert G. *Crecimiento radical y colonización micorrítica en maíz bajo labranza convencional y siembra directa.* In: VII Congreso Nacional de Maíz, Pergamino, Buenos Aires, 2001 (in Spanish).

Daft, MJ; Spencer, D; Thomas, GE. Infectctivity of vesicular-arbuscular mycorrhizal inocula after storage under various envirorонmental conditions. *Transactions of the British Mycological Society,* 1987, 88, 21–27.

Daniell, TJ; Husband, R; Fitter, AH; Young, JPW. Molecular diversity of arbuscular mycorrhizal fungi colonizing arable crops. *FEMS Microbiology Ecology*, 2001, 36, 203–209.

Douds, DD; Galvez, L; Franke-Snyder, M; Reider, C; Drinkwater, LE. Effect of compost addition and crop rotation point upon VAM fungi. *Agriculture, Ecosystems & Environment*, 1997, 65, 257–266.

De Souza, FA; Dalpé, Y; Declerck, S; de la Providencia, IE; Séjalon-Delmas, N. Life History Strategies in Gigasporaceae: Insight from Monoxenic Culture. In: Declerck S, Strullu DG, Fortin, JA, editors. *In Vitro Culture of Mycorrhizas*. Springer; 2005; 4, 73–91.

De Souza, FA; Silva, ICL; Berbara, RLL. Fungos micorrízicos arbusculares: muito mais diversos do que se imaginava. In: Moreira FMS, Siqueira JO, Brussaard L, editors. *Biodiversidade do solo em ecossistemas brasileiros*. Editora UFLA, Lavras; 2008 (in Portuguese).

Doran JW. Soil microbiological and biochemical changes associated with reduced tillage. *Soil Science Society of America Journal*, 1980, 44,765–771.

Echeverría, H; García, F. Guía para la fertilización fosfatada de trigo, maíz, girasol y soja. Boletín Técnico No. 149. EEA INTA Balcarce, 1998 (in Spanish).

Egerton-Waburton, LM; Allen, EB. Shifts in arbuscular mycorrhizal communities along an anthropogenic nitrogen deposition gradient. *Ecological Applications*, 2000, 10, 484–496.

Evans, DG; Miller, MH. The role of the external mycelial network in the effect of soil disturbance upon vesicular-arbuscular mycorrhizal colonization of maize. *New Phytologist*, 1990, 114:65–71.

Fearnside, PM. Soybean cultivation as a threat to the environment in Brazil. *Environmental Conservation,* 2001, 28,1, 23–38.

Franke-Snyder, M; Douds, DD; Galvez, L; Phillips, JG; Wagoner, P; Drinkwater, L; Morton, JB. Diversity of communities of arbuscular mycorrhizal (AM) fungi present in conventional versus low-input agricultural sites in eastern Pennsylvania, USA. *Applied Soil Ecology*, 2001, 16, 35–48.

Galdino, IASA; Lima, DS; Miyauchi, MYH. *Effect of transgenic soybean on the functional groups of microorganisms in the rhizosphere*. In: 6th International Conference on Mycorrhiza ICOM6, 2009, Belo Horizonte. Abstracts ICOM6. Viçosa: editors, 1, 94.

Gamper, H; Leuchtmann, A. Taxon-specific PCR primers to detect two inconspicuous arbuscular mycorrhizal fungi from temperate agricultural grassland. *Mycorrhiza*, 2007, 17,145–152.

Geue, H; Hock, B. Determination of *Acaulospora longula* and *Glomus* subgroup Aa in plant roots from grassland using new primers against the large subunit ribosomal DNA. *Mycological Research*, 2004, 108, 76–83.

Giovannetti, M. Spore germination and pre-symbiotic mycelial growth. In: Kapulnick Y, Douds DD, editors. *Arbuscular mycorrhizas: physiology and function*. Dordrecht, the Netherlands, Kluwer Academic Publishers; 2000; 47–68.

Giovannetti, M; Avio, LX. Biotechnology of Arbuscular Mycorrhizas. In: Khachatourians GG, Arora DK, editors. *Applied mycology and biotechnology*, *Agriculture and food production*. Amsterdam, the Netherlands, Elsevier; 2000; 275–310.

Goss, MJ; de Varennes, A. Soil disturbance reduces the efficacy of mycorrhizal associations for early soybean growth and N_2 fixation. *Soil Biology and Biochemistry*, 2002, 34, 1167–1173.

Gosling, P; Hodge, A; Goodlass, G; Bending, GD. Arbuscular mycorrhizal fungi and organic farming. *Agriculture, Ecosystems & Environment*, 2006, 113, 17–35.

Gryndler, M; Leština, J; Moravec, V; Přikryl, Z; Lipavský, J. Colonization of maize roots by VAM-fungi under conditions of long-term fertilization of varying intensity. *Agriculture, Ecosystems & Environment*, 1990, 29, 1-4, 183–186.

Helgason, T; Daniell, TJ; Husband, R; Fitter, AH; Young, JPW. Ploughing up the wood-wide web? *Nature*, 1998, 394, 431.

Hijri, M; Hosny, M; van Tuinen, D; Dulieu, H. Intraspecific ITS polymorphism in *Scutellospora castanea* (Glomales, Zygomycota) is structured within multinucleate spores. *Fungal Genetics and Biology,* 1999, 26, 141–151.

Hosny, M; Hijri, M; Passerieux, E; Dulieu, H. rDNA units are highly polymorphic in *Scutellospora castanea* (Glomales, Zygomycetes). *Gene,* 1999, 226, 61–71.

Hungria, M; Vargas, MAT; Campo, RJ; Chueire, LMO; Andrade, DS. The Brazilian Experience with the Soybean (*Glycine max*) and Common Bean (*Phaseolus vulgaris*) Symbioses. In: Pedrosa FO, Hungria M, Yates G, Newton WE, editors. *Nitrogen Fixation: From Molecules to Crop Productivity*. Current Plant Science and Biotechnology in Agriculture, Springer Netherlands; 2002; 515–518.

Jansa, J; Mozafar, A; Anken, T; Ruh, R; Sanders, IR; Frossard, E. Diversity and structure of AMF communities as affected by tillage in a temperate soil *Mycorrhiza*, 2002, 12, 225–234.

Johnson, NC; Copeland, PJ; Crookston RK; Pfleger FL. Mycorrhizae: Possible explanation for yield decline with continuous corn and soybean. *Agronomy Journal*, 1992, 84, 387–390.

Johnson, NC. Can fertilization of soil select less mutualistic mycorrhizae? *Ecological Applications*, 1993, 3, 749–757.

Johnson, NC; Wolf, J; Koch, GW. Interactions among mycorrhizae, atmospheric CO_2 and soil N impact plant community composition. *Ecology Letters*, 2003, 6, 532–540.

Karlen, DL; Varvel, GE; Bullock, DG; Cruse, RM. Crop Rotations for the 21st Century. *Advances in Agronomy,* 1994, 53, 1-45.

Kil, YJ; Lee, JE; Shin, HC. *Diversity of arbuscular mycorrhizal fungi and glomalin.* In: 6th International Conference on Mycorrhiza ICOM6, 2009, Belo Horizonte. Abstracts ICOM6, Viçosa editors, 1, 40.

Kjøller, R; Rosendahl, S. Molecular diversity of glomalean (arbuscular mycorrhizal) fungi determined as distinct *Glomus* specific DNA sequences from roots of field grown peas. *Mycological Research*, 2001,105, 9, 102–1032.

Koide, R; Li, M; Lewis, J; Irby, C. Role of mycorrhizal infection in the growth and reproduction of wild vs. cultivated plants. I. Wild vs. cultivated oats. *Oecologia,* 1988, 7, 7, 537–543.

Kuhn, G; Hijri, M; Sanders, IR. Evidence for the evolution of multiple genomes in arbuscular mycorrhizal fungi. *Nature,* 2001, 414,745–748.

Kurle, JE; Pfleger, FL. The effects of cultural practices and pesticides on VAM fungi. In: Pfleger FL, Linderman RG, editors. *Mycorrhizae and plant health*. American Phytopathological Society, St. Paul, Minn.; 1994; 101–132.

Lima, RLFA; Salcedo, IH; Fraga, VS. Propagules of arbuscular mycorrhizae in P deficient soils under different land uses, in semiarid NE Brazil. *Revista Brasileira de Ciência do Solo*, 2007, 31,257–268 (in Portuguese).

Liu, A; Hamel, C; Hamilton, RI; Ma, BL; Smith, DL. Acquisition of Cu, Zn, Mn and Fe by mycorrhizal maize (*Zea mays* L.) grown in soil at different P and micronutrient levels. *Mycorrhiza*, 2000, 9, 331–336.

Ma, WK; Siciliano, SD; Germida, JJ. A PCR-DGGE method for detecting arbuscular mycorrhizal fungi in cultivated soils. *Soil Biology & Biochemistry,* 2005, 37, 1589–1597.

Manns, HR; Maxwell, C; Emery, RJN. The effect of ground cover or initial organic carbon on soil fungi, aggregation, moisture and organic carbon in one season with oat (*Avena sativa*) plots. *Soil & Tillage Research*, 2007, 96, 83–94.

Martinez, A; Obertello, M; Pardo, A; Ocampo, JA; Godeas, A. Interactions between *Trichoderma pseudokoningii* strains and the arbuscular mycorrhizal fungi *Glomus mosseae* and *Gigaspora rosea*. *Mycorrhiza*, 2004,14, 2, 79–84.

Menéndez, AB; Scervino, JM; Godeas, AM. Arbuscular mycorrhizal populations associated with natural and cultivated vegetation on a site of Buenos Aires province, Argentina. *Biology and Fertility of Soils,* 2001, 33, 373–381.

Mercy, L; Gianinazzi-Pearson, V; Wipf, D. *IBG: a tool for promoting ecosystem services in plant production systems*. In: 6th International Conference on Mycorrhiza ICOM6, 2009, Belo Horizonte. Abstracts ICOM6. Viçosa editors, 1, 85.

Mergulhão, CESA; Silva, MV; Figueiredo, VBM; Burity, AH, Maia, LC. Characterisation and identification of arbuscular mycorrhizal fungi species by PCR/RFLP analysis of the rDNA internal transcribed spacer (ITS). *Annals of Microbiology*, 2008, 58, 341–344.

Merryweather, JW; Fitter, AH. The arbuscular mycorrhizal fungi of *Hyacinthoides* non-scripta II. Seasonal and spatial patterns of fungal populations. *New Phytologist,* 1998, 138, 131–142.

Miranda, JCC; Vilela, L; de Miranda, LN. Dynamics and contribution of arbuscular mycorrhiza in culture systems with crop rotation. *Pesquisa Agropecuária Brasileira*, 2005, 40, 1005–1014 (in Portuguese).

Miranda, JCC; Miranda, LN. Micorriza arbuscular e uso de adubos verdes em solos do Bioma Cerrado. In: Carvalho AM; Amabile RF, editors. Cerrado: adubação verde. Planaltina, DF, 2006, 69–123. (in Portuguese).

Miranda, JCC. Cerrado, micorriza arbuscular, ocorrência e manejo. Embrapa Cerrados, Planaltina, 2008, 169p. (in Portuguese).

Miranda, JCC. Porque a micorriza é importante para a produção agrícola, frutífera e florestal; 2010; Available at: www.agrosoft.org.br/agropag/20721.htm, (in Portuguese).

Miransari, M; Bahrami, HA; Rejali, F; Malakouti, MJ; Torabi, H. Using arbuscular mycorrhiza to reduce the stressful effects of soil compaction on corn (*Zea mays* L.) growth. *Soil Biology & Biochemistry*, 2007, 39, 8, 2014–2026.

Miransari, M; Bahrami, HA; Rejali, F; Malakouti. MJ. Effects of soil compaction and arbuscular mycorrhiza on corn (*Zea mays* L.) nutrient uptake. *Soil & Tillage Research*, 2009a, 103, 282–290.

Miransari, M; Bahrami, HA; Rejali, F; Malakouti, MJ. Effects of arbuscular mycorrhiza, soil sterilization, and soil compaction on wheat (*Triticum aestivum* L.) nutrients uptake. *Soil & Tillage Research*, 2009b, 104, 48–55.

Munkvold, L; Kjøller, R; Vestberg, M; Rosendahl, S; Jakobsen, I. High functional diversity within species of arbuscular mycorrhizal fungi. *New Phytologist,* 2004, 164, 357–364.

Muthukumar, T; Prakash S. Arbuscular mycorrhizal morphology in crops and associated weeds in tropical agro-ecosystems. *Mycoscience*, 2009, 50, 233–239.

Oehl, F; Sieverding, K; Ineichen, K; Mader, P; Boller, T; Wiemken, A. Impact of Land Use Intensity on the Species Diversity of Arbuscular Mycorrhizal Fungi in Agroecosystems of Central Europe. *Applied and Environmental Microbiology*, 2003, 69, 2816–2824.

Oehl, F; Sieverding, K; Ineichen, K; Ris, E; Boller, T; Wiemken, A. Community structure of arbuscular mycorrhizal fungi at different soil depths in extensively and intensively managed agroecosystems. *New Phytologist,* 2005, 165, 1, 273–283.

Oliveira, CA; Sa, NMH, Gomes, EA, Marriel, IE, Scotti,MR; Guimarães, CT; Schaffert, RE; Alves, VMC. Assessment of the mycorrhizal community in the rhizosphere of maize (*Zea mays* L.) genotypes contrasting for phosphorus efficiency in the acid savannas of Brazil using denaturing gradient gel electrophoresis (DGGE). *Applied Soil Ecology*, 2009, 41, 249–258.

Öpik, M; Moora, M; Liira, J; Zobel, M. Composition of root-colonizing arbuscular mycorrhizal fungal communities in different ecosystems around the globe. *Journal of Ecology*, 2006, 94, 778–790.

Pagano MC; Cabello MN; Bellote AF; Sa NM; Scotti MR. Intercropping system of tropical leguminous species and *Eucalyptus camaldulensis*, inoculated with rhizobia and/or mycorrhizal fungi in semiarid Brazil. *Agroforestry Systems*, 2008, 74, 231–242.

Pagano, MC; Scotti, MR; Cabello, MN. Effect of the inoculation and distribution of mycorrhizae in *Plathymenia reticulata* Benth under monoculture and mixed plantation in Brazil. *New Forests,* 2009, 38,197–214.

Pagano, MC; Cabello, MN; Scotti MR. Agroforestry in dry forest, Brazil: Mycorrhizal Fungi Potential. In: Kellymore LR, editor. *Handbook on Agroforestry: Management Practices and Environmental Impact.* Nova Science Publishers, New York; 2010; 367–388.

Pagano, MC. Soil Tillage in Agroforestry and Agroecosystems: Mycorrhizal Benefits. In: *Soil Tillage and Microbial Activities Research.* Signpost Publications, India. Miransari M, editor, in press.

Parniske, M. Plant–Fungal Associations: Cue for the branching connection. *Nature*, 2005, 435, 750–751.

Prado, RM. Base saturation and corn hybrids under no-tillage system. *Scientia Agricola*, 2001, 58, 2, 391–394 (in Portuguese).

Pringle, A; Moncalvo, JM; Vilgalys, R. High levels of variation in ribosomal DNA sequences within and among spores of a natural population of the arbuscular mycorrhizal fungus *Acaulospora colossica*. *Mycologia*, 2000, 92, 259–268.

Purin, S; Klauberg, O; Stürmer, SL. Mycorrhizae activity and diversity in conventional and organic apple orchards from Brazil. *Soil Biology & Biochemistry*, 2006, 38, 1831–1839.

Qin, R; Stamp, P; Richner, W. Impact of tillage on root systems of winter wheat. *Agronomy Journal*, 2004, 96, 1523–1530.

Quadros, VJ; Antoniolli, ZI; Denega, GL; Casali, CA; Pujol, S; Weber, MA. *Associações biológicas e atributos econômicos em sistemas de cultivo orgânico de batata, soja, feijão e milho.* Available from: www.ufsm.br/ppgcs/congressos/Fertbio2004/Mb112.pdf. (in Portuguese).

Rao, AV; Tarafdar, JC; Sharma, SK; Praveen-Kumar; Aggarwal, RK. Influence of cropping systems on soil biochemical properties in an arid rain-fed environment. *Journal of Arid Environments*, 1995, 31, 237–244.

Redecker, D; Morton, JB; Bruns, TD. Ancestral lineages of arbuscular mycorrhizal fungi (Glomales). *Molecular Phylogenetics and Evolution*, 2000, 14, 276–284.

Rillig, MC; Wright, SF; Nichols, KA; Schmidt, WF; Torn, MS. Large contribution of arbuscular mycorrhizal fungi to soil carbon pools in tropical forest soils. *Plant and Soil*, 2001, 233, 167–177.

Rillig, MC; Ramsey, PW; Morris, S; Paul, EA. Glomalin, an arbuscular-mycorrhizal fungal soil protein, responds to land-use change. *Plant and Soil,* 2003, 253, 293–299.

Rillig, MC. Arbuscular mycorrhizae, glomalin, and soil aggregation. *Canadian Journal of Soil Science,* 2004, 84, 355–363.

Saiki, RK; Gelfand, DH; Stoffel, S; Scharf, SJ; Higuchi, R; Horn, GT; Mullis, KB; Erich, HA. Primer-directed enzyme amplification of DNA with a thermostable DNA polymerase. *Science*, 1988, 239, 487–491.

Sanders, IR; Alt, M; Groppe, K; Boller, T; Wiemken, A. Identification of ribosomal DNA polymorphisms among and within spores of the Glomales: application to studies on the genetic diversity of arbuscular mycorrhizal fungal communities. *New Phytologist*, 1995, 130, 419–427.

Santos, IPA; Pinto, JC; Siqueira, JO; Morais, AR; Santos, CL. Influência do fósforo, micorriza e nitrogênio no conteúdo de minerais de *Brachiaria brizantha* e *Arachis pintoi* consorciados. *Revista Brasileira de Zootecnia*, 2002, 31, 2, 605–616 (in Portuguese).

Schwab, SM; Reeves, FB.The role of endomycorrhizae in revegetation practices in the semi-arid west. III. Vertical distribution of vesicular-arbuscular (VA) mycorrhiza inoculum potential. *American Journal of Botany*, 1981, 68, 1293–1297.

Schalamuk, S; Velázquez, S; Chidichimo, H; Cabello, M. Fungal spore diversity of arbuscular mycorrhizal fungi associated with spring wheat: effects of tillage. *Mycologia*, 2006, 98,1, 16–22.

Schalamuk, S; Cabello, MN. Effect of Tillage Systems on the Arbuscular Mycorrhizal Fungi (AMF) Propagule Bank in Soils. In: *Management of Fungal Plant Pathogens*, Arya A, Perelló AE, editors, CAB International, 2010a, 162–170.

Schalamuk, S; Cabello, MN. Arbuscular mycorrhizal fungal propagules from tillage and no-tillage systems: possible effects on Glomeromycota diversity. *Mycologia*, 2010b, 102, 2, 261–268.

Schüßler, A; Gehrig, H; Schwarzott, D; Walker, C. Analysis of partial Glomales SSU rRNA genes: implications for primer design and phylogeny. *Mycological Research*, 2001,105, 5–15.

Siddiqui, ZA; Akhtar, MS; Futai K. (editors) *Mycorrhizae: Sustainable Agriculture and Forestry*, Springer; 2008.

Siddiqui, ZA; Pichtel, J. Mycorrhizae: an overview. In: Mycorrhizae: Sustainable Agriculture and Forestry (Siddiqui ZA, Akhtar MS, Futai K, editors. Springer, New Delhi; 2008; 1–35.

Sieverding, E. Vesicular-arbuscular mycorrhiza management in tropical agrosystems. Eschborn: Deutsche Gesellschaft für Technische Zusammenarbeit (GTZ) GmnBH, Esch boran; 1991.

Silva, CF; Pereira, MG; Silva, EMR; Correia, MEF; Saggin-Júnior, OJ. Fungos micorrízicos arbusculares em áreas no entorno do parque estadual da serra do mar em Ubatuba (SP). *Revista Caatinga*, 2006, 19, 1, 1–10 (in Portuguese).

Simon, L; Lalonde, M; Bruns, TD. Specific amplification of 18S fungal ribosomal genes from vesicular-arbuscular endomycorrhizal fungi colonizing roots. *Applied and Environmental Microbiology*, 1992, 58, 291–295.

Simon, L; Levesque, RC; Lalonde, M. Identification of endomycorrhizal fungi colonising roots by fluorescent single-strand conformation polymorphism-polymerase chain reaction. *Applied and Environmental Microbiology*, 1993, 59, 4211– 4215.

Simon L, Lalonde M. DNA probes for the detection of arbuscular endomycorrhizal fungi. United States Patent 5434048. 1995. Available from: http://www.freepatentsonline.com /5434048.html

Siqueira, JO; Colozzi-filho, A; de Oliveira, E; Fernandes, AB; Florence, ML. Vesicular-Arbuscular Mycorrhizae in Coffee Seedlings in the Southern Region of Minas-Gerais State, Brazil. *Pesquisa Agropecuária Brasileira*, 1987, 22, 31–38 (in Portuguese).

Siqueira, JO; Colozzi-filho, A; Oliveira, E. Occurrence of Vesicular-Arbuscular Mycorrhizae in Agroecosystems and Natural Ecosystems of Minas-Gerais State. *Pesquisa Agropecuária Brasileira*, 1989, 24, 1499–1506 (in Portuguese).

Siqueira, JO; Lambais, MR; Stürmer, SL. Fungos micorrízicos arbusculares. Características, associação simbiótica e aplicação na agricultura. *Biotecnologia Ciência & Desenvolvimento*, 2002, 25, 12–21 (in Portuguese).

Siqueira, JO; Saggin-Júnior, OJ; Flores-Ayles, WW; Guimarães, PTG. Arbuscular mycorrhizal inoculation and superphosphate application influence plant development and yield of coffee in Brazil. *Mycorrhiza*, 1998, 7, 293–300.

Siqueira, JO; Saggin-Júnior, OJ. Dependency on arbuscular mycorrhizal fungi and responsiveness of some Brazilian native woody species. *Mycorrhiza*, 2001, 11, 245–255.

Smith, SE; Read DJ. *Mycorrhizal Symbiosis*. New York: Elsevier; 2008.

Smith, SE; Facelli E; Pope S; Smith FA. Plant performance in stressful environments: interpreting new and established knowledge of the roles of arbuscular mycorrhizas. *Plant Soil*, 2010, 326,3–20.

Solbrig, OT. The Dilemma of Biodiversity Conservation. Revista: Harvard Review of Latin America. Flora and Fauna. *Nature in Latin America*, 2005.

Stukenbrock, EH; Rosendahl, S. Clonal diversity and population genetic structure of arbuscular mycorrhizal fungi (*Glomus* spp.) studied by multilocus genotyping of single spores. *Molecular Ecology*, 2005, 14, 743–752.

Stürmer, SL; Siqueira, JO. Diversity of arbuscular mycorrhizal fungi in Brazilian ecosystems. In: Moreira, FMS, Siqueira, JO, Brussard, L, editors. *Soil biodiversity in Amazonian and other Brazilian ecosystems.* Wallingford: CABI-Publications, 2006, 206–236.

Tchabi, A; Coyne, D; Hountondji, F; Lawouin, L; Wiemken, A; Oehl, F. Arbuscular mycorrhizal fungal communities in sub-Saharan Savannas of Benin, West Africa, as affected by agricultural land use intensity and ecological zone. *Mycorrhiza*, 2008, 18,181–195.

Tchabi, A; Burger, S; Coyne, D; Hountondji, F; Lawouin, L; Wiemken, A; Oehl F. Promiscuous arbuscular mycorrhizal symbiosis of yam (*Dioscorea* spp.), a key staple crop in West Africa. *Mycorrhiza*, 2009, 19, 375–392.

Urquiaga, S; Jantalia, CP; Resende, AS; Alves, BJR; Boddey, RM. Contribuição da fixação biológica de nitrogênio na produtividade dos sistemas agrícolas na América Latina. In: *Processos biológicos no sistema solo-planta - Ferramentas para uma agricultura sustentável.* Aquino AM, Assis RL, editors, Embrapa, Brasília, 2005, 181–220 (in Portuguese).

Velázquez, MS; Cabello, MN. Mycobization as a Biotechnological Tool: A Challenge, In: Thangadurai D, Busso CA, Hijri M, editors. *Mycorrhizal Biotechnology*, 2010.

Walker, C; Trappe, JM. Names and epithets in the Glomales and Endogonales. *Mycological Research*, 1993, 97, 339–344.

Wenke, L; Lianfeng, D. Interactions between Bt transgenic crops and arbuscular mycorrhizal fungi: a new urgent issue of soil ecology in agroecosystems. *Acta Agriculturae Scandinavica Section B - Soil and Plant Science*, 2008, 58, 187–192.

Wright, SF; Green, VS; Cavigelli, MA. Glomalin in aggregate size classes from three different farming systems. *Soil & Tillage Research*, 2007, 94, 546–549.

In: Mycorrhizal Fungi
Editor: Susanne M. Fulton

ISBN 978-1-61122-659-1
© 2011 Nova Science Publishers, Inc.

Chapter 3

Plant Coexistence and Diversity Mediated Belowground: The Importance of Mycorrhizal Networks

Arsene Sanon[1], Fatou Ndoye[2], Ezekiel Baudoin[1], Yves Prin[3], Antoine Galiana[3] and Robin Duponnois[1,*]

[1]IRD. UMR 113 CIRAD/INRA/IRD/SUP-AGRO/UM2. Laboratoire des Symbioses Tropicales et Méditerranéennes. TA A-82/J, Campus International de Baillarguet. Montpellier Cedex 5, France

[2]Université Cheikh Anta Diop. Laboratoire Commun de Microbiologie IRD/ISRA/UCAD. Centre de Recherche de Bel Air. BP 1386. Dakar, Sénégal

[3]CIRAD. UMR 113 CIRAD/INRA/IRD/SUP-AGRO/UM2. Laboratoire des Symbioses Tropicales et Méditerranéennes. TA A-82/J, Campus International de Baillarguet. Montpellier Cedex 5, France

Abstract

The acceleration of species' lost is such that understanding relationships between biodiversity and ecosystem function is a crucial aspect of ecology. Plant community composition and distribution has been thought, for a long time, to be a matter of aboveground interactions such as plant – plant feedbacks, plant – herbivores and/or parasites relationships, disturbance creating new patches for plant colonization, … Due to relevant advances in our understanding of plant and soil ecology, particularly in the investigation of rhizosphere (that is the soil biologically active compartment where root – root and root – microbes communications occurred), several findings provide evidence that plant – soil microbes feedbacks, particularly those with mycorrhizal fungi, are of great importance in determining plant species competitive performance and, could thus

* Corresponding author. Present address: IRD. Laboratoire Commun de Microbiologie IRD / ISRA / UCAD. Centre de Recherche de Bel Air. BP 1386. Dakar. Sénégal. Tel: + 221 33 849 33 22; Fax : + 221 33 849 33 02; Email: Robin.Duponnois@ird.fr

contribute to regulate plant community structure. Indeed, mycorrhizal symbioses are involved in plant nutrient mobilization and acquisition strategies, in plant health promotion, in interactions between plant and other biotic and/or abiotic factors, ... As plants differ in their specificity with mycorrhizal symbionts, plant host – mycorrhizal symbiont feedback would result in plant species' differential fitness and abilities when colonizing soil patches. Moreover, mycorrhizal hyphal network is known to interconnect conspecific as well as interspecific plants, and nutrient exchange may occur *via* this route to support subordinates. Consequently, mycorrhizal associations may be considered as an important biological mediator for plant coexistence and diversity.

This chapter reviews some of the recent research work implemented on the involvement of soil microbial community in regulating plant community composition and distribution. We particularly focus on the implication of mycorrhizal symbioses in promoting multi-species assemblages in plant communities.

Keywords: vegetation diversity; plant coexistence; belowground interactions; soil microflora; mycorrhizae

INTRODUCTION

An area of continued study in ecology is the investigation of mechanisms that allow for many species of plants to coexist in contravention of a simple interpretation of the principle of competitive exclusion. Ecologists have formerly emphasized physical heterogeneity that changes competitive hierarchies (Tilman, 1980), but theory does suggest that other above- and belowground biotic factors can also interfere and act as resources. Among belowground organisms that have been proposed to mediate coexistence of plant are mycorrhizal fungi (van der Heijden et al., 1998 a,b; Hart et al., 2003; Sanon et al., 2010). The influence of mycorrhiza in controlling plant diversity may be vitally important in ecosystems, where communities can be subjected to dramatic disturbances, and may rely on high diversity to maintain stability (Grime, 1997).

Mycorrhizas are symbiotic associations between fungi and the roots of numerous plant species in which soil resources accessed by fungi are exchanged for photosynthetic carbon produced by plants (Smith & Read, 2008). Associations with mycorrhizal fungi not only influence the performance of individual plants but also alter plant community structure, plant productivity, and nutrient cycling (van der Heijden et al., 1998a; Reynolds et al., 2003; Cardoso & Kuyper, 2006; Smith & Read, 2008) and, evidence suggests that mycorrhizal effect on the makeup of plant communities may derive directly or indirectly through multitrophic interactions (Figure 1.; Janos, 1980; Eom et al., 2000; Bever, 2002b; Bidartondo et al., 2002; Duponnois et al., 2008; Gehring & Bennett, 2009).

While mycorrhizal associations, particularly those with AM fungi, have been thought to be largely non-specific, there is growing evidence of host-specific differences in plant response to fungi and in fungal response to plants. The extent of plant growth promotion by mycorrhizal fungi depends upon the specific plant and fungal combinations (Streitwolf-Engel et al., 1997; van der Heijden et al., 1998 a,b; Hart et al., 2003). Conversely, the growth response and development of mycosymbionts also depends on the associated plant host species (Eom et al., 2000). Thus, host-specific changes in the mycorrhizal fungal community could lead to positive or negative feedbacks on the plant (Bever et al., 1997; Bever, 1999; Bever, 2002 a,b). These two dynamics lead to very different predictions for the community in terms of competing plant species' ability to coexist.

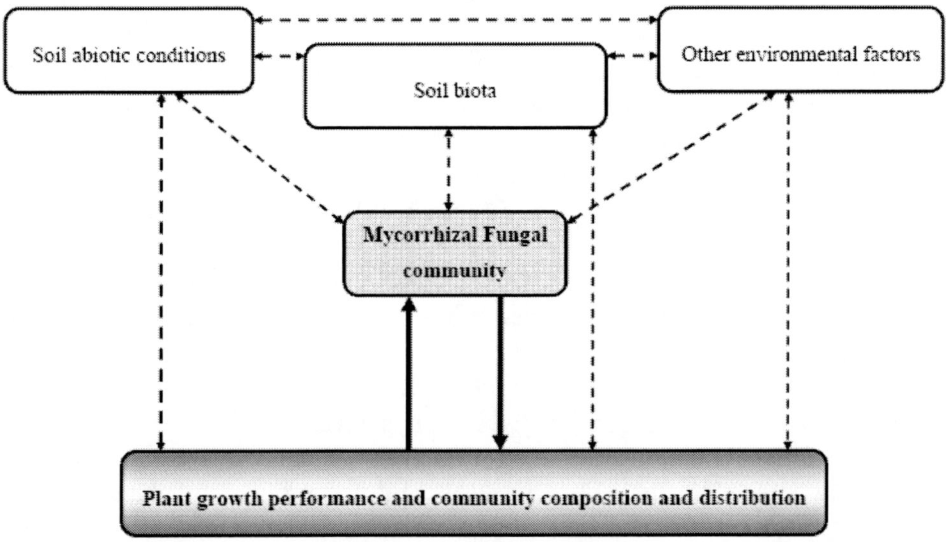

Figure 1. Conceptual diagram illustrating key interrelations susceptible to mediate plant competitive performance and community structure. Full arrows represent interrelations that are discussed in this chapter.

The effects of mycorrhizal fungi diversity and species composition are thought to arise through differential effects of different mycorrhizal fungi taxa, i.e. different mycorrhizal fungi-plant host combinations that may evolved, on the growth of individual plant species (van der Heijden et al., 1998 a,b; Eom, 2000; Vandenkoornhuyse et al., 2002, 2003). Such varying effects on the growth performance of plant species, in an environment characterized by high heterogeneity in resource availabilities (Tilman, 1980; Ryel et al., 1996), may constitute a strong driving force for the regulation of plant species coexistence and community composition and distribution.

Also, in plant communities, the influence of mycorrhizae on individual plant growth is affected by interactions between the individual plants. An established mycorrhizal mycelium is a potential resource and co-occurring plant species may differ in their ability to compete for this resource despite showing individual responsiveness to mycorrhizal association when grown separately in pots (Newman et al., 1992).

This chapter aims to review some of the recent advances in the understanding of the implication of mycorrhizal symbioses in the maintenance of multi-species assemblages in plant communities. We will specifically focus on the two most widespread types of mycorrhizal fungal associations: the Arbuscular Mycorrhizal (AM) fungi and the Ectomycorrhizal (EM) fungi (Smith & Read, 2008). We highlight whether mycorrhizal community diversity and abundance could evolve and whether the symbiotic associates community influences plant growth and competitive performance. Finally, we address the consequence of the 'multi-facets' plant host – fungal symbiont feedbacks on plant community composition and distribution.

1. Regulation Forces Selecting for Mycorrhizal Fungal Community Composition and Distribution

Whether high diversity of mycorrhizal fungi is maintained in a single community is steadily debated by specialists in mycorrhizae studies (Eom et al., 2000; Bever et al., 2001; Richard et al., 2005; Tedersoo et al., 2010). When trying to address this question, Bever et al. (2001) documented two hypotheses to support high diversity within their study site. The first hypothesis related to ecological equivalence of all fungal species; that means they are competitively equivalent within the cortical cells of plant roots and in this case, diversity is sustained by random drift processes. This hypothesis of functional redundancy suggested that the fungi are in some ways equivalent or redundant, given that a wide range of fungal species can colonize a particular plant species. Nevertheless, fungi have been clearly reported to differ in their effects on plant hosts (Bever et al., 1996; Hart et al., 2003). In addition, the distribution of fungal species within the study area strongly suggested that the local isolates of the species might differ in ecologically traits (Bever et al., 2001). These inconsistencies thus encourage the authors to formulate a second hypothesis postulating that fungal species are ecologically distinct and occupy different niches. That is, individual fungi would therefore be competitively superior in their specific niche, and the presence of multiple niches in a habitat results in the active maintenance of a speciose fungal community (Bever et al., 2001; Hart et al., 2003). The fungal diversity pattern discovered in the study of Bever et al. (2001), displaying different fungi predominating in various trap cultures under different environmental conditions, thus supports the second hypothesis described above. More specifically, this second hypothesis could be investigated by addressing the importance of individual environmental variables on which mycorrhizal symbionts differentiate, namely plant hosts, seasonality, and edaphic factors.

1.1. Host Specificity

Mycorrhizal fungi are thought to have evolved at the same time as land plants, allowing the latter to colonize and radiate on land (Brundrett, 2002). Identifying the fungi that form mycorrhizas with a plant species and the specificity of these relationships are important for understanding how plants interact with their environment. Knowledge of fungal identity and specificity can be used to determine how susceptible plants are to habitat alterations and to assess how the conservation of threatened plant species should be approached. Knowing the identity of the mycorrhizal fungi provides insight into nutritional pathways of the host plant while the specificity of the relationship often elucidates plant species distribution patterns, rarity and growth requirements (Bougoure et al., 2009). For instance, a 'successful' plant species, that is, one that is widely distributed amongst a variety of habitat types, will associate with one or more fungal species with collectively high tolerances to a broader range of particular environmental variables, such as nutrients availability, temperature, aridity etc. In contrast, 'susceptible' plant species, that is, those with limited distribution in specific habitat types, will form mycorrhizae with only one or few fungal species that have specific growth requirements and limited distribution (Selosse et al., 2002; Bidartondo et al., 2004).

AM fungi are documented to display low specificities of association with their plant host species, but these findings almost exclusively followed from experiments carried out separately with individual isolates of species, apart from competitive interactions. When fungi are examined as a community, evidence suggests that host plant species composition strongly influences AM fungal species composition, underlying that '*plant host preferences*' (Bever, 1996; Eom, 2000; Vandenkoornhuyse et al., 2002, 2003) or '*alteration in mycorrhizal fungal community*' (e.g. by phytotoxic mechanisms; Mummey et al., 2005; Mummey & Rillig, 2006) may exist. For instance, in an experiment in which AM fungi were trapped on different plant hosts, isolates of different fungal species sporulated differentially, with the relative dominance of fungal species being reversed, depending on the plant species with which they were associated (Bever et al., 1996). Similarly, Eom et al. (2000) conducted a greenhouse study in which different host plant species were grown in similar tallgrass prairie soil. After 4 months of growth, AM fungal species composition was significantly different beneath each host species and, the authors concluded that AM fungi show degree of host-specificity and are not randomly distributed in tallgrass prairies. Importantly, Mummey et al. (2005) and Mummey & Rillig (2006) documented that the presence of the exotic plant species, *Centaurea maculosa*, may alter AM fungi communities indirectly *via* phytotoxic mechanisms as *C. maculosa* is known to secrete catechin, which exhibits antimicrobial properties (Bais et al., 2002, 2003).

Host specificity in ectomycorrhizal communities may also be a rule (Bruns et al., 2002; Richard et al., 2005; Morris et al., 2009; Tedersoo et al., 2010). In their study, Richard et al. (2005) analyzed the ectomycorrhizal (EM) fungal diversity in a Mediterranean old-growth *Quercus ilex* forest stand from Corsica (France) where *Arbutus unedo* was the only other EM host. Regarding the distribution of the EM fungi community among the two host species, 46 RFLP (Restriction Fragment Length Polymorphism) were found on *A. unedo* and 112 on *Q. ilex* with some taxa (12.9% of the taxa recorded) infected both plant hosts (Richard et al., 2005). The authors also reported that senescent (170-yr-old) plants of *Q. ilex* harbored more rare species than seedlings (1-yr-old) and saplings (10-yr-old) of the same plant species. These findings thus strongly confirm the hypothesis that the hosts contribute to mycorhizal fungal diversity and also support the conclusion that established conspecific seedlings recruit EM symbionts in an opportunistic way among mycobionts colonizing the old surrounding trees (Jonsson et al., 1999; Kranabetter & Friensen, 2002).

Also, Diédhiou et al. (2010) investigated the diversity of EM fungi on root system of adult trees and seedlings of five plant species in a tropical rain forest of Guinea by sequencing the rDNA Internal Transcribed Spacer (ITS) region. Thirty-nine EM fungal taxa were recorded of which 19 were multi-hosts, 9 single-hosts and 11 singletons (Diédhiou et al., 2010). The same authors found that the multi-host fungi represented 92% of the total abundance. In addition, except for one plant species, namely *Anthonotha fragrans*, the adults of the host species displayed significant differentiation for their fungal communities whereas their seedlings harbored a similar fungal community. Their results thus indicated variation in EM fungal diversity with respect to the seedling *vs* adult development stage of trees. However, Diédhiou et al. (2010) documented low specificity of EM associates, owing to the dominance of multi-host fungi over single-host in their studied area and, similar observation has also been recorded in other area (Cullings et al., 2000). Importantly, difference could exist in the colonization pattern of multi-host fungal associates when taking into account the distribution of the fungal species on the host root system (Morris et al., 2009; Tedersoo et al.,

2010), and such differential colonization pattern might reveal potential specificity between a host and its fungal associates.

1.2. Seasonality

Evidence exists in the literature highlighting seasonality in the occurrence and abundance of mycorrhizal fungal propagules in the rhizosphere. Schultz et al. (1999) found that AM fungi differ in their seasonality, with some fungi sporulating in late spring and others doing so at the end of summer. Likely, Pringle & Bever (2002) found that AM fungi maintained different and contrasting seasonalities and they concluded that contrasting seasonal and spatial niches may facilitate the maintenance of a diverse community of AM fungi.

It has been stated that if mycorrhizal infection is high in a large number of individual plants in a community, then potentially, a significant part of the photosynthate produced by those individuals could be directed to the fungal symbionts, i.e. the carbon cost to the hosts would be high. Nevertheless, in a situation of high cost to the plant host, selection pressure should act to reduce infection, unless there are corresponding benefits, thus decreasing the level of infection within the community (Fitter, 1991). Also, in a survey of AM fungi in temperate deciduous woodland, Brundrett & Kendrick (1990) attributed variation in infection levels to periods of root growth and senescence. Sanders & Fitter (1992) documented variation in AM fungi infection between plant species and seasonal changes in infection under field conditions. Interspecific differences in mycorrhizal structures colonizing the root might indicate that aspects of the fungus/plant relationship are different in different plant species (Sanders & Fitter, 1992). More recently, DNA-based methods have been used to accurately explore the dynamic of AM fungi colonizing the root systems of different species (Husband et al., 2002; Vandenkoornhuyse et al., 2002; Rosendahl & Stukenbrock, 2004; Santos et al., 2006; Santos-González et al. (2007). All together, those studies did suggest seasonal variations in the composition of AM fungi communities across the growing season. For instance, in the experimental work implemented by Santos-González et al. (2007), though results indicated no significant seasonal changes in the species compositions of the AM fungi communities as a whole, the two studied plant species hosted significantly different AM fungi communities. In addition, *Prunella vulgaris* hosted a rich AM fungi community throughout the entire growing season and the presence of AM fungi in *Antennaria dioica* decreased dramatically in autumn, while an increased presence of *Ascomycetes* species was recorded.

Similar pattern in seasonal dynamics was also documented for the occurrence of ectomycorrhizal fungus assemblages. In this respect, the abundance and diversity of ectomycorrhizal fungi have been assessed based on the collection of basidiocarps during 12 months in southern Brazil. The results indicated that fruiting patterns of EM fungi differed with host and season, and host specificity was apparent in some (Giachini et al., 2004). Consistently, Walker et al. (2008) compared EM fungus distributions on root systems of out-planted oak seedlings at two sites in mixed southeastern Appalachian Mountain forests in North Carolina (USA) from samples taken in mid-July and early September. Seventy-four EM fungal ITS were recorded, most of which occurred only in the midsummer or early-fall samples. Abundance and relative frequency of EM fungal types, except *Cenococcum geophilum*, were significantly higher in the July samples, while *C. geophilum* was significantly more frequent and abundant in September (Walker et al., 2008). Globally, though their study is based on a single growing season, it did suggest that changes in abundance and frequency, detection of significant indicator species, and the apparent systematic affinities of shifting EM types support the potential for seasonal variability in EM associations in their system

1.3. Abiotic Factors

Mycorrhizal symbioses are also known to respond to a set of abiotic factors, which can strongly affect mycorrhizal fungi community abundance and composition. The mineral status of the soil, by influencing plant nutrients budget and availability, may affect the need for the plant to associate with mycorrhizal symbionts for nutrients mobilization and acquisition. Indeed, it is well established that mycorrhizal symbioses are more efficient and beneficial in nutrient-deficient soil and that plants may control their infection by mycorrhizal fungi when the cost mycosymbionts impose to the plant extends the benefit they provide (Fitter, 1991; Smith & Read, 2008). It is generally assumed that large amount of nutrients, especially phosphorus, negatively affects plants infection by mycosymbionts as well as the development of mycorrhizal propagules in soil (Cardoso & Kuyper, 2006; Smith & Read, 2008). Nevertheless, individual fungi could show opposite associations with certain soil parameters. In some instances, the distribution and abundance of *Acaulospora colossica* was negatively associated with soil phosphorus concentration, while the reverse was true for *Gigaspora gigantea* (Schultz, 1996). Several other parameters, including soil pH, drought, soil temperature, soil degradation status, soil compaction, tillage, pesticides, salinity, pollution, ... are also documented to have ramifications on mycorrhizal fungi development (Furlan & Fortin, 1973; Daniels & Trappe, 1986; Cairney & Meharg, 1999; Cardoso & Kuyper, 2006; Smith & Read, 2008).

1.4. Additional Factors

Other factors, such as difference in life history characters, including duration of dormancy, germination and sporulation requirements are documented to affect mycorrhizal fungi community (Bever et al., 2001). Additionally, these fungi may also differ in their palatability and resistance to grazing by belowground herbivores. Consistently, species-specific differences in palatability have been explicitly reported for ectomycorrhizal hyphae (Schultz, 1991), and similar patterns might likely occur among the arbuscular mycorrhizal fungi as well (Moore, 1985; Klironomos & Ursic, 1998; Bever et al., 2001).

2. Mycorrhizal Symbiosis Effects on Plant Growth and Competitive Performance

Though parasitic interactions have, sometimes, been documented regarding mycorrhizal partners (see Johnson et al., 1997; Purin & Rillig, 2008), mycorrhizal fungi have generally been found as essential components of sustainable soil – plant systems and consistently, the successful establishment of certain plant species has been conditioned by the presence of a suitable mycorrhizal symbiont (Janos, 1980; Richardson et al., 1994). In this respect, major benefits of mycorrhizae include an improved mobilization and acquisition of nutrients by their hosts and, this has been documented for a large range of macro- and micronutrients (Caris et al., 1998; Pare, 2000; He & Nara, 2007; Lambers et al., 2008; Smith & Read, 2008). Increased access to nutrients by plant hosts may result from a greater soil volume, more efficiently, explored by external hyphal network (Sylvia, 1988; Smith & Read, 2008) but also, from active release of nutrients from mineral particles and rock surfaces through weathering (Landeweert et al., 2001; Finlay, 2004, 2008) and from breaking down complex organic matter by the production of extracellular hydrolytic enzymes (Hodge et al., 2001; Cappellazzo et al., 2008; Lambers et al., 2008; Smith & Read, 2008). Additional benefits of mycorrhizae are related to an enhanced plant resistance to pathogens and others environmental stresses (i.e. metal and organic pollution, salinity, acidity, soil compaction, …), an improved water relations and, an improved formation and stability of soil aggregates that could significantly impact on plant growth and sustain productivity (St-Arnaud et al., 1997; Auge, 2001; Joner & Leyval, 2003; Cardoso & Kuyper, 2006; Rillig & Mummey, 2006; Smith & Read, 2008; Miransari, 2010).

Furthermore, interactions of mycorrhizal fungi with other soil microorganisms are diverse and, stimulatory, inhibitory or no effect on either partner could be reported (Founoune et al., 2002; André et al., 2005; Frey-Klett et al., 2005; Smith & Read, 2008). Mycorrhizal fungi and rhizosphere microorganisms like the plant growth promoting fungi (PGPF) or rhizobacteria (PGPR) have equally good potential in both plant growth promotion and plant disease control and, since both are beneficial microorganisms, their synergistic or additive effects could be more valuable than their individual effects (Hyakumachi & Kubota, 2004; Lesueur & Sarr, 2008; Siddiqui & Pichtel, 2008; Bonfante et al., 2009). Also, certain microorganisms have been found to stimulate the development of mycorrhizae in the root system of the plant hosts, which resultantly affects mycorrhizal efficiency on the growth of the hosts (Founoune et al., 2002; Duponnois & Plenchette, 2003).

Importantly, individual mycorrhizal fungal species are thought to have a multiplicity of effects on different hosts, promoting growth in one host while showing neutral effect or, in extreme case, inhibiting growth in other hosts (Streitwolf-Engel et al., 1997; van der Heijden et al., 1998a; Klironomos, 2003). Moreover, when fungi are examined as a community, they are expected to differentially impact on plant species with respect to plant host-fungal species preference and the generated feedback on plant host (Bever, 1996; Eom, 2000; Hart et al., 2003; Vandenkoornhuyse et al., 2002, 2003). Thereby, differential plant host-fungal species feedbacks may strongly alter and direct competitive relationships between plants (Hart et al., 2001). Importantly, van der Heijden et al. (1998b) and Stampe & Daehler (2003) demonstrated that the identity of AM fungi present could influence the outcome of plant competitive relationships as much as whether AM fungi are present or absent. In addition, evidence did suggest that mycorrhizae presence can significantly mediate resource acquisition by plants or alter resource distribution between competing plant species, thus affecting the growth performance of co-occurring plant species (Zabinski et al., 2002; Reynolds et al., 2003; Simard & Durall, 2004; Selosse et al., 2006; Sanon et al., 2010).

Table 1. Potential mechanisms mediated by mycorrhizae and though to affect coexistence and diversity within plant communities

Potential mechanism of action responsible for the promotion of plant coexistence	Type of mycorrhiza	Reference
Specificity/Preference in the plant host-fungal symbiont relation	AMF; EMF	van der Heijden *et al.* (1998a,b); Eom *et al.* (2000); Klironomos, 2000; Vandenkoornhuyse *et al.* (2002,2003); Richard *et al.* (2005).
Resource niches specialization	AMF; EMF	Bever *et al.*, 2001; Facelli & Facelli (2002); Hartnett & Wilson (2002); Reynolds *et al.* (2003).
Inter-canopy resource exchanges through mycorrhizal networks	AMF; EMF	Grime *et al.* (1987); Newman (1988); Bidartondo *et al.* (2002); Leake (2004); Simard & Durall (2004).
Alleviation of dominant canopy allelopathic/inhibitory effect	AMF	Sanon *et al.* (2006); Kisa *et al.* (2007); Sanon *et al.* (2009); Barto *et al.* (2010).
Multitrophic interactions with other organisms	AMF; EMF	Gange (2001); Gehring & Bennett (2009).
Successional dynamics within mycorrhizal communities	Non-mycorrhizal *vs* mycorrhizal plants; AMF *vs* EMF plants	Gorham *et al.* (1979); Janos (1980); Boemer *et al.* (1996); Smith & Read (2008).
Negative feedback within the mutualism	AMF	Bever. (2002b).

AMF, Arbuscular Mycorrhizal Fungi; EMF, Ectomycorrhizal Fungi.

3. Implications of Mycorrhizal Symbiosis Feedbacks for Plant Diversity and Community Distribution

Mycorrhizal symbioses and their feedbacks on plant growth performance might have strong ramifications on the relative abundance of plant species and their distribution within a community and these interrelations would be translated both in terms of promotion of multi-species assemblages and in terms of species replacement as succession proceeds (Table 1).

3.1. Maintenance of Multi-species Assemblages in Plant Community

With most plants possessing similar nutritional requirements, competition is a key factor in their interactions and only best competitors are expected to establish successfully thereby reducing diversity in communities. Nevertheless, multi-species assemblages are common within plant communities indicating that driving forces, counterbalancing species exclusion and encouraging diversity in plant communities, may exist. '*Agent-mediated coexistence*' is a non-interaction theory that has been proposed as a mechanism for maintaining multi-species assemblages in plant communities (Pacala & Crawley, 1992). Among the agents able to promote plant coexistence, mycorrhizal fungi have been proposed as efficient drivers for the maintenance of biodiversity within plant communities (Janos, 1980; Allen & Allen, 1990; Zobel & Moora, 1995; van der Heijden et al., 1998a,b; Renker et al., 2004; Simard & Durall, 2004; Kisa et al., 2007; Smith & Read, 2008). It is clearly demonstrated from various studies that mycorrhizae have large influences on plant community structure and are an important factor for the stability of plant species composition, as evidenced by the large magnitude of changes in plant communities reported in response to the presence *vs* absence of mycorrhizal symbionts or to the particular species of mycorrhizal fungi present (Grime et al., 1987; van der Heijden et al., 1998 a,b; Hart et al., 2003).

Spatial heterogeneity of mycorrhizal fungal infectivity may increase plant species diversity, allowing non-mycotrophic and mycotrophic species to coexist in patches of low and high mycorrhizal soil potential, respectively (Allen, 1991). Hartnett & Wilson (2002) further extend Allen's hypothesis and suggest that, in habitats with significant spatial heterogeneity in mycorrhizal soil infectivity, there will be a positive relationship between interspecific variability in host plant mycotrophy and plant species diversity. A related hypothesis is that spatial heterogeneity in mycorrhizal soil infectivity, coupled with variation in host species mycorrhizal dependency for colonization and establishment (e.g., regeneration niche differences) may enhance species diversity in plant communities (Hartnett et al., 1994).

Importantly, plant host specificity, in combination with mycorrhizal fungal diversity, is tightly involved in the maintenance of diversity in plant communities. Indeed, because different species of mycorrhizal fungi have different effects on the growth of particular plant species, variation in mycorrhizal fungus species composition could cause changes in the strength of plant-plant interactions. For instance, in settings by van der Heijden et al. (1998a), increasing fungal diversity resulted in greater plant species diversity and higher productivity, suggesting that changes in belowground diversity of mycorrhizal symbionts can drive changes in aboveground diversity and productivity. The mechanism behind these effects is likely to be differential effects of specific plant-fungus combinations on the growth of different plant species. With respect to this, some authors argued that higher AM fungi diversity could lead to higher plant coexistence simply by increasing the probability of individual plant species to associate with a compatible and effective AM fungi partner (Hart et al., 2003). Therefore, if the addition of new fungal species leads to increases in the survival and vigor of more plant species that are responsive to mycorrhizal colonization, there may be a positive feedback of the mycorrhizal fungi, leading to more efficient resource utilization and increases in overall productivity (Finlay, 2004, 2008). Nevertheless, these effects may be context dependent as Vogelsang et al. (2006) suggested that plant diversity and productivity are more responsive to AM fungi identity than to AM fungi diversity *per se*, and that AM fungal identity and P environment can interact in complex ways to alter community-level properties. Whatever, increased AM fungi species richness is argued to be beneficial both in terms of host compatibility (Sanders & Fitter, 1992; Bever et al., 1996; van der Heijden et al., 1998b; Eom et al., 2000; Klironomos, 2000; Klironomos et al., 2000) and in terms of multifunctionality of AM fungi (Newsham et al., 1995; Klironomos, 2000). However, as previously underlined by van der Heijden & Scheublin (2007), defining functionally distinct AM fungal groups is essential to fully understand the interactions between plant and AM fungal communities in agricultural and natural ecosystems.

Also, if mycorrhizal fungi are capable of direct access to pools of soil nutrients (e.g. N and P) not necessarily or really inaccessible to non-mycorrhizal plants (Lambers et al., 2008; Smith & Read, 2008), two hypotheses could emerged: the first assumes that mycorrhized plants could colonize certain soil patches that could be unsuitable for non-mycorrhized plants and secondly, it may be possible to predict for mycorrhized plants that different fungal species might facilitate differential access of plant species to these pools. The evolvement of these two hypotheses might thus promote plant species coexistence through nutritional niche partitioning (Reynolds et al., 2003). Later, Reynolds et al. (2006) reported that nutrient partitioning might be a less advantageous strategy than an ability to take advantage of multiple forms of nitrogen or phosphorus, especially for sessile organisms in a temporally and spatially heterogeneous world. Thereby, partitioning along any number of other niche dimensions (Tilman & Pacala, 1993) or competitive equivalence (Hubbell, 2001; Silvertown, 2004; Tilman, 2004) must account for the coexistence of largely co-dominant plant species, reflecting the 'fundamental' but not the 'realized' niches of these species (Reynolds et al., 2006). Additionally, plant mycorrhizal dependence and their position in the local dominance hierarchy (Urcelay & Diaz, 2003) in combination with AM fungal presence and abundance in soil may strongly influence plant community dynamics. In this context, it has been suggested that if an otherwise less competitive plant species is infected by more AM fungi than is a highly competitive plant species, then AM fungi should promote coexistence by increasing the ability of less competitive species to access nutrients (Zobel & Moora, 1995; Moora & Zobel, 1996). Consistent with this, Facelli & Facelli (2002) documented that mycorrhizal symbiosis has the potential to strongly influence plant community structure by favoring coexistence of mycorrhizal plants when soil nutrient distribution is heterogeneous because it promotes pre-emption of limiting resources. Contrastingly, one may note that if a highly competitive plant species is also more infected by mycorrhizal fungi, then mycosymbionts could simply reinforce competitive dominance by that species (West, 1996). Accordingly, Hartnett & Wilson (1999) have shown that suppression of mycorrhizal fungi, through a mycorrhizal fungal-specific fungicide, resulted in an increase in floristic diversity in a tallgrass prairie, likely because the dominant C4 grasses in that system are more strongly responsive to mycorrhizal colonization than the other species present. Therein, the competitive dominance of a mycorrhized plant may be related to his mycorrhizal dependence and his position in the local dominance hierarchy (Urcelay & Diaz, 2003).

Different plant species can be compatible with the same species of mycorrhizal fungi and be connected to one another by a common mycelium and to a large extent, plants within communities can be interconnected and form a common mycorrhizal network based on their shared mycorrhizal associates (Francis & Read, 1984; Simard et al., 1997; Selosse et al., 2006; Richard et al., 2009; Diédhiou et al., 2010). Common mycorrhizal networks either could originate from fungal genets colonizing neighboring roots during their growth, but also from hyphal fusions uniting previously separated mycelia (Selosse et al., 2006). Transfer of carbon, nitrogen, and phosphorus (Finlay & Read, 1986; Newman & Eason, 1993; Simard & Durall, 2004; Selosse et al., 2006) are well documented between interconnected plants and, exchanges may occur either with conspecific and interspecific plants. One important ecological consequence of nutrient transfer through common mycorrhizal networks is that adult plants could favor early establishment of conspecific or interspecific seedlings (Newman, 1988; Simard & Durall, 2004) but also, subordinate species could be maintained in plant community due to nutrients uptake from mycorrhizal mycelia guilds (Hoeksema & Kummel, 2003; Smith & Read, 2008). Over 400 non-photosynthetic species from 10 families of vascular plants obtain their carbon from fungi and are thus defined as myco-heterotrophs (Leake, 1994). Many of these plants are epiparasitic on green plants from which they obtain carbon by likely '*cheating*' shared mycorrhizal fungi (Selosse et al., 2002; Leake, 2004). Most epiparasitic plants examined have been shown to depend on ectomycorrhizal fungi networks for carbon transfer and exhibit exceptional specificity for these fungi. Recently, Bidartondo et al. (2002) show that non-photosynthetic plants associate with AM fungi and can display the characteristic specificity of epiparasites.

Also, Grime et al. (1987) hypothesized that their detected increase in plant species diversity in turfgrass microcosms in response to the presence of mycorrhizae was due to extensive mycelial networks facilitating interplant resource transfer *via* hyphal connections. They postulated that, allowed small, suppressed plants to obtain carbohydrates from the larger, dominant species *via* shared mycorrhizal hyphae, ultimately increased the equitability of species abundances. These same authors then suggested that the "export of assimilate from 'source' (canopy dominants) to 'sink' (understory components) through a common mycelial network may be an important mechanism for the maintenance of multi-species assemblages in infertile soils". Nevertheless, Pfeffer et al. (2004) found no evidence to support the movement of carbon between interconnected roots of AM plants.

Contradictorily, it is crucial to consider that mycorrhizal networks may, in certain situations, lead to lower diversity if one species in the network is the dominant sink for nutrients (Connell & Lowman, 1989; Allen & Allen, 1990). This case is well illustrated by the ability of some nonnative invasive plant species to divert to their advantage carbon and nutrients (P) transfers through the common mycelial network, thus severely reducing the growth performance of native plants (Zabinski et al., 2002; Carey et al., 2004). In the same line of ideas, some findings did document that the common mycorrhizal networks may maintain tropical monodominance. Tropical rainforest is known to harbor a hotspot of tree diversity; however, tree diversity is not uniformly diverse, and the existence of tropical monodominance, i.e. where a single tree species dominates the canopy, is one example of the extreme variation found in rainforests (Torti et al., 2001; Leigh et al., 2004). Indeed, McGuire (2007) reported a positive distance-dependant distribution and survival with respect to conspecific adults, suggesting that the negative distance-dependant mechanisms at the seedling stage thought to maintain tropical rain forest diversity (Janzen, 1970; Condit et al., 1994; Harms et al., 2000) are reversed for ectomycorrhized seedlings, which experience positive feedbacks from the EM network. The incorporation of seedling roots to the common EM network served as an effective mechanism for higher seedling survival in the monodominant forest, potentially providing seedlings with photosynthate from overstory individuals of the same species (McGuire, 2007).

The results of these various experiments indicate the potential that the movement of plant resources through hyphal interconnections may be an important mechanism influencing plant species interactions and community structure in grasslands and other plant communities. However, results are equivocal and much further studies are needed to determine whether this mechanism plays a significant role in patterns of species abundances and diversity in natural communities. In addition, the methodological limitations and difficulties in measuring the patterns and consequences of this phenomenon in the field will be the greatest challenge (Hartnett & Wilson, 2002).

Mycorrhizal associations have recently been reported to counterbalance the allelopathic effects of exotic fast growing trees against native herbaceous plant species community adjacent to stands of exotic plants. In experiments carried out in greenhouse in Sahelian conditions, Sanon et al. (2006) and Kisa et al. (2007) observed that previous inoculation of exotic trees, *Gmelina arborea* or *Eucalyptus camaldulensis*, with the AM fungus *Glomus intraradices* significantly increased the growth of native herb species in the mesocosms where the trees were cultivated; thus favoring coexistence of exotic trees and native understory herbaceous plants. In the same lines of ideas, from experiments carried out in greenhouse conditions, we observed that the growth of native *Acacia* species was severely reduced in the soil invaded by *Amaranthus viridis*, an annual weed native from Central America (Sanon et al., 2009). Interestingly, the inoculation of *G. intraradices* was highly beneficial to the growth and nodulation of *Acacia* species. From both studies (inoculation of exotic fast growing trees to alleviate their allelopathic effect against endogenous communities and inoculation of native *Acacia* species to favor their reestablishment in soils displaying severe alteration of chemical and microbial characteristics due to invasion of *A. viridis*), the authors postulated that the beneficial effect of AM fungus inoculation may result from either the well-developed mycelial network owing equalization of distribution of soil resources among competitively dominant and subdominant plant species (Grime et al., 1987; Wirsel, 2004) or allelochemical-mediating effects from AM fungi which altogether with their mycorrhizosphere microbial communities are known to inactivate or catabolize toxic compounds (Pellissier & Souto, 1999; Blum et al., 2000). Similar results regarding restoration of native plant diversity by utilizing AM fungi have also been reported in southern California by Vogelsang et al. (2004). In these ecosystems, most native weed plant species were dependent on AM symbioses for optimal growth. Conversely, growth of many of the pernicious exotic weedy species was not improved, and could be reduced by these fungi. Then, inoculation of AM fungi enabled speeded establishment of vigorous locally adapted vegetation in these areas, which will ultimately reduced exposure of natural communities to the forces of erosion.

In combination to their direct effects on their host plants, mycorrhizal fungi may influence plant communities indirectly through their effects on interactions between plants and their herbivores, pathogens, pollinators, and other microbial patners (Finlay &Soderstrom, 1989; Wurst et al., 2004 ; Cahill et al., 2008 ; Gehring & Bennett, 2009). In turn, these interactions can indirectly and, occasionally directly influence mycorrhizal fungal communities and their functions (Eom et al., 2001; Gehring & Bennett, 2009). Accumulating evidence suggests that indirect effects of mycorrhizae may be of great relevance in plant communities' composition and dynamics and should not be ignored (Hartnett & Wilson, 2002). Such complex, multitrophic interactions still remain not enough documented and may strongly vary in time and space. In addition to plant P and N content, other plant compounds, such as secondary metabolites (Gange & West, 1994; Koide, 2000), are affected by AM fungi root colonization. Changes in foliar chemistry may influence plant herbivore interactions and, herbivore performance has been reported to be positively affected (Gange & West, 1994; Gange et al., 1999) or negatively (Gange & West, 1994; Gange, 2001) by AM fungus colonization, depending on plant, herbivore and fungal species present. Ecosystem multitrophic interactions may result ultimately in subsequent effects on plant fitness, community composition and distribution. These ecological data underscore the importance of above-and belowground linkages and indicate that alterations in mycorrhizal and rhizosphere processes can have large indirect effects on plant communities through their effects on plant responses to above- and belowground consumers.

3.2. Species Replacement in Plant Community

Species replacement could be considered as a mechanism to ensure plant diversity mainly in the situation where some individuals of a dominant species are progressively replaced by individuals of a single one or several other species. We explicitly avoid extreme situations such as those in which for instance, a species B completely displaces a species A as in this case we assist to species replacement during succession rather than long-term coexistence (Van der Putten et al., 1993).

Plant species ability to specifically associate with particular symbionts could be of crucial importance in mediating species replacement during succession. Mycorrhizal associations are relatively ubiquitous and act to increase access to soil nutrients. The availability of P and other nutrients often decrease in late succession (Walker & Syers, 1976; Lambers et al., 2008). In this concern, Reeves et al., (1979) and Janos (1980) envisioned a shift from nonmycorrhizal to obligately mycorrhizal plants from early to late secondary succession, with low levels of facultatively mycorrhizal associations throughout. Contrarily to these previous observations, more recent studies rather suggest that very mycotrophic plants species would be pioneers in early successions in certain environmental conditions, acting as '*nurse plants*' as termed by Carrillo-García et al. (1999) to promote '*fertility islands*' (Garner & Steinberger, 1989) or '*resource islands*' (Reynolds et al., 1990; Schlesinger et al., 1996) where facilitation and replacement among plants species may be highly fostered (Callaway, 1995, 1997). Again, Smith & Read (2008) also reported in sand dune and many other successional communities a shift later in succession from herbaceous plant species involved in obligate symbioses with AM fungi to woody species involved in obligate symbioses with ecto- or ericoid mycorrhizae, coincident with a shift in predominance of inorganic *vs* organic N (Read, 1993). Consistently, some authors postulated that higher host specificity of ecto- compared to endomycorrhizae leads to dominance by ectomycorrhizal species (Connell & Lowman, 1989). Nevertheless, positive feedback, operating alone, would be expected to lead to monocultures, or even, in the absence of soil building or temporal shifts in forms of nutrients (e.g., inorganic to organic N), arrested successions (Reynolds et al., 2003).

Resource partitioning mediated by soil microbes might play a role in species replacements if the forms of available nutrients (N or P) change through succession. Gorham et al. (1979) implicated fungal symbionts in such sequential partitioning. Indeed, the example of a shift from arbuscular to ecto- or ericoid mycorrhizal plant species with change in inorganic to organic forms of N, previously reported in our discussion, well illustrated this process. Others shifts in forms of nutrients over succession (e.g., nitrate to ammonium, or protein to chitin) would also provide opportunity for additional species replacements on the basis of soil resource specialization (Reynolds et al., 2003). Again, Reynolds et al. (2003) also document that as the diversity of microbes and nutrient inputs increase over succession, the opportunity for microbially mediated differentiation in resource use would increase, promoting increase plant community diversity over succession. Supporting this idea, ectomycorrhizal infectiveness and diversity has been found to increase over a successional gradient (Boerner et al., 1996). In addition, mycorrhizal fungi associations typically exhibit higher levels of specificity as successions proceeds and, some authors envisioned the same pattern for the majority of microorganisms involved in mediating resources differentiation, thus predicting that greater opportunities for species coexistence (i.e., greater specialization of resource niches) may arise as succession proceeds (Reynolds et al., 2003, Smith & Read, 2008).

Negative feedback dynamics have been reported on plant growth through changes in the composition of their mutualistic fungal symbionts, AM fungi, deriving from plant host-mycorrhizal symbiont specificity (Bever, 2002b). Negative feedback results from asymmetries in the delivery of benefit between plant and AM fungal species in which the AM fungus that grows best with the plant *Plantago lanceolata* is a poor growth promoter for *Plantago*. Growth of *Plantago* is, instead, best promoted by the AM fungal species that accumulate with a second plant species, *Panicum sphaerocarpon*, which is not inhibited in its soil of culture. The resulting community dynamic leads to a decline in mutualistic benefit received by *Plantago* and that effect contributes to the coexistence of these two competing plant species (Bever, 2002b).

CONCLUSION

Overall, evidence suggests that the presence or absence of mycorrhizal fungi, the growth responses (nutrient uptake improvement but certainly mycorrhizal mediation of other biotic interactions such as plant-herbivore or plant-pathogen interactions also) or mycorrhizal dependency of host plant species, differential plant species responses to particular species of mycorrhizal fungi may be potentially important factors shaping the performance and relative abundance of plant species within local communities.

Nevertheless, complementary experimental works remain needed for a better understanding of whether mycorrhizal community density may impact on plant community composition and dynamic. For instance, at a high concentration of mycorrhizal inoculum, infection by the symbiotic fungi is thought to become detrimental rather than beneficial because heavily infected plants might experience a large carbon removal that outweighs any benefit (Gange & Ayres, 1999; Hart et al., 2003). Experimental works must then be undertaken to fully address the outcome of plant – plant relationships in a context of high mycorrhizal soil infectivity.

REFERENCES

Allen, E. B. & Allen, M. F. (1990). The mediation of competition by mycorrhizae in successional and patchy environments. In : J. B. Grace, & G. D. Tilman (Eds), *Perspectives on plant competition* (pp. 367–389). New York, USA : Academic Press.

Allen, M. F. (1991). *The ecology of mycorrhizae*. Cambridge, UK : Cambridge University Press.

André, S., Galiana, A., Le Roux, C., Prin, Y., Neyra, M. & Duponnois R. (2005). Ectomycorrhizal symbiosis enhanced the efficiency of two *Bradyrhizobium* inoculated on *Acacia holosericea* plant growth. *Mycorrhiza, 15*, 357–364.

Auge, R.M. (2001). Water relations, drought and vesicular-arbuscular mycorrhizal symbiosis. *Mycorrhiza, 11*, 3-42.

Bais, H. P., Walker, T. S., Stermitz, F. R., Hufbauer, R. A. & Vivanco, J. M. (2002). Enantiomeric-dependent phytotoxic and antimicrobial activity of (±) catechin. A rhizosecreted racemic mixture from spotted knapweed. *Plant Physiology, 128*, 1173–1179.

Bais, H. P., Walker, T. W. S., Kennan, A. J., Stermitz, F. R. & Vivanco, J. M. (2003). Structure-dependent phytotoxicity of catechins and other flavonoids: Flavonoid conversions by cell-free protein extracts of *Centaurea maculosa* (Spotted Knapweed) roots. *Journal of Agricultural & Food Chemistry, 51*, 897–901.

Barto, K., Friese, C. & Cipollini, D. (2010). Arbuscular mycorrhizal fungi protect a native plant from allelopathic effects of an invader. *Journal of Chemical Ecology, 36*, 351-360.

Bever, J. D. (1999). Dynamics within mutualism and the maintenance of diversity: inference from a model of interguild frequency dependence. *Ecology Letters, 2*, 52–61.

Bever, J. D. (2002a) Host-specificity of AM fungal population growth rates can generate feedback on plant growth. *Plant & Soil , 244*, 281–290.

Bever, J. D. (2002b). Negative feedback within a mutualism: host-specific growth of mycorrhizal fungi reduces plant benefit. *Proceedings of the Royal Society of London B, 269,* 2595-2601.

Bever, J. D., Morton, J. B., Antonovics, J. & Schultz, P. A. (1996). Host-dependent sporulation and species diversity of arbuscular mycorrhizal fungi in a mown grassland. *Journal of Ecology, 84,* 71-82.

Bever, J. D., Westover, K. M. & Antonovics, J. (1997). Incorporating the soil community into plant population dynamics: the utility of the feedback approach. *Journal of Ecology, 85*, 561–573.

Bever, J. D., Schultz, P., Pringle, A. & Morton, J. (2001). Arbuscular mycorrhizal fungi: more diverse than meets the eye and the ecological tale of why. *BioScience, 51,* 923-931.

Bidartondo, M. I., Redecker, D., Hijri, I., Wiemken, A., Bruns, T. D., Dominguez, L., Sérsic, A., Leake, J. R. & Read, D. J. (2002). Epiparasitic plants specialized on arbuscular mycorrhizal fungi. *Nature, 419,* 339-392.

Bidartondo, M. I., Burghardt, B., Gebauer, G., Bruns, T. D. & Read, D. J. (2004). Changing partners in the dark: isotopic and molecular evidence of ectomycorrhizal liaisons between forest orchids and trees. *Proceedings of the Royal Society of London, 271,* 1799–1806.

Blum, U., Statman, K. L., Flint, L. J. & Shaefer, S. R. (2000). Induction and/or selection of phenolic acid-utilizing bulk-soil and rhizospheric bacteria and their influence on phenolic acid phytotoxicity. *Jounal of Chemical Ecology, 26,* 2059-2078.

Boerner, R. E. J., DeMars B. G. & Leicht, P. N. (1996). Spatial patterns of mycorrhizal infectiveness of soils along a successional chronosequence. *Mycorrhiza, 6,* 79-90.

Bonfante, P. & Anca, I-A. (2009). Plants, Mycorhizal Fungi, and Bacteria : A Network of Interactions. *Annual Review of Microbiology. 63*, 363-383.

Bougoure, J., Ludwig, M., Brundrett, M. & Grierson, P. (2009). Identity and specificity of the fungi forming mycorrhizas with the rare mycoheterotrophic orchid *Rhizanthella gardneri. Mycological Research, 113*, 1097-1106.

Brundrett, M. C. (2002). Coevolution of roots and mycorrhizas of land plants. *New Phytologist, 154*, 275–304.

Brundrett, M. & Kendrick, B. (1990). The roots and mycorrhizas of herbaceous woodland plants. I. Quantitative aspects of morphology. *New Phytologist, 114*, 457-468.

Bruns, T. D., Bidartondo, M. I. & Taylor, D. L. (2002). Host specificity in ectomycorrhizal communities: what do the exceptions tell us? *Integrative and Comparative Biology, 42*, 352-359.

Cahill, J. F., Elle, E., Smith, G. R. & Shore, B. H. (2008). Disruption of a belowground mutualism alters interactions between plants and their floral visitors. *Ecology, 89,* 1791-1801.

Cairney, J. W. G. & Meharg, A. A. (1999). Influences of anthropogenic pollution on mycorrhizal fungal communities. *Environmental Pollution, 106*, 169-182.

Callaway, R. M. (1995). Positive interactions among plants. *Botanical Review, 61*, 306–349.

Callaway, R. M. (1997). Positive interactions in plant communities and the individualistic-continuum concept. *Oecologia, 112,* 143-149.

Cappellazzo, G., Lanfranco, L., Fitz, M., Wipf, D. & Bonfante, P. (2008). Characterization of an amino acid permease from the endomycorrhizal fungus *Glomus mosseae. Plant Physiology, 147*, 429-437.

Cardoso, I. M. & Kuyper, T. W. (2006). Mycorrhizas and tropical soil fertility. *Agriculture, Ecosystems & Environment, 116*, 72-84.

Carey, E. V., Marler, M. J. & Callaway, R. M. (2004). Mycorrhizae transfer carbon from a native grass to an invasive weed: evidence from stable isotopes and physiology. *Plant Ecology, 172,* 133-141.

Caris, C., Hördt, W., Hawkins, H-J., Römheld, V. & George, E. (1998). Studies of iron transport by arbuscular mycorrhizal hyphae from soil to peanut and sorghum plants. *Mycorrhiza, 8,* 35-39.

Carrillo-García, A., León de la Luz, J. L., Bashan, Y. & Bethlenfalvay, G. J. (1999). Nurse plants, mycorrhizae, and plant establishment in a disturbed area of the Sonoran desert. *Restoration Ecology, 7,* 321-335.

Condit, R., Hubbell, S. P. & Foster, R. B. (1994). Density dependence in two understory tree species in neotropical forest. *Ecology, 75*, 671– 680.

Connell, J. H. & Lowman, M. D. (1989). Low-diversity tropical rainforests: some possible mechanisms for their existence. *American Naturalist, 134*, 88–119.

Cullings, K. W., Vogler, D. R., & Finley, S. K. (2000). Ectomycorrhizal specificity patterns in a mixed *Pinus contorta* and *Picea engelmannii* forest in Yellowstone National Park. *Applied & Environmental Microbiology, 66*, 4988-4991.

Daniels, B. A. & Trappe, J. M. (1986). Factors affecting spore germination of the vesicular-arbuscular mycorrhizal fungus, *Glomus epigeus*. *Mycologia, 72*, 457-463.

Diédhiou, A. G., Selosse, M-A., Galiana, A., Diabaté, M., Dreyfus, B., Bâ, A. M., Miana de Faria, S. & Béna, G. (2010). Multi-host ectomycorrhizal fungi are predominant in a Guinean tropical rainforest and shared between canopy trees and seedlings. *Environmental Microbiology* (*in press*.). doi: 10.1111/j.1462-2920.2010.02183.x.

Duponnois, R., Galiana, A. & Prin, Y. (2008). The mycorrhizosphere effect: a multitrophic interaction complex improves mycorrhizal symbiosis and plant growth. In: Z. A. Siddiqui, M. S. Akhtar & K. Futai (Eds), *Mycorrhizae: Sustainable Agriculture and Forestry* (pp: 227-238). Dordrecht, The Netherlands: Springer.

Duponnois R. & Plenchette, C. (2003). A mycorrhiza helper bacterium (MHB) enhances ectomycorrhizal and endomycorrhizal symbiosis of Australian *Acacia* species. *Mycorrhiza*, *13*, 85-91.

Eom, A. H., Hartnett, D. C. & Wilson, G. W. T. (2000). Host plant species effects on arbuscular mycorrhizal fungal communities in tall grass prairie. *Oecologia, 122*, 435-444.

Eom, A. H., Wilson, G. W. T. & Hartnett, D. C. (2001). Effects of ungulate grazers on arbuscular mycorrhizal symbiosis and fungal community composition in tallgrass prairie. *Mycologia, 93*, 233-242.

Facelli, E. & Facelli, J. M. (2002). Soil phosphorus heterogeneity and mycorrhizal symbiosis regulate plant intra-specific competition and size distribution. *Oecologia, 133*, 54-61.

Finlay, R. D. (2004). Mycorrhizal fungi and their multifunctional roles. *Mycologist, 18,* 91-96.

Finlay, R. D. (2008). Ecological aspects of mycorrhizal symbiosis: with special emphasis on the functional diversity of interactions involving the extraradical mycelium. *Journal of Experimental Botany, 59,* 1115-1126.

Finlay, R. & Read, D. J. (1986). The structure and function of the vegetative mycelium of ectomycorrhizal plants. I. Translocation of ^{14}C-labeled carbon between plants interconnected by a common mycelium. *New Phytologist, 103*, 143-156.

Finlay, R. D., & Soderstrom, B. (1989) Mycorrhizal mycelia and their role in soil and plant communities. In M. Clarholm, & L. Bergstrom (Eds), *Ecology of Arable Land* (pp. 139-148). Amsterdam : Kluwer.

Fitter, A. H. (1991). Costs and benefits of mycorrhizas : implications for functioning under natural conditions. *Experientia, 47*, 350-355.

Founoune, H., Duponnois, R., Meyer, J. M., Thioulouse, J., Masse, D., Chotte, J. L. & Neyra, M. (2002). Interactions between ectomycorrhizal symbiosis and fluorescent pseudomonads on *Acacia holosericea*: isolation of Mycorrhiza Helper Bacteria (MHB) from a Soudano-Sahelian soil. *FEMS Microbiology Ecology*. *41*, 37-46.

Francis, R. & Read, D. (1994). The contributions of mycorrhizal fungi to the determination of plant community structure. *Plant & Soil, 159*, 11-25.

Frey-Klett, P., Chavatte, M., Clausse, M. L., Courrier, S., Le Roux, C., Raaijmakers, J., Martinotti, M. G., Pierrat, J. C. & Garbaye, J. (2005). Ectomycorrhizal symbiosis affects functional diversity of rhizosphere fluorescent pseudomonads. *New Phytologist, 165*, 317-328.

Furlan, V. & Fortin, J. A. (1973). Formation of endomycorrhiza by *Endogone calospora* on *Allium cepa* under three temperature regimes. *Naturalist Canadian, 100*, 589-593.

Gange, A. C. (2001). Species-specific responses of a root- and shoot-feeding insect to arbuscular mycorrhizal colonization of its host plant. *New Phytologist, 150*, 611-618.

Gange, A. & West, H. M. (1994). Interactions between arbuscular mycorrhizal fungi and foliar-feeding insects in *Plantago lanceolata* L. *New Phytologist, 128*, 79-87.

Gange, A. C. & Ayres, R. L. (1999). On the relation between arbuscular mycorrhizal colonization and plant "benefit". *Oikos, 87*, 615-621.

Gange, A. C., Bower, E. & Brown, V. K. (1999). Positive effects of an arbuscular mycorrhizal fungus on aphid life history traits. *Oecologia, 120*, 123-131.

Garner, W. & Steinberger, Y. (1989). A proposed mechanism for the formation of fertile islands in the desert ecosystem. *Journal of Arid Environments, 16*, 257-262.

Gehring, C. & Bennett, A. (2009). Mycorrhizal fungal-plant-insect interactions: the importance of a community approach. *Environmental Entomology, 38*, 93-102.

Gianchini, A. J., Souza, L. A. B. & Oliveira, V. L. (2004). Species richness and seasonal abundance of ectomycorrhizal fungi in plantations of *Eucalyptus dunnii* and *Pinus taeda* in southern Brazil. *Mycorrhiza, 14*, 375-381.

Gorham, E., Vitousek P. M. & Reiners, W. A. (1979). The regulation of chemical budgets over the course of terrestrial ecosystem succession. *Annual Review of Ecological Systems, 10*, 53-84.

Grime, J. P. (1997). Biodiversity and ecosystem function: the debate deepens. *Science, 277*, 1260–1261.

Grime, J. P., Mackey, J. M. L., Hillier, S. H. & Read, D. J. (1987). Floristic diversity in a model system using experimental microcosms. *Nature, 328*, 420-422.

Harms, K. E., Wright, S. J., Calderon, O., Hernandez, A. & Herre, E. A. (2000). Pervasive density-dependent recruitment enhances seedling diversity in a tropical forest. *Nature, 404*, 493-495.

Hart, M. M., Reader, R. J. & Klironomos, J. N. (2001). Life-history strategies of arbuscular mycorrhizal fungi in relation to their successional dynamics. *Mycologia, 93*, 1186-1194.

Hart, M. M., Reader, R. J. & Klironomos, J. N. (2003). Plant coexistence mediated by arbuscular mycorrhizal fungi. *Trends in Ecology & Evolution, 18*, 418-423.

Hartnett, D. C., Samenus R. J., Fischer, L. E & Hetrick B. A. D. (1994). Plant demographic responses to mycorrhizal symbiosis in tallgrass prairie. *Oecologia, 99*, 21-26.

Hartnett, D. & Wilson, G. (1999). Mycorrhizae influence plant community structure and diversity in tallgrass prairie. *Ecology, 80*, 1747-1761.

Hartnett, D. C. & Wilson, G. W. T. (2002). The role of mycorrhizas in plant community structure and dynamics : lessons from grasslands. *Plant & Soil, 244*, 319-331.

He, X. & Nara, K. (2007). Element biofortification: can mycorrhizas potentially offer a more effective and sustainable pathway to curb human malnutrition? *Trends in Plant Science, 12*, 331-333.

Hodge, A., Campbell, C. D. & Fitter, A. H. (2001). An arbuscular mycorrhizal fungus accelerates decomposition and acquires nitrogen directly from organic material. *Nature, 413*, 297-299.

Hoeksema, J. D. & Kummel, M. (2003). Ecological persistence of the plant-mycorrhizal mutualism: a hypothesis from species coexistence theory. *American Naturalist, 162*, 40-50.

Hubbell, S. P. (2001). *The unified neutral theory of biodiversity and biogeography*. Monographs in population biology. Princeton: Princeton University Press.

Husband, R., Herre, E. A. & Young, J. P. W. (2002). Temporal variation in the arbuscular mycorrhizal communities colonising seedlings in a tropical forest. *FEMS Microbiology Ecology, 42*, 131–136.

Hyakumachi, M. & Kubota, M. (2004). Fungi as plant growth promoter and disease suppressor. In: DK Arora (Ed.), *Fungal Biotechnology in Agricultural, Food, and Environmental Applications* (pp. 101-110). New York: Marcel Dekker,.

Janos, D. P. (1980). Mycorrhizae influence tropical succession. *Biotropica, 12*, 56-64.

Janzen, D. H. (1970). Herbivores and the number of tree species in tropical forests. *American Naturalist, 102*, 592-595.

Johnson, N. C., Graham, J. H. & Smith, F. A. (1997). Functioning of mycorrhizal associations along the mutualism-parasitism continuum. *New Phytologist*, *135*, 575-585.

Joner, E. J. & Leyval, C. (2003). Rhizosphere gradients of polycyclic aromatic hydrocarbon (PAH) dissipation in two industrial soils and the impact of arbuscular mycorrhiza. *Environmental Science & Technology*, *37*, 2371-2375.

Jonsson, L., Dahlberg, A., Nilsson, M-C., Kåren, O. & Zackrisson, O. (1999). Continuity of ectomycorrhizal fungi in self-regenerating boreal *Pinus sylvestris* forests studied by comparing mycobiont diversity on seedlings and mature trees. *New Phytologist, 142*, 151–162.

Kisa, M., Sanon, A., Thioulouse, J., Assigbetse, K., Sylla, S., Spichiger, R., Dieng, L., Berthelin, J., Prin, Y., Galiana, A., Lepage, M. & Duponnois, R. (2007). Arbuscular mycorrhizal symbiosis can counterbalance the negative influence of the exotic tree species *Eucalyptus camaldulensis* on the structure and functioning of soil microbial communities in a sahelian soil. *FEMS Microbiology Ecology, 62*, 32-44.

Klironomos, J. N. (2000). Host-specificity and functional diversity among arbuscular mycorrhizal fungi. In C. R. Bell, Brylinsky, M., & Johnson-Green, P. (Eds), *Microbial Biosystems : New Frontiers*, In Proceedings of the 8th International Symposium on Microbial Ecology (pp. 845-851). Halifax: Atlantic Canada Society for Microbial Ecology.

Klironomos, J. N. & Ursic, M. (1998). Density-dependent grazing on the extraradical hyphal network of the arbuscular mycorrhizal fungus, *Glomus intraradices*, by the collembolan, *Folsomia candida. Biology & Fertility of Soils, 26*, 250-253.

Klironomos, J. K., McCune, J., Hart, M. & Neville, J. (2000). The influence of arbuscular mycorrhizae on the relationship between plant diversity and productivity. *Ecology Letters, 3*, 137-147.

Klironomos, J. N. (2003). Variation in plant response to native and exotic arbuscular mycorrhizal fungi. *Ecology, 84*, 2292-2301.

Koide, R. T. (2000). Mycorrhizal symbiosis and plant reproduction. In Y. Kapulnik, & D. D. Douds (Eds), *Arbuscular Mycorrhizas: Physiology and Function* (pp. 19–46). Dordrecht: Kluwer Academic Publishers.

Kranabetter, J. M. & Friesen, J. (2002). Ectomycorrhizal community structure on western hemlock (*Tsuga heterophylla*) seedlings transplanted from forests into openings. *Canadian Journal of Botany, 80*, 861–868.

Lambers, H., Raven, J. A., Shaver, G. R., Smith, S. E. (2008). Plant nutrient-acquisition strategies change with soil age. *Trends in Ecology & Evolution, 23*, 95-103.

Landeweert, R., Hofflund, E., Finlay, R. D. & van Breemen, N. (2001). Linking plants to rocks: Ectomycorrhizal fungi mobilize nutrients from minerals. *Trends in Ecology & Evolution, 16,* 248-254.

Leake, J. R. (1994) The biology of myco-heterotrophic ('saprophytic') plants. *New Phytologist, 127*, 171–216.

Leake, J. R. (2004). Myco-heterotroph/epiparasitic plant interactions with ectomycorrhizal and arbuscular mycorrhizal fungi. *Current Opinion in Plant Biology, 7*, 422-428.

Leigh, E.G., Davidar, P., Dick, C.W., Puyravaud, J., Terborgh, J., ter Steege, H., and Wright, S.J., 2004, Why do some tropical forests have so many species of trees? *Biotropica,* 36, 445–473.

Lesueur, D. & Sarr, A. (2008). Effects of single and dual inoculation with selected microsymbionts (rhizobia and arbuscular mycorrhizal fungi) on field growth and nitrogen fixation of *Calliandra calothyrsus* Meissn. *Agroforestry Systems*, *73*, 37-45.

McGuire, K. L. (2007). Common ectomycorrhizal networks may maintain monodominance in a tropical rain forest. *Ecology, 88*, 567-574.

Miransari, M. (2010). Contribution of arbuscular mycorrhizal symbiosis to plant growth under different types of soil stress. *Plant Biology* (*in press*). doi: 10.1111/j.1438-8677.2009.00308.x.

Moora, M. & Zobel, M. (1996). Effect of arbuscular mycorrhiza and inter- and intraspecific competition of two grassland species. *Oecologia, 108*, 79-84.

Moore, J. C., John, T. V. St. & Coleman, D. C. (1985). Ingestion of vesicular-arbuscular mycorrhizal hyphae and spores by soil microarthropods. *Ecology, 66*, 1979-1981.

Morris, M. H., Pérez-Pérez, M. A., Smith, M. E. & Bledsoe, C. S. (2009). Influence of host species on ectomycorrhizal communities associated with two co-occuring oaks (*Quercus* spp.) in a tropical cloud forest. *FEMS Microbiology Ecology, 69*, 274-287.

Mummey, D. L., Rillig, M. C. & Holben, W. E. (2005). Neighboring plant influences on arbuscular mycorrhizal fungal community composition as assessed by T-RFLP analysis. *Plant & Soil, 271*, 83-90.

Mummey, D. L., Rillig, M. C. (2006). The invasive plant species *Centaurea maculosa* alters arbuscular mycorrhizal fungal communities in the field. *Plant & Soil, 288*, 81-90.

Newman, E. I. (1988). Mycorrhizal links between plants: their functioning and ecological significance. *Advances in Ecological Research, 18*, 243-270.

Newman, E. I., Eason, W. R., Eissenstat, D. M. & Ramos, I. R. F. (1992). Interactions between plants: the role of mycorrhizae. *Mycorrhiza, 1*, 47–53.

Newman, E. I. & Eason, W. R. (1993). Rates of phosphorus transfer within and between ryegrass (*Lolium perenne*) plants. *Functional Ecology, 7*, 242-248.

Newsham, K., Fitter, A. H. & Watkinson, A. R. (1995). Arbuscular mycorrhiza protect an annual grass from root pathogenic fungi in the field. *Trends in Ecology & Evolution, 83*, 991-1000.

Pacala, S. & Crawley, M. J. (1992). Herbivores and plant diversity. *American Naturalist, 155*, 435-453.

Pare, T., Gregorich, E. G. & Nelson, S. D. (2000). Mineralization of nitrogen from crop residues and N recovery by maize inoculated with vesicular arbuscular mycorrhizal fungi. *Plant & Soil, 218*, 11-20.

Pellissier, F. & Souto, X. (1999). Allelopathy in Northern temperate and boreal semi-natural woodland. *Critical Reviews in Plant Sciences 18*, 637-652.

Pfeffer, P. E., Douds, D. D., Bücking, H., Schwartz, D. P. & Shachar-Hill, Y. (2004). The fungus does not transfer carbon to or between roots in an arbuscular mycorrhizal symbiosis. *New Phytologist, 163*, 617-627.

Pringle, A. & Bever, J. D. (2002). Divergent phonologies may facilitate the coexistence of arbuscular mycorrhizal fungi in a North Carolina grassland. *Ameriacn Journal of Botany, 89*, 1439-1446.

Purin, S. & Rillig, M. C. (2008). Parasitism of arbuscular mycorrhizal fungi : reviewing the evidence. *FEMS Microbiology Ecology*, *279*, 8-14.

Read, D. J. (1993). Mycorrhiza in plant communities. *Advances in Plant Pathology, 9*, 1-31.

Reeves, F. B., Wagner, D., Moorman, T. & Kiel, J. (1979). The role of endomycorrhizae in revegetation practices in the semi-arid Est. I. A comparison of incidence of mycorrhizae in severely disturbed *vs* natural environments. *American Journal of Botany, 66*, 6-13.

Renker, C., Zobel, M., Öpik, M., Allen, M. F., Allen, E. B., Vosátka, M., Rydlová, J. & Buscot, F. (2004). *Structure, Dynamics, and Restoration of Plant Communities: Do Arbuscular Mycorrhizae Matter?* In : V. M. Temperton, R. J. Hobbs, T. Nuttle, S. Halle (Eds.), *Assembly Rules and Restoration Ecology – Bridging the Gap between Theory and Practice* (pp. 189-229). Washington DC, USA : Island Press.

Reynolds, J. F., Virginia, R. A. & Cornelius, J. M. (1990). Resource island formation associated with the desert shrubs, creosote bush (*Larrea tridentata*) and mesquite (*Prosopis glandulosa*) and its role in the stability of desert ecosystems: a simulation model. *Supplemental Bulletin of the Ecological Society of America, 70*, 299-300.

Reynolds, H.L., Packer, A., Bever, J.D. & Clay, K. (2003). Grassroots ecology: plant-microbe-soil interactions as drivers of plant community structure and dynamics. *Ecology, 84*, 2281-2291.

Reynolds, H. L., Vogelsang, K. M., Hartley, A. E., Bever, J. D. & Schultz, P. A. (2006). Variable response of old-field perennials to arbuscular mycorrhizal fungi and phosphorus source. *Oecologia, 147*, 348-358.

Richard, F., Millot, S., Gardes, M. & Selosse, M-A. (2005). Diversity and specificity of ectomycorrhizal fungi retrieved from an old-growth Mediterranean forest dominated by *Quercus ilex*. *New Phytologist, 166*, 1011-1023.

Richard, F., Selosse, M. A. & Gardes, M. (2009). Facilitated establishment of *Quercus ilex* in shrub-dominated communities within a mediterranean ecosystem: do mycorrhizal partners matter ? *FEMS Microbiology Ecology, 68*, 14-24.

Richardson, D. M., Williams, P. A. & Hobbs, R. J. (1994). Pine invasions in the Southern Hemisphere-determinants of spread and invadability. *Journal of Biogeography, 21*, 511-527.

Rillig, M.C. & Mummey, D. L. (2006). Mycorrhizas and soil structure. *New Phytologist.171*, 41-53.

Rosendahl, S. & Stukenbrock E. H. (2004). Community structure of arbuscular mycorrhizal fungi in undisturbed vegetation revealed by analyses of LSU rDNA sequences. *Molecular Ecology, 13*, 3179–3186.

Ryel, R. J., Caldwell, M. M. & Manwaring, J. H. (1996). Temporal dynamics of soil spatial heterogeneity in sagebrush–wheatgrass steppe during a growing season. *Plant & Soil, 184*, 299-309.

Sanders, I. R. & Fitter, A. H. (1992). The ecology and functioning of vesicular-arbuscular mycorrhizas in co-existing grassland species. I. Seasonal patterns of mycorrhizal occurrence and morphology. *New Phytologist, 120*, 517-524.

Sanon, A., Martin, P., Thioulouse, J., Plenchette, C., Spichiger, R., Lepage, M. & Duponnois, R. (2006). Displacement of an herbaceous plant species community by mycorrhizal and non-mycorrhizal *Gmelina arborea*, an exotic tree, grown in a microcosm experiment. *Mycorrhiza, 16*, 125-132.

Sanon, A., Beguiristain, T., Cébron, A., Berthelin, J., Ndoye, I., Leyval, C., Sylla, S. & Duponnois, R. (2009). Changes in soil diversity and global activities following invasions of the exotic invasive plant, *Amaranthus viridis* L., decrease the growth of native Sahelian *Acacia* species. *FEMS Microbiology Ecology, 70*, 118-131.

Sanon, A., Beguiristain, T., Cébron, A., Berthelin, J., Ndoye, I., Leyval, C., Prin, Y., Galiana, A., Baudoin, E. & Duponnois, R. (2010). Towards the influence of plant-soil-microbes feedbacks on plant biodiversity, grassland variability and productivity. In: J. Runas & T. Dahlgren (Eds.), *Grassland Biodiversity: Habitat types, Ecological Processes and Environmental impacts* (pp: 267-301) Hauppauge, New York: Nova Sciences Publishers.

Santos, J. C., Finlay R. D., & Tehler A. (2006). Molecular analysis of arbuscular mycorrhizal fungi colonising a semi-natural grassland along a fertilisation gradient. *New Phytologist, 172*, 159–168.

Santos-González, J. C., Finlay, R. D. & Tehler, A. (2007). Seasonal dynamics of arbuscular mycorrhizal fungal communities in roots in a seminatural grassland. *Applied & Environmental Microbiology, 73,* 5613-5623.

Schlesinger, W. H., Raikes, J. A., Hartley, A. E. & Cross, A. F. (1996). On the spatial pattern of soil nutrients in desert ecosystems. *Ecology, 77*, 364-374.

Schultz, P. A. (1991). Grazing preferences of two collembolan species, *Folsomia candida* and *Proisotoma minuta*, for ectomycorrhizal fungi. *Pedobiologia, 35*, 313–325.

Schultz, P. A. (1996). Arbuscular mycorrhizal species diversity and distribution in an old field community. PhD dissertation. Duke University, Durham, NC.

Schultz, P. A., Bever, J. D. & Morton, J. (1999). *Acaulospora colossica* sp. nov. from an old field in North Carolina and morphological comparisons with similar species, *A. laevis* and *A. koskei*. *Mycologia, 91*, 676-683.

Selosse, M. A., Weiss, M., Jany, J. L. & Tillier, A. (2002). Communities and populations of sebacinoid basidiomycetes associated with the achlorophyllous orchid *Neottia nidus-avis* (L.) LCM Rich. and neighbouring tree ectomycorrhizae. *Molecular Ecology, 11*, 1831–1844.

Selosse, M. A., Richard, F., He, X. & Simard, S.W. (2006). Mycorrhizal networks : des liaisons dangeureuses ? *Trends in Ecology & Evolution, 21*, 621-628.

Siddiqui, Z. A. & Pichtel, J. (2008). Mycorrhizae : an overview. In : Z. A. Siddiqui, M. S. Akhtar & K. Futai (Eds.), *Mycorrhizae : Sustainable Agriculture and Forestry* (pp. 1-35). Dordrecht, The Netherlands : Springer.

Silvertown, J. (2004). Plant coexistence and the niche. *Trends in Ecology & Evolution, 19*, 604-611.

Simard, S. W., Perry, D. A., Jones, M. D., Myrold, D. D., Daniel M. Durall, D. M. & Molina, R. (1997). Net transfer of carbon between ectomycorrhizal tree species in the field. *Nature, 388*, 579-582.

Simard, S. W. & Durall, D. M. (2004). Mycorrhizal networks: a review of their extent, function, and importance. *Canadian Journal of Botany, 82*, 1140-1165.

Smith, S. E. & Read, D. J. (2008). *Mycorrhizal Symbiosis*. London : Academic Press.

Stampe, E. D. & Daehler, C. C. (2003). Mycorrhizal species identity affects plant community structure and invasion: A microcosm study. *Oikos, 100*, 362-372.

St-Arnaud, M., Hamel, C., Vimard, B., Caron, M. & Fortin, J. A. (1997). Inhibition of *Fusarium oxysporum* f.sp. *dianthi* in the non-VAM species *Dianthus caryophyllus* by co-culture with *Tagetes patula* companion plants colonized by *Glomus intraradices*. *Canadian Journal of Botany. 75*, 998-1005.

Streitwolf-Engel, R., Boller, T., Wiemken, A. & Sanders, I. R. (1997). Clonal growth traits of two *Prunella* species are determined by co-occurring arbuscular mycorrhizal fungi from a calcareous grassland. *Journal of Ecology, 85*, 181-191.

Sylvia, D.M. (1988). Activity of external hyphae of vesicular-arbuscular **mycorrhizal** fungi. *Soil Biology & Biochemistry, 20*, 39-43.

Tedersoo, L., Sadam, A., Zambrano, M., Valencia, R. & Bahram, M. (2010). Low diversity and high host preference of ectomycorrhizal fungi in Western Amazonia, a neotropical biodiversity hotspot. *The ISME Journal, 4*, 465-471.

Tilman, D. (1980). Resources: a graphical-mechanistic approach to competition and predation. *American Naturalist, 116*, 362–393.

Tilman, D. & Pacala, S. (1993). The maintenance of species richness in plant communities. In R. E. Ricklefs, & D. Schluter (Eds.), *Species diversity in ecological communities : historical and geographical perspectives* (pp. 13-25). Chicago: University of Chicago Press.

Tilman, D. (2004). Niche tradeoffs, neutrality, and community structure : a stochastic theory of resource competition, invasion, and community assembly. *Proceedings of National Academy of Sciences USA, 101*, 10854-10861.

Torti, S.D., Coley, P.D. & Kursar, T.A. (2001). Causes and consequences of monodominance in tropical lowland forests. *American Natur*alist, *157*, 141–153.

Urcelay, C. & Diaz, S. (2003). The mycorrhizal dependence of surbordinates determines the effect of arbuscular mycorrhizal fungi on plant diversity. *Ecology Letters, 6*, 388-391.

van der Heijden, M. G. A., Klironomos, J. N., Ursic, M., Moutoglis, P., Streitwolf-Engel, R., Boller, T., Wiemken, A. & Sanders, I. R. (1998a) Mycorrhizal fungal biodiversity determine plant biodiversity, ecosystem variability and productivity. *Nature, 396*, 69-72.

van der Heijden, M. G. A., Boller, T., Wiemken, A. & Sanders I. R. (1998b). Different arbuscular mycorrhizal fungal species are potential determinants of plant community structure. *Ecology, 79*, 2082-2091.

van der Heijden, M. G. A. & Scheublin, T. R. (2007). Functional traits in mycorrhizal ecology: their use for predicting the impact of arbuscular mycorrhizal fungal communities on plant growth and ecosystem functioning. *New Phytologist, 174*, 244-250.

Van der Putten, W. H., Van Dijk, C. & Peters, B. A. M. (1993). Plant-specific soil-borne diseases contribute to succession in foredune vegetation. *Nature, 362*, 53-56.

Vandenkoornhuyse, P., Husband, R., Daniell, T. J., Watson, I. J., Duck, M., Fitter, A. H. & Young, J. P. W. (2002). Arbuscular mycorrhizal community composition associated with two plant species in a grassland ecosystem. *Molecular Ecology, 11*, 1555–1564.

Vandenkoornhuyse, P., Ridgway, K. P., Watson, I. J., Duck, M., Fitter, A. H. & Young, J. P. W. (2003). Co-existing grass species have distinctive arbuscular mycorrhizal communities. *Molecular Ecology, 12*, 3085–3095.

Vogelsang, K. M., Bever, J. D., Griswold, M. & Schulz, P. A. (2004). The use of mycorrhizal fungi in erosion control applications. Final report for Caltrans. California Department of Transportation Contract no. 65A0070, Sacremento (California).

Vogelsang, K. M., Reynolds, H. L. & Bever, J. D. (2006). Mycorrhizal fungal identity and richness determine the diversity and productivity of a tallgrass prairie system. *New Phytologist, 172*, 554-562.

Walker, J. F., Miller, Jr. O. K. & Horton, J. L. (2008). Seasonal dynamics of ectomycorrhizal fungus assemblages on oak seedlings in the southeastern Appalachian Mountains. *Mycorrhiza, 18*, 123-132.

Walker, T. W. & Syers, J. K. (1976). The fate of P during pedogenesis. *Geoderma, 15*, 1-19.

West, H. M. (1996). Influence of arbuscular mycorrhizal infection on competition between *Holcus lanatus* and *Dactylis glomerata*. *Journal of Ecology, 84*, 429-438.

Wirsel, S. G. R. (2004). Homogenous stands of wetland grass harbour dicerse consortia of arbuscular mycorrhizal fungi. *FEMS Microbiology Ecology, 48*, 129-138.

Wurst, S., Dugassa-Gobena, D., Langel, R., Bonkowski, M. & Scheu, S. (2004). Combined effects of earthworns and vesicular-arbuscular mycorrhizas on plant and aphid performance. *New Phytologist, 163*, 169-176.

Zabinki, C. A., Quinn, L. & Callaway, R. M. (2002). Phosphorus uptake, not carbon transfer, explains arbuscular mycorrhizal enhancement of *Centaurea maculosa* in the presence of native grassland species. *Functional Ecology, 16*, 758-765.

Zobel, M. & Moora, M. (1995). Interspecific competition and arbuscular mycorrhiza : importance for the coexistence of two calcareous grassland species. *Folia Geobotanica Phytotaxon, 30,* 223-230.

In: Mycorrhizal Fungi
Editor: Susanne M. Fulton
ISBN 978-1-61122-659-1
© 2011 Nova Science Publishers, Inc.

Chapter 4

MYCORRHIZAL FUNGI AND ECOSYSTEM EFFICIENCY

Mohammad Miransari*
Department of Soil Science, College of Agricultural Sciences, Shahed University, Tehran, Qom Highway, Tehran, 18151/159, Iran

ABSTRACT

Arbuscular mycorrhizal fungi are soil fungi developing symbiotic association with most terrestrial plants as their usually non-specific hosts. In their symbioses the fungi provide their host plant with water and nutrients in the exchange for carbon. The beneficiary effects of mycorrhizal fungi on the ecosystem have been previously indicated by numerous number of research work. Their contribution to the ecosystem is by enhancing the efficiency of environment including soil under different conditions including stress, which eventually increases the production of crop yield. In greater details such implications include: 1) enhanced water and nutrient uptake by plant, 2) alleviation of different soil stresses on plant growth including drought, salinity, compaction, and pathogens, 3) their interactions with other soil microbes, and 4) their likely use as bio-fertilizers. In this article the most recent advances regarding such implications, which eventually contribute to the enhanced ecosystem efficiency, are reviewed. Some likely approaches, which can improve the efficiency of mycorrhizal fungi under different conditions, are also highlighted.

Keywords: Arbuscular mycorrhizal fungi, plant growth, plant hormones, soil microbes, stress

* E-mail: Miransari1@gmail.com, Miransari@Shahed.ac.ir

INTRODUCTION

Arbuscular mycorrhiza are soil fungi, developing symbiotic association with most terrestrial plants. AM fungi are obligate biothrophs and the presence of the host plant for the establishment and development of symbiosis is necessary, although the fungal spore is able to germinate in the absence of the host plant (Rosendahl, 2008; Johnson, 2010). In this symbiosis plants provide the fungi with carbon in the exchange of water and nutrients. The following stages are usually common before the onset of symbiosis between the fungi and the host plant (Smith and Read, 2008).

There are some communications between the fungi and the host plant resulting in the realization of the two symbionts by each side and hence successful establishment of symbiosis (Tikhonovich and Provorov, 2009; Ivanov et al., 2010). The germinated hyphae grow toward the plant root and eventually the extensive network of hypha is formed between the plant roots and the soil interface. AM hyphae grow inside the root cortex cells and develop some organelles called arbuscule and vesicle. Arbuscules are branched like hyphal structures, acting as the exchanging interface for nutrients between the plant root and the soil environment. Vesicles are also storage tissues storing different organic and mineral compounds (Balestrini and Lanfranco, 2006).

AM hyphae represent a major part of microbial biomass as well as organic carbon (C) in the soil (Wilson, 2009). Carbon, nitrogen (N) and phosphorus (P) are the three most important nutrients controlling the level of AM symbiosis and their contribution determine the community structure of plant and AM fungi (Johnson, 2010). The balance between these three nutrients determines the establishment and development of AM symbiosis. Carbon is necessary for plant food production in the ecosystem by photosynthesis process. Nitroegen is a necessary nutrient for chlorophyll formation and hence chloroplast activity during photosynthesis process and P is the major component for energy storing molecules affecting the related biochemical pathways. As the plant host is necessary for the establishment and development of AM symbiosis, the performance of AM symbiosis is dependent on the availability and balance between these three nutrients (Miller et al., 2002). With respect to the availability of soil nutrients, plant can adjust their dependency on AM symbiosis (Graham and Eissenstat, 1994). High nutrient availability in the soil can significantly decrease plant dependency on AM symbiosis and vice versa. In other words AM fungi develop the highest level of symbiosis with their host plants in soils with low levels of nutrients including P (Marschner and Dell, 1994; Balestrini and Lanfranco, 2006; Miransari, 2010).

AM fungi can affect plant fitness or in other words plant adaptation with the environmental parameters (Koide and Dickie, 2002), the structural combination of plant community (Hartnett and Wilson, 2002) and the flow of matter and energy in the ecosystem (van der Heijden et al., 2008). AM fungi have also some unique specifications including: 1) production of the extensive hyphal network, which can significantly enhance the host plant ability for water and nutrient uptake by increasing the solubility and availability of different nutrients such as phosphorus (P) 2) production of plant hormones, 3) controlling pathogens, 4) improving soil structure, and 5) alleviating the unfavorable effects of soil stresses on plant growth. Accordingly, AM fungi can significantly contribute to the enhanced ecosystem efficiency.

HYPHAL NETWORK

AM fungi are able to produce a very extensive network of hypha, which can grow in the soil even in zones further than the root development zone and absorb water and nutrients for the host plant. The hyphal network can significantly increase the surface area of the plant root resulting in the significant increase of water and nutrient uptake by plant. The hypha are thin and can grow in fine soil pores, where even the finest root hairs are not able to grow, and absorb water and nutrients (Rillig and Mummey, 2006). Although, the symbiosis between the fungi and the host plant are non-specific, there are different parameters affecting the symbiosis specificity and efficiency. Plant and fungal species, soil and climate properties are among the most important factors affecting the AM-host plant symbiosis (Daei et al., 2009).

In some cases the fungi may perform more efficiently, for example the morphological and physiological properties of the host plant can considerably influence the symbiotic efficiency. These include: 1) the structure of the host plant root, 2) the properties of the rhizosphere, which is initially affected by the host plant root, 3) root biochemical properties including the production of different enzymes, organic acids, sugars and other organic products, 4) plant physiological properties, 5) plant chemotactic response to AM presence, and 6) rhizodeposition and carbon flux (Drew et al., 2003; Finlay, 2008).

ENHANCED WATER AND NUTRIENT UPTAKE

The symbiosis development between arbuscular mycorrhizal fungi and the host plant can significantly increase the uptake of water and nutrients by the host plant. The extensive hyphal network is able to grow beyond the zones where the host plant root can grow and absorb water and nutrients. It is because the AM hypha are thin and can grow inside even the finest micropores where the thinnest root hairs are not able to grow. The considerable volume of hyphae can also contribute to the enhanced water and nutrient uptake. Enhanced water and nutrient uptake by mycorrhizal plants can contribute to the alleviation of the unfavorite effects of stresses on the host plant growth (Rillig and Mummey, 2006; Smith and Read, 2008).

More details regarding the molecular aspects affecting the process of nutrient uptake by mycorrhizal plants are being elucidated (Harrison, 2005; Javot et al., 2007; Landis et al., 2008). There are much more that must be known about the fluxes of nutrients between the two symbionts. The rate of C utilized by AM fungi and rhizobium ranges between 4 to 16%. The enhanced nutrient uptake by AM fungi as well as the increased rate of plant photosynthetic and crop yield can compensate for the energy used by the host plant to produce C necessary for fungal utilization. For example, AM symbiosis increased the rate of leaf P and photosynthates by 6 and 14%, respectively, in the host plant (Kaschuk et al., 2009).

It has also been indicated that AM fungi can contribute to the uptake of ammonium and nitrate, which is assimilated into the structure of arginine in the AM hyphae and transported to the plant, however at very little amoutns, not comparable with the amounts absorbed by ectomycorrhizal fungi (Govindarajulu et al., 2005; Olsson et al., 2005; Smith and Read, 2008). AM symbiosis can also result in the significant absorbion of mineral and organic N and transfer them to the host plant (George et al., 1995; Leigh et al., 2008).

Among the important parameters affecting AM efficiency are AM species and their inoculation rate. This has recently become the center of attention by researchers (Lekberg et al., 2007; Maherali and Klironomos; 2007). There is a high rate of competition among different AM species as well as the isolates of the same species when inoculating the host plant roots. Plant species, growth stage, and soil and climate properties can affect the rate of inoculation by AM species singly or in combination (Johnson et al., 1997; Munkvold et al., 2004; Jansa et al., 2008). It is common, especially under field conditions, that more than one species of AM fungi inoculate the host plant roots (Merryweather and Fitter, 1998). Hence, it is important to indicate the contribution of each species for the indication of appropriate strategies, which can enhance ecosystem efficiency. Among such strategies is the use of AM species as bioinoculants.

Using real-time PCR method, which has recently been used to determine the identity of AM species in a mixture (Alkan et al., 2004; Isayenkov et al., 2004), Jansa et al. (2008) tested and indicated the contribution of three different species of AM fungi including *Glomus mosseae*, *G. claroideum* and *G. intraradices* singly or combined on the host plants medic (*Medicago trancatula*) and leek (*Allium porrum*). These fungi, with different strategies to absorb P, were isolated from the field (Jansa et al., 2002). Their differences, which are related to the distance from which they absorb P and their P uptake efficiency from soil (P absorption per unit of hyphal length), were determined using labeled P. The hypha of *G. claroideum* absorbed P from distances less than 6 cm, while for *G. mosseae* and *G. intraradices* the related distance was around 15 cm (Jansa et al., 2005). They found the contribution of each species by indicating the combination of the ribosomal RNA genes and mentioned that *G. mosseae* was the dominant species and the mixture of *G. claroideum* and *G. intraradices* inoculated the leek roots. *G. mosseae* colonized the roots at the fastest rate followed by *G. intraradices* and *G. claroideum*. They all reached their maximum colonization rate at week 4.

In cases where the AM fungi species can be complementary in their functioning, for example by absorbing nutrients from different resources, their mixed inoculation can be superior to their single inoculation (Koide et al., 2000; Alkan et al., 2006; Gustafson and Casper, 2006). Specifically mentioning, the N and P concentration in the plant can determine the combination of AM species on the roots. It is, however, important to clearly indicate the identity of each AM species (Robinson-boyer et al., 2009) as well as its functioning before performing the experiments related to the functional diversity and specific functioning of each species when testing mixed AM species. The rate of plant C utilization by AM fungi can affect plant growth, the higher the C utilization rate, the lower the plant growth (Rosendahl, 2008).

The enhanced uptake of nutrients by AM fungi alter plant metabolism in a way, so that the host plant can handle the absorbed nutrients and incorporate it into structural organic compounds. Accordingly, under stress AM fungi are able to modify plant physiology so that it can cope with the stress (Miransari et al., 2008). Mycorrhizal plants transfer photosynthates to area where P is absorbed at higher amounts (Fitter, 2006). Although C moves in the plant in form of sucrose, it is translocated to AM fungi in form of hexoses, which is eventually turned into tryacylglycerides and trehaloses to be utilized by fungi (Pfeffer et al., 2004).

In addition to the effects of root morphology on plant nutrient uptake, root exudates can also affect the solubility and hence availability of nutrients in the soil. Production of organic acids such as tricarboxylic acid anions by plant roots can affect plant P uptake (Lambers and Poot, 2003; Lambers et al., 2008a,b; Smith et al., 2010)., AM symbiosis usually increase the growth of the host plant by enhancing nutrient uptake. However, there are situations where AM symbiosis may not stimulate plant growth or it may even decrease plant growth, indicating the importance of functional diversity in the performance of AM fungi (Johnson et al., 1997; Jakobsen et al., 2002; van der Heijden et al., 2008; van der Heijden and Horton, 2009; Smith et al., 2010).

The following parameters influence AM responsiveness and hence plant growth: 1) plant genotype affecting plant growth including roots and hence uptake of nutrients such as P, 2) fungal genotype affecting fungal growth including hypha, and 3) the genomic interactions between AM hypha and plant roots affecting the rate of root colonization and P uptake (Jakobsen et al., 2002; Smith and Read, 2008; Smith et al., 2010).

The two pathways affecting P uptake by plant including the plant direct P uptake and the AM pathway are significantly different. Plant absorbs most of its necessary P by the AM pathway, which considerably absorbs higher P than the plant pathway (Smith et al., 2010). If plant responds positively to P uptake, the two P pathways act additively, however when plant growth is not affected or adversely affected the two P pathways are not additive. Although, P behavior in the soil is a determining factor in its uptake by plant, apparently the sensitivity of P transporters in the two pathways is different. In other words the fungal P transporters have a much higher affinity for P and hence resulting in the higher P uptake by AM fungi. Such expression of P transporter genes is indicated by the accumulation of RNA (Paszkowski et al., 2002; Glassop et al., 2005; Smith et al., 2010).

P uptake is much affected by the expression of P transporters genes in mycorrhizal plants (Poulsen et al., 2005; Balestrini and Lanfranco, 2006; Grunwald et al., 2009). Plant uses different strategies to compensate for N and P deficiency, among which alteration in root growth and morphological properties as well as symbiotic development with AM fungi are the most important ones. However, interestingly with aging the soil, the methods of plant nutrient uptake may also alter as soil properties result in the development of different biological and chemical processes. During the early years of soil development usually AM fungi develop symbiotic with their host plants, however as the soil ages, other kinds of mycorrhizal symbioses including ectomycorrhizal, ericoid mycorrhizal as well as non-symbiotic mycorrhizal develop, respectively. Accordingly, soil total P decreases with time, however total N increases to a pick and hence diminishes. Under nutrient deficient conditions plant produces cluster roots. Cluster roots are able to produce high amounts of carboxylates and phosphatases, which enhance the solubility of P in the soil. In addition cluster growth of roots, increases root density per soil area unit and hence increases the rate of nutrient uptake by plant. Regarding Pi uptake, Pi movement in the soil and in the plant determines its rate of uptake by plant. P moves faster in the plant than in the soil. Hence, plant must be able to produce roots at a higher rate compared to the production of proteins, which transfer P in the plant (Clarkson, 1985; Lambers et al., 2008b). MicroRNAs (single stranded RNA with almost 20 nucleotides, activated by binding to RNAs) regulate gene expression in plant cells. MicroRNAs are able to regulate nutrient homeostasis in plant under stress by regulating the related genes (Chiou et al., 2006).

Production of Plant Hormones

Plant hormones including auxin (IAA), gibberellins (GA), abscisic acid (ABA), ethylene and jasmonic acid (JA) can regulate plant growth and development by controlling different mechanisms such as plant cell division and growth (auxin), seed germination (GA), stomatal activity (ABA), fruit ripening (ethylene) and induced systemic resistance (JA). The way plant hormones can affect microbial interaction with plant is mostly related to triggering plant resistance (Van Wees et al., 2008). There is also a new class of plant phytohormones, strigolactones, which have been recently identified (Akiyama et al., 2005). These hormones are able to increase hyphal branching and also trigger the germination of *Striga* seed, a parasitic weed. Interestingly, mycorrhizal plants are able to inhibit the germination and growth of such parasitic weeds. Researchers have indicated that Jasmonic acid (JA) can regulate AM-host symbiosis (Herrera-Medina et al., 2008). In mutant plants, due to the absence of JA pathway, the rate of AM inoculation was not controlled. In other words, the autoregulation of AM symbiosis in mutant plants was adversely affected. It has been speculated that plant hormones including IAA, ABA and JA are able to autoregulate AM formation in the host plant. Exogenous application of JA, decreased AM inoculation by affecting P metabolism and arbuscule formation (Meixner et al., 2008). The role of ethylene in the regulation of AM symbiosis has also been indicated affecting plant related genes including chitinases and β-1,3-glucanases during the establishment of AM symbiosis (Zsögön et al., 2008).

Controlling Plant Pathogens

Plants develop both symbiotic and pathogenic association with soil microbes including arbuscular mycorrhizal fungi, N-fixing bacteria and plant pathogens. There is some kind of signal communications between the host plant and soil microbes, determining the type of plant-microbe interactions. Soil microbes elicit some molecules, which are perceived by plant cellular receptors including kinases (Zhao and Qi, 2008). Plants perceive pathogens in soil as a kind of stress and hence respond accordingly. Plant pathogens result in the expression of genes related to plant systemic resistance. As a result, some controlling processes are activated to inhibit the adverse effects of pathogenic invasion on plant growth. Reactive oxygen species are also produced during the development of AM-host plant symbiosis. This has been related to the activation of NADPH oxidase located in the cellular plasma membrane. More details in this regard have yet to be indicated. The interesting point is the differences, present among the activation of reactive oxygen species during the pathogenic invasion and the symbiotic interactions (Nanda et al., 2010).

Under stress reactive oxygen species are produced, which are not favorable to plant growth and activity. Plants are able to produce a range of antioxidants degrading the unfavorable produced products. The higher the plant ability to produce antioxidants, the higher the chance of inhibiting or alleviating the adverse effects of pathogen on plant growth.

It has been known for some time that AM fungi are able to enhance plant tolerance to pathogens in soil. The following mechanisms have been suggested for increased plant tolerance to pathogens by AM fungi: 1) enhanced nutrient uptake, 2) the competition between

AM fungi and plant pathogens in soil for root colonization as well as for resources, 3) higher root growth and altered root morphology, 4) the effects of AM fungi on root exudates influencing soil microbes in the plant rhizosphere including plant pathogens, and 5) induced plant resistance to pathogens by increasing the concentration of plant resistance molecules in the root cortex cells (Graham, 2001; Avis et al., 2008; Smith et al., 2010; Wehner et al., 2010). Fungal diversity as well as their single or combined use may also influence their alleviating effects on pathogenic invasion. In other words, in cases of synergistic interactions between AM species, their combined use can enhance plant resistance to pathogens, compared with their single use (Wehner et al., 2010).

The use of new techniques, such as proteomic studies, has helped researchers to identify the molecular alterations taking place during plant microbe interactions including pathogenic invasions. Expression of different genes and the subsequent production of proteins indicated by such methods can be useful to evaluate the response of host plant during different interactions (Mathesius, 2009). This may help to develop strategies that can enhance plant resistance to pathogens. According to Valot et al. (2005; 2006) the proteome analysis of proteins in the cellular membranes of mycorrhizal roots of *M. trancatula* (infected with *G. intraradices*) indicated that with respect to plant species and AM species as well as the amounts of P, different genes and hence proteins (23) were expressed. The proteome analysis of prearbuscular membrane of roots inoculated with *G. intraradices* indicated the presence of 78 proteins in the plasma membrane of uninfected roots, however two proteins were altered during the infection. Carlsen et al. (2008) analyzed the effects of a pathogenic fungi *Phytium ultimum* and two species of AM fungi including *G. mosseae* and *G. claroideum* on the flavonoids levels in the roots of white clover (*Trifolium repens* L.). They found that while AM species adversely affected the growth of *P. ultimum*, by significantly decreasing its biomass, they were also able to prevent the fungi from infecting the host root in some cases. The type of fungi (pathogenic or beneficial), fungal species as well as plant variety determined the level of flavonoids produced. *G. claroideum* produced significantly higher amounts of coumestrol or kaempferol than *G. mosseae*, which was cultivar dependent. P. *ultimum* increased the amounts of formononetrin and daidzein in the roots of white clover. Such products are usually produced in plant in response to stress at the time of pathogenic invasion.

The type of interaction (pathogenic, parasitic and mutalistic) or in other words the trophic level between soil microbes and the host plant is determined by different parameters including plant, microbe and environmental ones. Different environmental stressors can limit the availability of different resources such as water and nutrients to both the microbes and the host plant. Accordingly, the related signals may differ affecting plant response the microbial presence. For example, some pathogens can enhance the production of salicylic acid in plant, which can adversely affect the production of other plant hormone JA. JA can increase plant resistance to necrothrophic pathogens and insect herbivors (Newton et al., 2010).

IMPROVING SOIL STRUCTURE

Hyphal length is different under different conditions, ranging from 1 to 111m/g of soil; however the common length is in the medium range, in comparison with the <10 μm length of plant roots (Miller et al., 1995; Rillig and Mummey, 2006). The average diameter for plant root and root hairs is 300 and 10 μm, respectively, while for the fungal hyphae ranges from 2 to 20 μm. Plant roots influence the soil environment at the highest distance of 0.3 mm; however for AM hyphae the distance is equal to 25 cm (Smith et al., 2010). AM produces high amounts of C in the soil beyond the root growth zone affecting the activities of other soil microorganisms. AM hyphae are also important in stabilizing soil structure by binding soil particles, especially in the range of 20-200 μm diameter (Miller and Justrow, 2002).

It should also be mentioned that the tremendously thin diameter of AM hyphae is extremely important in the soil as the hyphae is able to grow in micropores, where even the finest root hairs are not able to grow (Drew et al., 2003). This is important for the uptake of water and nutrients by mycorrhizal roots under different conditions including stresses like soil drought and soil compaction (Auge, 2001; Miransari et al., 2007; 2008). In a compacted soil the size of soil pores decreases significantly affecting plant root growth and hence water and nutrient uptake. However, as mentioned, AM hyphae are able to grow in even the finest soil micropores and absorb water and nutrients and hence considerably alleviate the unfavorable effects of soil compaction on plant growth and yield production (Mitransari 2009a; b). As previously mentioned, AM fungi are also able to significantly increase water uptake by plant, because of their extensive hyphal network and their growth beyond the root growth zone in the finest soil micropores.

AM hypha are able to bind soil particles and hence improve soil structure. Therefore, soils with higher mycorrhizal population are of more improved soil structure and hence higher efficiency. In addition, AM fungi are able to produce a glycoprotein called glomalin, which is able to bind soil particles. Glomalin is also able to act as a source of C for soil microbes affecting their growth and population in soil (Rillig and Mummey, 2006).

ALLEVIATING SOIL STRESSES

Under stress, plant is able to alleviate the unfavorable effects of stress by inducing morphological and physiological changes. However, interestingly stress response differs among different plant species, as some of which are more tolerant to stress due to the development of different mechanisms. Tolerant plants are able to alleviate the stress more efficiently than non tolerant plants. This indicates the importance of selecting appropriate plant species, with regard to the present conditions. In other words, under stress, tolerant plants must be selected so that they can perform more efficiently (Munns and James, 2003).

There are stress genes in plant, which are expressed during the stress resulting in some mechanisms, alleviating the stress up to some extent. For example, under drought plant is able to produce proteins such as prolin, adjusting cellular water potential. They may also allocate more carbon to their roots, searching for water, for a more developed growth in the soil. Under drought plant adjust their growth in accordance with the present moisture. Plant leaf may roll, and the related angle may change. Plant hormones including ABA adjust the stomatal activity under drought resulting in decreased transpiration (Zinselmeier et al., 1995; Boomsma and Vyn, 2008).

When subjected to salinity stress, plants may store the extra ions in their vacuoles or they may exudate extra salt into the rhizosphere through their roots or accumulate salt on their leaf. Under nutrient deficiency, plant roots produce organic acids in the rhizosphere, enhancing the solubility of nutrients. They may also translocate nutrients from their old leaf to their young leaf. Under acidic conditions plants try to adjust their cytoplasmic pH, by controlling the movement of different ions across their cellular membrane, specifically hydrogen (H^+) ions or they may store acidic ions such as aluminum in their vacuoles (Marschner, 1995). Soil compaction may result in cluster growth of roots enhancing root efficiency to absorb water and nutrients under compaction stress. Furtherer more, under compaction stress, plant roots send signals such as plant hormones to their shoot adjusting plant growth (Miransari et al., 2006; 2007; 2008).

Numerous experimental data have indicated the alleviating effects of AM fungi on plant growth and crop production under stress. AM fungi own some unique properties, which can make the host plant survive under stress. Most of the experiments regarding the alleviating effects of AM fungi on plant growth under stress have been performed under greenhouse conditions and there are little data under field conditions (Miransari et al., 2006; Daei et al., 2009). In the following the effects of AM fungi on plant growth under different kinds of stress with some details are mentioned. As previously metntioned, under stress AM fungi are able to alter plant physiology in a way, so that the plant can handle the stress (Miransari et al., 2008).

Drought and AM Fungi

Soil moisture deficiency is usual in different parts of the world resulting in the reduction of plant growth and yield production. Water is necessary for different plant development and physiological processes including cellular and tissue growth and development, nutrient movement, uptake and translocation, evapo-transpiration, photosynthesis, temperature adjustment, enzymatic activities, etc. In the arid and semi arid areas of the world drought is common affecting plant growth and yield production. In addition to the mechanisms utilized by plant, as previously mentioned, AM fungi are also able to enhance plant tolerance to drought stress (Boomsam and Vyn, 2009).

The extensive network of hypha is able to absorb high amounts of water and nutrients, which can enhance plant growth under different conditions including stress. The growth of hypha into the soil micropores where even plant the finest plant roots hairs are not able to grow, and in the zones beyond the root zone growth can substantially increase the amount of water and nutrients absorbed by the host plant (Auge, 2001). Different experiments have indicated the positive effects of AM fungi on plant growth under stress.

Under drought, AM fungi enhance plant uptake of mobile nutrients such as nitrogen and relatively immobile nutrients including phosphorous, potassium, calcium, magnesium, sulfur, iron, manganese, zinc, and copper (Marschner, 1995; Liu et al., 2007). AM fungi result in the following physiological alterations in plant under drought stress. AM fungi affect the influx and efflux of water influencing plant water potential and the related physiological processes and hence the process of transpiration and gas exchange. As a result of higher root surface and growth of AM hypha into the soil micropores, AM plants dry the soil faster relative to non-AM plants. Altered production of carbohydrates and antioxidants by AM fungi are also among their other effects on plant growth under drought stress (Sylvia et al., 1993; Ruiz-Lozano, 2003).

There are some mycorrhizal effects, enhancing plant tolerance to drought stress including enhanced plant water uptake and hydraulic activity as well as positively affecting water behavior in soil. For example, improved soil structure, extensive hyphal network and their growth into the soil micropores and hence increased available water, as well as the alteration of plant signals (IAA and cytokinins production) are among the most important effects of AM fungi on plant growth under stress (Auge, 2001).

In brief, Auge (2001) indicated the benefits of AM symbiosis on plant water characters by the followings: enhanced nutrient uptake, increased plant tolerance to drought, considerably higher water uptake by mycorrhizal plants, greater stomatal activities and transpiration, higher nutrients diffusion and alleviating drought stress by reducing ABA concentration in the xylem of mycorrhizal plants.

Under drought stress mycorrhizal plants are able to absorb higher rate of nutrients including P (Auge, 2001; Neumann and George, 2004). It is because, under such conditions, the AM pathway is more active than the plant pathway. It must also be mentioned that P is effective on root growth (Miransari et al., 2007; 2008) and enhanced P uptake under drought by mycorrhizal plants can significantly contribute to plant tolerance to stress. P can also influence root hydraulic activity as under P deficient conditions root permeability and hence water uptake by plant decreases (Carvajal et al., 1996; Reinbott and Blevins, 1999). It is because the membrane water channels (aquaporins), controlling water movement in the plant are produced by the expression of genes, which are responsive to different environmental parameters including nutrients (Luu and Maurel, 2005). Hence, under P sufficient conditions, aquaporins production can enhance root hydraulic activity and hence water uptake. However, genes similar to P transporter genes, affecting aquaporins production by AM symbiosis have not yet been recognized. However, the expression of aquaporins genes (PIP1 and PIP2) have been observed in the plasma membrane of mycorrhizal Medicago *truncatula* not under drought stress (Journet et al. 2002; Aroca et al., 2007).

With respect to the significance of AM fungi for plant growth and yield production under different conditions including stress the followings are suggested to enhance the persistence and efficiency of AM fungi in soil: 1) reduced soil tillage, preserving fungal hypha in soil, 2) proper rotation including the use of cover plants, intercropping and continuous cropping, which provide the necessary C as well as host plant for AM growth and development, 3) appropriate P fertilization maintaining AM fungal growth and activity in soil, especially when in symbiosis with the host plant, and 4) decreased use of fungicides (Johnson and Pfleger, 1992; Plenchette et al., 2005; Boomsma and Vyn, 2009).

Salinity and AM Fungi

Salinity stress is common in different parts of the world, which is mainly due to low precipitation and high evaporation. As a result, high amounts of salt are accumulated in soil decreasing plant growth and crop yield. The adverse effects of salinity on plant growth and yield production are through the excess concentration of cations and anions in soil and hence plant (Cornillon and Palloix, 1997; Munns and James, 2003; Miransari and Smith, 2007) affecting plant functioning by the followings.

1) Extra amounts of ions in the soil increases soil water tension and hence decrease soil water availability to plant. In other words although the amounts of water in a While mycorrhizal associations, particularly those with AM fungi, have been thought to be largely non-specific, there is growing evidence of host-specific differences in plant response to fungi and in fungal response to plants. The extent of plant growth promotion by mycorrhizal fungi depends upon the specific plant and fungal combinations (Streitwolf-Engel et al., 1997; van der Heijden et al., 1998 a,b; Hart et al., 2003). Conversely, the growth response and development of mycosymbionts also depends on the associated plant host species (Eom et al., 2000). Thus, host-specific changes in the mycorrhizal fungal community could lead to positive or negative feedbacks on the plant (Bever et al., 1997; Bever, 1999; Bever, 2002 a,b). These two dynamics lead to very different predictions for the community in terms of competing plant species' ability to coexist.

saline soil may be adequate, ions can significantly influence water availability by binding to water molecules. The same is also true in plant as excess absorbed ions in plant can influence water cellular behavior and its availability. 2) In addition, extra amounts of salts in plant can adversely affect plant growth by interfering with the functions of other ions. For example, the similar chemical properties of sodium (Na^+) and potassium (K^+) and of chloride (Cl^-) and nitrate (NO_3^-) decreases the uptake of K^+ and NO_3 in a salty soil resulting in the appearance of toxic effects of salinity in plant (Munns and James, 2003). Accordingly, it is suggested that higher rate of N and K fertilization be applied in a saline soil.

To alleviate the unfavorable effects of salinity on plant growth in a salty soil different strategies are used including the use of tolerant plants and leaching the soil by calculating the leaching requirement, which is the ratio of soil salinity to irrigation water salinity. In addition, identification of genes, expressed under salinity stress can be useful for the development of tolerant crop plants under salinity (Evelin et al., 2009).

Mycorrhizal plants have also demonstrated their superior tolerance to salinity, relative to non mycorrhizal plants. AM fungi are able to significantly increase water and nutrient uptake by the host plant, which can be effective in alleviating the stress (The presence of vesicles with high number of vacuoles can also store high concentration of ions in AM vacuoles). Eventually, through adjusting the K/Na ratio in the host plant, AM fungi can alleviate the adverse effects of salinity on plant growth and yield production (Mukerji and Chamol, 2003; Scheloske et al., 2004; Giri and Mukerji, 2004; Al-Karaki, 2006; Daei et al., 2009).

Inoculation of lettuce plants with AM fungi, at salinity levels of 2, 4, 8, 12 dS/m increased plant growth by 3.4, 8.2, 11.7 and 29.3%, respectively, relative to non-mycorrhizal plants (Cantrell and Linderman, 2001). Plant growth and nutrient uptake of P, K, Zn, Cu and Fe in mycorrhizal tomato was significantly higher than non mycorrhizal plants under salinity stress. Tomato fresh weight in plants inoculated with *G. mosseae* was increased by 60 and 29% under saline and non-saline conditions, respectively (Al-Karaki and Hammad, 2001; Al-Karaki, 2006). These results indicate the higher effectiveness of AM species on plant growth at higher levels of stress (Miransari et al., 2007; 2008).

Under field conditions, Daei et al. (2009) indicated that when using soil and irrigation salinity of 7.41 and 13.87 dS/m, AM species increased wheat dry weight and grain yield by 42.08 and 38.3%, respectively. *G. etunicatum* was the most efficient species compared with *G. mosseae* and *G. intradices*, indicating that the right combination of AM species and plant host may result in the highest performance of mycorrhizal plant (Jansa et al., 2008) under stress. In addition, AM species resulted in the lowest concentration of Na^+ and Cl^-. Their results also verified that genetical modification of the host plant may enhance the efficiency of mycorrhizal plant under stress. Under salinity stress, tolerant plant species are able to absorb higher rates of nutrients resulted by higher colonization rate indicating that there is interaction between AM species and host plant. Usually with increasing the salinity level AM species become more capable to alleviate the stress (Audet and Charest, 2006; Subramanian et al., 2006; Miransari et al., 2007, 2008).

In brief, the followings are the mechanisms, used by arbuscular mycorrhizal species to alleviate the stress of salinity on plant growth (Evelin et al., 2009). 1) enhanced nutrient uptake, 2) adjustment of K/Na ratio, 3) biochemical adaptations including the production of different osmolytes such as polyamines, proline, betaines, carbohydrates and antioxidants, 4) physiological alterations such as photosythetic adjustment, cellular permeability and water potential, production of ABA, and nodule formation and N fixation, 5) molecular adjustment including the expression of genes, and 6) morphological changes.

Soil Compaction and AM Fungi

Agricultural soils are subjected to compaction as a result of extensive use of agricultural machinery in the field. Use of different machines in the field has become very common for the implementation of different practices. Although there are multipurpose machines, soils are always subjected to compaction, especially when using machines at high moisture. Different mechanical techniques such as using subsoilers may temporarily alleviate the stress; however such a method is not economically and environmentally recommendable. Economically, it is of high expenses and environmentally although it may remove the hard pan, its adverse effects on soil structure and hence environment cannot be favored.

Accordingly, researchers have suggested and tested the use of biological methods, which are economically and environmentally superior to mechanical methods. For example, Miransari et al. (2006; 2007; 2008; 2009a, b) tested the effects of different arbuscular mycorrhizal species on corn and wheat growth, yield and nutrient uptake under compaction stress both under field and greenhouse conditions. The results are promising indicating that AM species are able to alleviate the stress of soil compaction on plant growth and yield under compaction. Hence, the field application of AM fungi can be beneficial under different conditions including stress.

Under field conditions using a John Deer tractor the field was compacted for two subsequent years. Corn seeds (cultivar 704) were planted and inoculated with four different species of arbuscular mycorrhiza including control, Iranian *G. mosseae*, Iranian *G. etunicatum*, and Canadian *G. mosseae*, received from GINCO (Glomales in vitro Collection), Canada. AM species were selected based on their Most Probable Number (MPN) test. Compared with other AM species *G. intraradices* resulted in the lowest root infection so was not used in the experiment.

Soil compaction resulted in the following adverse effects on plant growth: decreased plant height and growth, pale leaf color, and cluster and pan cake root growth. Use of AM species alleviated the stress; although with increasing the stress and at the highest level of stress (20-time tractor passing) the alleviating effects of AM fungi on corn growth and yield decreased (Miransari et al., 2006). Under high levels of stress, the symbiotic effects between AM fungi and the host plant may turn parasitic, as AM may not be able to function properly (Paszkowski, 2006).

Under greenhouse conditions 10-kg pots were compacted at four different levels using 2-kg weights, released from a 20 cm height. AM species with different origins significantly increased corn root growth. AM fungi are able to enhance the uptake of different nutrients including P. The extensive network of fungal hypha and the production of enzymes such as phosphatase can contribute to such enhanced nutrient availability and hence uptake by plant. P is an important macronutrient, which among its other functions in the plant can also significantly increase root growth (Miransari et al., 2007). The increase in the nutrient uptake of mycorrhizal plants (corn and wheat) under compaction has also been shown by the results of our experiments (Miransari et al., 2009a, b).

For the wheat experiments (Miransari et al., 2008), AM species increased root, shoot growth as well as grain yield. Interestingly, AM species increased root/shoot ratio relative to the control treatment. This is due to: 1) as just mentioned, under compaction stress AM species are able to significantly increase root growth, and 2) under different stresses including salinity (Miransari and Smith, 2007; 2009), suboptimal root temperature (Miransari and Smith 2008) and compaction (Miransari et al., 2007; 2008) plant allocate more carbon to their roots.

Canadian *G. mosseae* resulted in the highest uptake of P (60%) and Fe (58%) under compaction relative to the control treatments. This was also the case for Iranian *G. etunicatum*, as it increased corn nutrient uptake under compaction. Iranian *G. mosseae* was also effective, especially under the highest level of compaction, relative to the other species (Miransari et al., 2009a). For wheat, P uptake was also increased by AM species even under the highest level of compaction (Miransari et al., 2009b). For all greenhouse experiments both sterilized and unsterilized soils were tested. Although the tested soils were not biologically active, AM species did not seem to be very synergistic to the soil local microbes as in most cases the sterilized soils indicated to be more responsive under the highest level of compaction.

CONCLUSION

AM fungi are important components of ecosystem significantly affecting its efficiency. AM fungi are able to enhance plant growth and yield production by the followings: 1) enhanced water and nutrient uptake, 2) improving soil structure, 3) controlling pathogens, 4) alleviating soil stresses, and 5) interacting with other soil microbes. It is important to test their behavior under different conditions so that efficient strategies are used to enhance ecosystem efficiency. AM fungal diversity and their application (single or combined) can also affect plants performance as well as their community structure.

REFERENCES

Akiyama, K., Matsuzaki, K., Hayashi, H. 2005. Plant sesquiterpenes induce hyphal branching in arbuscular mycorrhizal fungi. *Nature* 435:824–827.

Alkan, N., Gadkar, V., Yarden, O., Kapulnik, Y. 2006. Analysis of quantitative interactions between two species of arbuscular mycorrhizal fungi, *Glomus mosseae* and *G. intraradices*, by real-time PCR. *Applied and Environmental Microbiology* 72: 4192–4199.

Al-Karaki, G.N., Hammad, R. 2001.Mycorrhizal influence on fruit yield and mineral content of tomato grown under salt stress. *Journal of Plant Nutrition* 24:1311–23.

Al-Karaki, G.N. 2006. Nursery inoculation of tomato with arbuscular mycorrhizal fungi and subsequent performance under irrigation with saline water. *Scientia Horticuture* 109:1–7.

Aroca, R., Porcel, R., Ruiz-Lozano, M. 2007. How does arbuscular mycorrhizal symbiosis regulate root hydraulic properties and plasma membrane aquaporins in *Phaseolus vulgaris* under drought, cold and salinity stresses? *New Phytologist* 173:808–816.

Audet P., Charest, C. 2006. Effects of AM colonizationon''wild tobacco''plants grown in zinc-contaminated soil. *Mycorrhiza* 16:277–83.

Augé, R.M. 2001. Water relations, drought and vesicular arbuscular mycorrhizal symbiosis. *Mycorrhiza* 11:3–42.

Avis, T., Gravel, V., Antoun, H., Tweddell, R. 2008. Multifaceted beneficial effects of rhizosphere microorganisms on plant health and productivity. *Soil Biology and Biochemistry 40*:1733-1740.

Balestrini, R. Lanfranco L. 2006. Fungal and plant gene expression in arbuscular mycorrhizal symbiosis. *Mycorrhiza* 16:509–524.

Boomsma, C., Vyn, T. 2008. Maize drought tolerance: Potential improvements through arbuscular mycorrhizal symbiosis? *Field Crops Research* 108: 14–31.

Cantrell, I.C., Linderman, R.G. 2001. Preinoculation of lettuce and onion with VA mycorrhizal fungi reduces deleterious effects of soil salinity. *Plant and Soil* 233:269–81.

Carlsen, S., Understrup, A., Fomsgaard, I., Mortensen, A., Ravnskov, S., 2008. Flavonoids in roots of white clover: interaction of arbuscular mycorrhizal fungi and a pathogenic fungus. *Plant and Soil* 302:33–43.

Carvajal, M., Cooke, D.T., Clarkson, D.T. 1996. Responses of wheat plants to nutrient deprivation may involve the regulation of water-channel function. *Planta* 199:372–381.

Chiou, T-J. et al. 2006. Regulation of phosphate homeostasis by MicroRNA in Arabidopsis. *Plant Cell* 18, 412–421.

Clarkson, D.T. 1985. Factors affecting mineral nutrient acquisition by plants. *Annual Reviews of Plant* Physiology 36:77–115.

Cornillon, P., Palloix, A. 1997. Influence of sodium chloride on the growth and mineral nutrition of pepper cultivars. *Journal of Plant Nutrition* 20:1085–94.

Daei, G., Ardakani, M., Rejali, F., Teimuri, S., Miransari, M. 2009. Alleviation of salinity stress on wheat yield, yield components, and nutrient uptake using arbuscular mycorrhizal fungi under field conditions. *Journal of Plant Physiology* 166: 617-625.

Drew, E.A., Murray, R.S., Smith, S.E. et al. 2003. Beyond the rhizosphere: growth and function of arbuscular mycorrhizal external hyphae in sands of varying pore sizes. *Plant, Cell and Environment.* 251:105–114.

Evelin, H., Kapoor, R., Giri, B. 2009. Arbuscular mycorrhizal fungi in alleviation of salt stress: a review. *Annals of Botany* 104: 1263–1280.

Finlay, R. 2008. Ecological aspects of mycorrhizal symbiosis: with special emphasis on the functional diversity of interactions involving the extraradical mycelium. *Journal of Experimental Botany* 59:1115–1126.

Fitter, A.H. 2006. What is the link between carbon and phosphorus fluxes in arbuscular mycorrhizas? A null hypothesis for symbiotic function. *New Phytologist* 172: 3–6.

George, E., Marschner, H., Jakobsen, I. 1995. Role of arbuscular mycorrhizal fungi in uptake of phosphorus and nitrogen from soil. *Critical Reviews in Biotechnology* 15: 257–270.

Giri, B., Mukerji, K.G. 2004. Mycorrhizal inoculants alleviates salt stress in *sesbania aegyptica* and *sesbania grandiflora* under field condition: evidence for reduced sodium and improved magnesium uptake. *Mycorrhiza* 14: 307–12.

Glassop, D., Smith, S.E., Smith, F.W. 2005. Cereal phosphate transporters associated with the mycorrhizal pathway of phosphate uptake into roots. *Planta* 222:688–698.

Govindarajulu, M., Pfeffer, P.E., Jin, H., Abudaker, J., Douds, D.D., Allen, J.A., Bucking, H., Lammers, P.J., Shachar-Hill, Y. 2005. Nitrogen transfer in the arbuscular mycorrhizal symbiosis. *Nature* 435: 819–823.

Graham, J.H., Eissenstat, D.M. 1994. Host genotype and the formation and function of VA mycorrhizae. *Plant and Soil* 159: 179–185.

Graham, J.H. 2001. What do root pathogens see in mycorrhizas? *New Phytologist* 149:357–359.

Grunwald, U., Guo, W. Kerstin Fischer, K. Isayenkov, S. Ludwig-Müller, J. Hause, B., Yan, X., Küster, H., Franken, P. 2009. Overlapping expression patterns and diVerential transcript levels of phosphate transporter genes in arbuscular mycorrhizal, Pi-fertilised and phytohormone-treated *Medicago truncatula* roots. *Planta* 229:1023–1034.

Gustafson, D.J., Casper, B.B. 2006. Differential host plant performance as a function of soil arbuscular mycorrhizal fungal communities: experimentally manipulating co-occurring *Glomus* species. *Plant Ecology* 183: 257–263.

Harrison, M.J. 2005. Signaling in the arbuscular mycorrhizal symbiosis. *Annual Review of Microbiology* 59:19–42.

Hartnett, D.C., Wilson, G.W.T. 2002. The role of mycorrhizas in plant community structure and dynamics: Lessons from grasslands. *Plant and Soil* 244: 319–331.

Herrera-Medina, M., Tamayo, M., Vierheilig, H., Ocampo, J., Garcıa-Garrido, J. 2008. The Jasmonic acid signalling pathway restricts the development of the arbuscular mycorrhizal association in tomato. *Journal of Plant Growth Regulation* 27:221–230.

van der Heijden, M.G.A., Bardgett, R.D., van Straalen, N.M. 2008. The unseen majority: Soil microbes as drivers of plant diversity and productivity in terrestrial ecosystems. *Ecology Letters* 11: 296–310.

van der Heijden, M., Horton, T. 2009. Socialism in soil? The importance of mycorrhizal fungal networks for facilitation in natural ecosystems. *Journal of Ecology* 97:1139–1150.

Ivanov, S., Fedorova, E., Bisseling, T. 2010. Intracellular plant microbe associations: secretory pathways and the formation of perimicrobial compartments *Current Opinion in Plant Biology* 13:1-6.

Isayenkov, S., Fester, T., Hause, B. 2004. Rapid determination of fungal colonization and arbuscule formation in roots of *Medicago truncatula* using real-time (RT) PCR. *Journal of Plant Physiology* 161: 1379–1383.

Jansa, J., Mozafar, A., Anken, T., Ruh, R., Sanders, I.R., Frossard, E. 2002. Diversity and structure of AMF communities as affected by tillage in a temperate soil. *Mycorrhiza* 12: 225–234.

Jansa, J., Mozafar, A., Frossard, E. 2005. Phosphorus acquisition strategies within arbuscular mycorrhizal fungal community of a single field site. *Plant and Soil* 276: 163–176.

Jansa, J., Smith, A., Smith, S. 2008. Are there benefits of simultaneous root colonization by different arbuscular mycorrhizal fungi? *New Phytologist* 177:779–789.

Javot, H., Penmetsa, R.V., Terzaghi, N., Cook, D.R., Harrison, M.J. 2007. A *Medicago truncatula* phosphate transporter indispensable for the arbuscular mycorrhizal symbiosis. *Proceedings of the National Academy of Sciences,* USA 104:1720–1725.

Jakobsen, I., Smith, S.E., Smith, F.A. 2002. Function and diversity of arbuscular mycorrhizae in carbon and mineral nutrition. In: van der Heijden, M.G.A, Sanders, I.R. (eds) *Mycorrhizal ecology.* Springer-Verlag, Berlin, Heidelberg, pp 75–92.

Johnson, N.C., Pfleger, F.L., 1992. Vesicular-arbuscular mycorrhizae and cultural stresses. In: Bethlenfalvay, J., Linderman, R.G. (Eds.), *Mycorrhizae in Sustainable Agriculture.* American Society of Agronomy, Crop Science Society of America, and Soil Science Society of America, Madison, pp. 71–99.

Johnson, N.C., Graham, J.H., Smith, F.A. 1997. Functioning of mycorrhizal associations along the mutualism-parasitism continuum. *New Phytologist* 135:575–586.

Johnson, N.C. 2010. Resource stoichiometry elucidates the structure and function of arbuscular mycorrhizas across scales. *New Phytologist* 185: 631–647.

Journet, E.P,, van Tuinen, D., Gouzy, J. et al. 2002. Exploring root symbiotic programs in the model legume Medicago truncatula using EST analysis. *Nucleic Acids Research* 30:5579–5592.

Kaschuk, G., Kuyper, T.W., Leffelaar, P.A., Hungria, M., Giller, K. 2009. Are the rates of photosynthesis stimulated by the carbon sink strength of rhizobial and arbuscular mycorrhizal symbioses? *Soil Biology and Biochemistry* 41:1233–1244.

Koide, R.T. 2000. Functional complementarity in the arbuscular mycorrhizal symbiosis. *New Phytologist* 147: 233–235.

Koide, R.T., Dickie, I.A. 2002. Effects of mycorrhizal fungi on plant populations. *Plant and Soil 244*: 307–317.

Lambers, H., Poot, P., (eds). 2003. *Structure and functioning of cluster roots and plant responses to phosphate deficiency.* Kluwer, Dordrecht, Boston, London.

Lambers H, Chapin FSI, Pons TL (2008a) *Plant Physiological Ecology.* Springer, New York

Lambers, H., Raven, J., Shaver, G, Smith, S. 2008b. Plant nutrient-acquisition strategies change with soil age. *Trends in Ecology and Evolution* 23: 95-103.

Landis, F.C., Fraser, L.H. 2008. A new model of carbon and phosphorus transfers in arbuscular mycorrhizas. *New Phytologist* 177:466–479.

Leigh, J., Hodge, A., Fitter, A.H. 2008. Arbuscular mycorrhizal fungi can transfer substantial amounts of nitrogen to their host plant from organic material. *New Phytologist* 181: 199–207.

Lekberg Y, Koide RT, Rohr JR, Aldrich-Wolfe L, Morton JB. 2007. Role of niche restrictions and dispersal in the composition of arbuscular mycorrhizal fungal communities. *Journal of Ecology* 95: 95–105.

Liu, A., Plenchette, C., Hamel, C., 2007. Soil nutrient and water providers: how arbuscular mycorrhizal mycelia support plant performance in a resourcelimited world. In: Hamel,

C., Plenchette, C. (Eds.), *Mycorrhizae in Crop Production. Haworth Food & Agricultural Products Press,* Binghamton, NY, pp. 37–66.

Luu, D-T., Maurel, C. 2005. AQPs in a challenging environment: molecular gears for adjusting plant water status. *Plant, Cell and Environment* 28:85–96.

Maherali, H., Klironomos, J.N. 2007. Influence of phylogeny on fungal community assembly and ecosystem functioning. *Science* 316: 1746–1748.

Marschner, H., Dell, B. 1994. Nutrient uptake in mycorrhizal symbiosis. *Plant and Soil* 159: 89–102.

Marschner, H., 1995. *Mineral Nutrition of Higher Plants*. Academic Press, San Diego.

Mathesius, U. 2009. Comparative proteomic studies of root–microbe interactions. *2009. Journal of Proteomics* 72:353-366.

Merryweather J, Fitter A. 1998. The arbuscular mycorrhizal fungi of *Hyacinthoides non-scripta* I. *Diversity of fungal taxa. New Phytologist* 138: 117–129.

Meixner, C., Ludwig-Müller, J., Miersch, O., Gresshoff, P., Staehelin, C. et al. 2008. Lack of mycorrhizal autoregulation and phytohormonal changes in the supernodulating soybean mutant. *Planta* 222:709-715.

Miller, R.M., Jastrow, J.D., Reinhardt, D.R. 1995. External hyphal production of vesicular-arbuscular mycorrhizal fungi in pasture and tallgrass prairie communities. *Oecologia* 103:17–23.

Miller, R.M., Miller, S.P., Jastrow, J.D., Rivetta, C.B. 2002. Mycorrhizal mediated feedbacks influence net carbon gain and nutrient uptake in *Andropogon gerardii. New Phytologist* 155: 149–162.

Miller, R.M., Jastrow, J.D. 2002. Mycorrhizal influence on soil structure. In: Kapulnik, Y., Douds, D.D. (eds) *Arbuscular mycorrhizae: molecular biology and physiology.* Kluwer Academic Publishers, Dordrecht, The Netherlands, pp 3–18.

Miransari, M, Smith, D.L. 2007. Overcoming the stressful effects of salinity and acidity on soybean[Glycine max (L.) Merr.] nodulation and yields using signal molecule genistein under field conditions. *Journal of Plant Nutrition* 30: 1967–92.

Miransari, M., Bahrami, H. A., Rejali, F. Malakouti, M. J. 2006. Evaluating the effects of arbuscular mycorrhizae on corn nutrient uptake and yield in a compacted soil. In Persian, Abstract in English (CAB Abstracts), *Iranian Journal of Soil and Water* 20:106-121.

Miransari, M., Smith, D.L., 2007. Overcoming the stressful effects of salinity and acidity on soybean [Glycine max (L.) Merr.] nodulation and yields using signal molecule genistein under field conditions. *Journal of Plant Nutrition* 30, 1967-1992.

Miransari, M., Smith, D.L., 2008. Using signal molecule genistein to alleviate the stress of suboptimal root zone temperature on soybean-Bradyrhizobium symbiosis under different soil textures. *Journal of Plant Interactions,* 3, 287-295.

Miransari M., Smith, D. 2009. Alleviating salt stress on soybean (*Glycine max* (L.) Merr.) - Bradyrhizobium japonicum symbiosis, using signal molecule genistein. *European Journal of Soil Biology,* 45, 146-152.

Miransari, M., H.A. Bahrami, F. Rejali, M. J. Malakouti, H. Torabi. 2007. Using arbuscular mycorrhiza to reduce the stressful effects of soil compaction on corn (*Zea mays* L.) growth. *Soil Biology and Biochemistry* 39:2014-2026.

Miransari, M., Bahrami, H.A., Rejali, F., Malakouti, M.J., 2008. Using arbuscular mycorrhiza to reduce the stressful effects of soil compaction on wheat (*Triticum aestivum* L.) growth. *Soil Biology and Biochemistry* 40:1197-1206.

Miransari M., Rejali F., Bahrami H.A., Malakouti, M.J., 2009a. Effects of soil compaction and arbuscular mycorrhiza on corn (*Zea mays* L.) nutrient uptake. *Soil and Tillage Research* 103:282-290.

Miransari M., Rejali F., Bahrami H.A., Malakouti, M.J., 2009b. Effects of arbuscular mycorrhiza, soil sterilization, and soil compaction on wheat (*Triticum aestivum* L.) nutrients uptake. *Soil and Tillage Research* 104:48-55.

Miransari, M.. Contribution of arbuscular mycorrhizal symbiosis to plant growth under different types of soil stresses. Review article. *Plant Biology,* 12, 563-569.

Mukerji, K.G., Chamol, B.P. 2003. *Compendium of mycorrhizal research.* New Delhi: A. P. H. Publisher. pp 310.

Munkvold, L., Kjøller, R., Vestberg, M., Rosendahl, S., Jakobsen, I. 2004. High functional diversity within species of arbuscular mycorrhizal fungi. *New Phytologist* 164: 357–364.

Munns, R., James, R.A. 2003. Screening methods for salinity tolerance: a case study with tetraploid wheat. *Plant Soil* 253:201–18.

Nanda, A., Andrio, E., Marino, D., Pauly, N., Dunand, C. 2010 Reactive Oxygen species during plant-microorganism early interactions. *Journal of Integrative Plant Biology* 52: 195-204.

Neumann, E, George, E. 2004. Colonisation with the arbuscular mycorrhizal fungus *Glomus mosseae* (Nicol. & Gerd.) enhanced phosphorus uptake from dry soil in Sorghum bicolor (L.). *Plant, Cell and Environment* 261:245–255.

Newton, A., Fitt, B., Atkins, S., Walters, D., Daniell, T. 2010. Pathogenesis, parasitism and mutualism in the trophic space of microbe–plant interactions. *Tends in Microbiology* 18: 65-373.

Olsson, P.A., Burleigh, S.H., van Aarle, I.M., 2005. The influence of external nitrogen on carbon allocation to Glomus intraradices in monoxenic arbuscular mycorrhiza. *New Phytologist* 168, 677–686.

Paszkowski, U. 2006. Mutualism and parasitism: the yin and yang of plant symbioses *Current Opinion in Plant Biology* 9:364–370.

Paszkowski, U., Kroken, S., Roux, C. et al. 2002. Rice phosphate transporters include an evolutionarily divergent gene specifically activated in arbuscular mycorrhizal symbiosis. *Proceedings of the National Academic of Sciences USA* 99:13324–13329.

Pfeffer, P.E., Douds, D.D., Bucking, H., Schwartz, D.P., Shachar-Hill, Y. 2004. The fungus does not transfer carbon to or between roots in an arbuscular mycorrhizal symbiosis. *New Phytologist* 163: 617–627.

Plenchette, C., Clermont-Dauphin, C., Meynard, J.M., Fortin, J.A., 2005. Managing arbuscular mycorrhizal fungi in cropping systems. *Canadian Journal of Plant Science* 85:31-40.

Poulsen, K.H., Nagy, R., Gao, L.L. et al. 2005. Physiological and molecular evidence for Pi uptake via the symbiotic pathway in a reduced mycorrhizal colonization mutant in tomato associated with a compatible fungus. *New Phytologist* 168:445–453.

Reinbott, T.M., Blevins, D.G. 1999. Phosphorus nutritional effects on root hydraulic conductance, xylem water flow and flux of magnesium and calcium in squash plants. *Plant, Cell and Environment* 209:263–273.

Rillig, M.C., Mummey, D.L. 2006. Mycorrhizas and soils structure. *New Phytologist,* 171, 41–53.

Robinson-boyer, L., Grzyb, I., Jeffies, P. 2009. Shifting the balance from qualitative to quantitative analysis of arbuscular mycorrhizal communities in field soils. *Fungal Ecology* 2:1 - 9.

Rosendahl, S. 2008. Communities, populations and individuals of arbuscular mycorrhizal fungi. *New Phytologist* 178: 253–266.

Ruiz-Lozano, J.M., 2003. Arbuscular mycorrhizal symbiosis and alleviation of osmotic stress. New perspectives for molecular studies. *Mycorrhiza* 13, 309–317.

Scheloske, S, Maetz, M., Schneider, T,, Hildenbrandt, U,, Bothe, H., Povh, B. 2004. Elemental distribution in mycorrhizal and non-mycorrhizal roots of the halophyte Aster tripo- lium determined by proton induced X-ray emission. *Protoplasma* 223:183–9.

Smith, S.E., Read, D.J., 2008. *Mycorrhizal Symbiosis*, third ed. Elsevier Ltd., London.

Smith, F.A., Grace, E.J., Smith, S.E. 2009. *More than a carbon economy: nutrient trade and ecological sustainability in facultative arbuscular mycorrhizal symbioses.* New Phytologist in press.

Smith, S., Facelli, E. Pope, S. 2010. Plant performance in stressful environments: interpreting new and established knowledge of the roles of arbuscular mycorrhizas. *Plant and Soil* 326:3–20.

Subramanian K., Santhanakrishnan P., Balasubramanian P. 2006. Responses of field grown tomato plants to arbuscular mycorrhizal fungal colonization under varying intensities of drought stress. *Scientia Horticulture* 107:245–53.

Sylvia, D.E., Hammond, L.C., Bennet, J.M., Hass, J.H., Linda, S.B., 1993. Field response of maize to a VAM fungus and water management. *Agron. J.* 85, 193–198.

Tikhonovich, I., Provorov, N. 2009. From plant–microbe interactions to symbiogenetics: a universal paradigm for the interspecies genetic integration. *Annals of Applied Biology* 154: 341-350.

Van Wees, S, Van der Ent, S., Pieterse, C. 2008. Plant immune responses triggered by beneficial microbes. *Current Opinion in Plant Biology.* 11:443–448.

Valot, B., Dieu, M., Recorbet, G., Raes, M., Gianinazzi, S., Dumas-Gaudot, E. 2005. Identification of membrane-associated proteins regulated by the arbuscular mycorrhizal symbiosis. *Plant Molecular Biology* 59:565–80.

Valot, B., Negroni, L., Zivy, M., Gianinazzi, S., Dumas-Gaudot, E. A. 2006. Mass spectrometric approach to identify arbuscular mycorrhiza-related proteins in root plasma membrane fractions. *Proteomics* 6:S145–55.

Wehner, J., Antunes, P., Powell, J., Mazukatow, J., Rillig, M. 2010. Plant pathogen protection by arbuscular mycorrhizas: A role for fungal diversity? *Pedobiologia* 53:197–201.

Wilson, G.W.T, Rice, C.W., Rillig, M.C., Springer, A., Hertnett, D.C. 2009. Soil aggregation and carbon sequestration are tightly correlated with the abundance of arbuscular mycorrhial fungi: Results from long-term field experiments. *Ecology Letters* 12: 452–461.

Zsögön, A., Lambais, M., Benedito, V., Vargas, A., Figueira, O., Peres, L. 2008. Reduced abuscular mycorrhizal colonization in tomato ethylene mutants. *Scientia Agriculture* 65:259-267.

Zhao, S., Qi, X. 2008. Signaling in Plant Disease Resistance and Symbiosis. *Journal of Integrative Plant Biology* 50: 799–807.

Zinselmeier, C., Westgate, M.E., Schussler, J.R., Jones, R.J., 1995. Low water potential disrupts carbohydrate metabolism in maize (Zea mays L.) ovaries. *Plant Physiology* 107: 385–391.

In: Mycorrhizal Fungi
Editor: Susanne M. Fulton

ISBN 978-1-61122-659-1
© 2011 Nova Science Publishers, Inc.

Chapter 5

INFLUENCE OF ORGANIC POLLUTANTS ON ARBUSCULAR MYCORRHIZAL FUNGI

Eswaranpillai Uma, Sarah Jaison and Thangavelu Muthukumar*
Root and Soil Biology Laboratory, Department of Botany, Bharathiar University, Coimbatore – 641 046, Tamilnadu, India

ABSTRACT

The numerous benefits contributed by arbuscular mycorrhizal (AM) fungi has lead to recognition of their benefits in sustainable agriculture and various natural environmental processes. Application of anthropogenic and other compounds intentionally in the form of pesticides, fungicides, fertilizers or unintentionally like polycyclic aromatic hydrocarbons (PAH) which affect soil fungi have shown varying effects on AM association. Not all organic pollutants are detrimental to AM fungi. Some fungicides such as Fludioxonil have a stimulatory effect on mycorrhizal formation and function. Rapid decrease in the metabolic activity of AM fungi, a decline in arbuscule production, which is the main site of nutrient exchange between the host plant and the fungus, reduction in the total root length colonized and slow colonization rates are the negative effects shown by certain fungicides. Pesticides generally inhibit AM colonization and spore numbers. Slow release mineral fertilizers though benefit AM fungi, their excessive use can result in suppression of the AM fungi. Certain PAH especially anthracene reduce the development of extra radical mycelium, sporulation, root colonization and spore germination. Different AM fungal species show differential sensitivity even within a particular class of organic pollutant. Caution should also be exercised while irrigating soils with effluents as these could affect AM fungi. A reduction in our dependency on conventional agricultural practices especially organic pesticides by using AM fungi as a biological tool which is possible only by maintaining low levels of organic pollutants in the soil. Minimal use of organic pollutants especially pesticides would enable the maintenance of a healthy AM fungal populations essential for sustainable plant production systems.

* Corresponding author, e-mail: tmkum@yahoo.com

INTRODUCTION

Organic pollutants have received much attention in the recent times. They are the major contaminants in soil and are of great concern worldwide because of their carcinogenic and /or mutagenic properties. Organic pollutants include biocides (fungicides, pesticides, herbicides, nematicides), hydrocarbons and fertilizers. Sewage arising from industrial and domestic point sources to agricultural run off and shipping activities are the major contributors to the anthropogenic organic load in the soil. Concern over the non-target effects of biocides employed in agriculture has shifted its focus from pests and disease control towards integrated management techniques that employ combinations of cultural practices, biological control and the use of phytochemicals to combat pests and pathogens (Schreiner and Bethlenfalvay, 1997a,b). Polycyclic hydrocarbons and other petroleum products are among the worst pollutants because of their toxicity, carcinogenicity, wide distribution and slow degradation in the environment. Indiscriminate use of fertilizers affects soil fauna. By nature, the rhizosphere has a complex ecology with numerous feedback loops that regulate microbial populations between the plant and its microflora. This controls the fate and degradation of organic contaminants in the soil. Some organic contaminants are more bioavailable and are more susceptible to microbial degradation because they can be used as a source of carbon, when present at non-toxic levels (Kirk et al., 2005).

Soil fungi have been virtually ignored despite being critical to ecosystem functioning, as decomposers, mutualists and a food-source for many other soil organisms. One such fungal group are the arbuscular mycorrhizal (AM) fungi, which form a permanent obligate symbiotic relationship with the roots of over 80% of all terestrial plant species (Wang and Qiu, 2006). AM fungi are responsible for enhanced uptake of macro and micro nutrients as well as access to nutrient pools typically not available to the root system of plants, such as nitrogen (N) (Hodge et al., 2001) and phosphorus (P) (Tarafdar and Marschner, 1994), directly from decomposing organic matter. In addition, AM fungi imparts resistance against drought (Auge, 2001), pathogenic organisms (Whipps, 2004), toxic organic compounds (Corgie et al., 2006) and heavy metals (Audet and Charest, 2006; Rivera-Beacerril et al., 2005). AM fungi in contaminated soil enhances plant establishment and reduce the toxic effects of pollutants (Joner and Leyval, 2001; Leyval and Binet, 1998; Olexa et al., 2000).

Overall levels of mycorrhizal colonization by AM fungi are known to be strongly influenced by the soil environment (Cairney and Meharg, 1999). The soil nutrient status (Smith and Read, 1997; Wallenda and Kottke, 1998), organic matter content (Jurgensen et al., 1977), and moisture (Parke et al., 1983; Haugen and Smith, 1992) are the key factors that influence the extent of root colonization by AM fungi. Levels of mycorrhizal colonization may also show variation in response to major soil disturbances like organic pollution. In such instances, reductions in level of AM colonization appear to reflect reductions in propagule density, establishment of mycelial systems (Allen et al., 1984; Jasper et al., 1987; Brundrett et al., 1996). Germination of AM fungal spores, root colonization, extent of arbuscule and vesicle formation which include a series of morpho-physiological and biochemical events are regulated by several environmental factors. It will be very interesting to evaluate how different organic pollutants affect the AM symbiosis and how AM fungi responds to these different soil stresses.

HYDROCARBONS

The response of AM fungi to exposure to hydrocarbons such as aromatic, polycyclic aromatic (PAH) and petroleum hydrocarbons (PHC) tend to vary greatly. Generally, the presence of these hydrocarbons in soils are detrimental to the formation or expression of AM fungal association (Davies et al., 2001; Cabello, 1997; Leyval and Binet, 1998). Spore germination and subsequent hyphal elongation are drastically reduced by phenanthrene (PHE). The reduction in spore germination may be greater than 90% in the presence of PHE. The toxic effects of PHE results from its hydrophobic nature and the ability of PHE to cover the major spore surface area because of its large particle size (Cerniglia, 1992). Further, the hydrophobic nature of PHE deters spore hydration and germination. Wu et al. (2009) found that root length colonization in mycorrhizal *Medicago sativa* decreased from 69.9% to 52.7% with increasing PHE concentrations. The spread of AM in rye grass roots raised in substrate spiked with fluorine and PHE generally increased with time and was 17.8% when assessed after 70 days of growth (Gao et al., 2010). The increase in plant biomass in PHE contaminated soil has been ascribed to the utilization of PHE as a carbon source by mycorrhizal roots (Corgie et al., 2006).

The effect of five ringed PAH Benzo[a]pyrene (BaP) whose metabolites are highly mutagenic tends to vary with AM fungal species. Increasing concentrations of BaP greatly reduces root colonization, extra radical hyphal length and sporulation in *Glomus intraradices* (Alarcon et al., 2006), *Glomus mosseae*, *Glomus ambisporum* and *Scutellospora heterogama* (Franco-Ramirez et al., 2007). In contrast, BaP does not adversely effect neither root colonization by *Gigaspora margarita* nor its efficiency to transfer phosphate (Tisserent et al., 1993). Furthermore, greater hyphal elongation occurs in G. intraradices spores germinating presence of 75 and 100 μg ml-1 of BaP (Alarcon et al., 2006). An increased activity of soil enzymes around AM fungal hyphae could enhance the degradation of BaP in the plant mycorhizosphere (Liu et al., 2004).

Leyval and Binet (1998) showed that AM colonization in some plant species was negatively affected by increasing additions of an industrial soil polluted with PAH, but not by the addition of similar concentrations of a single PAH like anthracene to the soil. Exposure of the AM fungus *G. intraradices* to anthracene reduces spore germination, extraradical mycelium development, root colonization and sporulation (Verdin et al., 2006). Joner et al. (2001) reported low levels of root colonization (0.5-5%) by *G. mosseae* in rye grass raised in soils spiked with anthracene, chrysene and dibenzo (a,h) anthracene. The positive influence of AM fungal inoculation on the dissipation of PAHs in soils have also been reported in several studies (Leyval and Binet, 1998; Binet et al., 2000; Joner et al., 2001, 2002; Verdin et al., 2006), suggesting a possible role of AM fungi in PAH degradation. But, there is no convincing evidence of the involvement of AM fungi in direct degradation of PAH. Binet et al. (2000) and Alarcon et al. (2006) failed to detect any dissipation or degradation of PAH in the mycorrhizospheres of *Lolium perenne* and *Echynochloa polystachya* respectively. However, AM fungi may enhance the tolerance of the host plants in PAH contaminated soils as shown for other contaminants. Binet et al. (2000) proposed that AM fungi might alleviate PAH tolerance in host plants by reducing the PAH uptake by plants although the mechanisms remain unclear. The limited dissipation of BaP in the mycorrhizosphere suggests that AM fungi can contribute to the stabilization or sequestration of PAH in contaminated soils through different mechanism (Alarcon et al., 2006). The first possible mechanism for PAH stabilization is the uptake and accumulation of PHE in AM fungal spores and extra radical mycelium (Gasper et al., 2002, Alarcon et al., 2006). The second mechanism is through the stimulation of microbial activity in the mycorrhizosphere leading to the degradation of hydrocarbons (Rilling and Steinberg, 2002). The third mechanism involves an indirect benefit of AM fungi through production of oxidative enzymes responsible for the degradation or transformation of PAH in mycorrhizal plants of contaminated soils (Leyval et al., 2002).

The reactivity of AM fungi to PAH tends to vary with the type of PAH and the AM fungal species involved. Aromatic hydrocarbon like xylene and particularly benzene are detrimental to root colonization and arbuscule formation by *G. mosseae*, while *Gigaspora rosea* was more sensitive to toluene and ethylbenzene (Cabello, 1997; Volante et al., 2005).

Contamination of soil with crude oil significantly reduces AM fungal propagule density (Cabello, 1997). Although not much information is available in literature about the effects of petroleum hydrocarbons on spore germination, changes in soil factors such as temperature, pH, soil moisture and root exudates in presence of organic chemicals may affect spore germination, development of the germ hyphae and hyphal branching in different AM fungal species (Tommerup and Briggs, 1981; Tommerup, 1984; Clark, 1997; Chiocchio et al., 2000). Several studies have shown the negative influence of PHC on AM fungal colonization (Cabello, 1997; Leyval and Binet, 1998; Gasper et al., 2002; Liu et al., 2004; Tang et al., 2009). Contrary to these observations, Geng et al. (2002) found that PHC had stimulatory influence on AM fungal colonization. Even as the rate of AM fungal colonization in maize roots declined with increasing concentration of diesel, the activity of antioxidant enzymes like superoxide dismutase (SOD), catalase (CAT) and peroxidase (POD) tend to increase (Tang et al., 2009). However, the presence of F2 diesel oil delays germ tube emergence and reduces the hyphal growth (Kirk et al., 2005). A moderate AM fungal colonization levels (63-77%) and spore numbers (715-912 per 100g of soil) have been reported in PHC contaminated soils (Franco-Ramirez et al., 2007). These evidences along with the formation of a functional AM in *Lolium multiflorum* by *G. intraradices* (Alarcon et al., 2008), suggests that AM fungal inocula can successfully establish in PHC contaminated soils.

EFFLUENTS

The use of domestic waste water and industrial effluents rich in organic pollutants to irrigate agricultural soils is practised at different scales. The use of refuse water to irrigate soil is a convienent solution in areas where the treatments of these waters are uneconomical or where these waters directly enter into water bodies. Long-term irrigation of wastewater or industrial effluents increases the concentration of nutrients and toxic elements in the soils (Ortega-Larrocea, 2001). Plants growing on wastewater irrigated soils have higher AM colonization levels, compared to plants growing on non-polluted soils (Srinivasan et al., 1995; Safir et al., 1990). Safir et al. (1990) indicated the absence of build up of any anti AM factors even after three years of continuous waste water irrigation. The response of AM formation to sewage irrigation depends on the associated plant species. For example, Kehri and Chandra (1990) studied the influence of sewage irrigation on 18 crop species. Sewage irrigation encouraged AM formation in five crops (Vigna sinensis, Spinacia oleracea, Coriandrum *sativum*, *Solanum tuberosum*, *Trigonella foenum-graecum*) but supressed colonization levels in *Saccharum officinarum*. Compared to wastewater, irrigation of pulp and paper mill effluent to agricultural soils, generally reduces AM formation, especially the root length with arbuscules in several crops and weedy species (Muthukumar and Vediyappan, 2010). The suppressive effect of the pulp and paper mill effluent has often been attributed to its organic nutrient richness. High nutrient inputs to soil or their buildup may diminish the colonization potential of AM fungi and also dependency of plants on AM fungi.

Sewage or effluent irrigation also affects the abundance and richness of AM fungal spores in the soils. A reduction in AM fungal spore numbers has been reported for soils irrigated with sugar mill effluent (Vani et al., 2005), sewage (Kehri and Chandra, 1990), paper mill effluent (Muthukumar and Vediyappan, 2010) and wastewater (Srinivasan et al., 1995; Ortega- Larrocea, 2001). This effect on spore numbers tend to change with soil types. In Mezquital Valley of Cental Mexico largest spore abundance was associated with Leptisols, while the lowest spore abundance occurred in Vertisols that were under long-term (90 years) and short-term wastewater irrigation (5 years) (Ortega-Larrocea, 2001). The effect of soil types on AM fungal spore abundance can vary with the frequency of irrigation and retention of the nutrients in the soil. In addition to spore density, effluent irrigation also reduces AM fungal diversity. Irrigation of soils with pulp and paper mill effluent reduced AM fungal species diversity by 50%. Effluent irrigation results in the elimination of certain AM fungal species. Though Ortega-Larrocea (2001) indicated that Glomus sinuosum (= Sclerocystis sinuosa) was less sensitive to eutrophication, this species tends to disappear from pulp and paper mill effluent irrigated soils and during conversion of natural soils to agriculture (Muthukumar and Vediyappan, 2010; Sieverding, 1989). Hence, this AM fungal species appears to be more sensitive to disturbance and nutrient enrichment. However, some AM fungal species like G. mosseae appear to be less sensitive to soil disturbances.

FUNGICIDES

Fungicides are extensively used in agricultural systems to control air borne, soil borne and seed borne fungal pathogens. Agricultural fungicides are not pathogen specific so they affect a wide range of non-target fungi (Bollen, 1979; Vyas, 1988). Several investigations have shown that fungicide application can affect AM fungi (Jalali and Domsch, 1975; Manjunath and Bagyaraj, 1984; Kough et al., 1987; Fitter and Nichols, 1988; Udaiyan et al., 1995; Wilson and Hartnett, 1998; Gupta and Borse, 1997; Pandey et al., 1996; Miller and Jackson, 1998; Parvathi et al., 1985). Fungicides have either a deleterious effect (Jalali and Domsch, 1975; De Bertoldi et al., 1977; Bailey and Safir, 1978; Boatman et al., 1978; Ocampo and Hayman, 1980; Kough et al., 1987; Fitter, 1986), no effect (Jalali and Domsch, 1975; Ocampo and Hayman, 1980) or stimulatory effect on the development of AM (Nemec, 1980; Jabaji-Hare and Kendrick, 1987).

Benomyl is a widely used fungicide and several investigations were carried out to understand the effect of benomyl/benlate on non-target fungi like AM fungi. Most of these studies indicate a deleterious effect of benomyl on AM fungi (Jalali and Domsch, 1975; Ocampo and Hayman, 1980, Kough et al., 1987; Fitter, 1986). Benomyl adversely affect the different stages of AM fungal development and function like spore germination, hyphal growth and sporulation (Trappe et al., 1984; Schreiner and Bethlenfalvay, 1997a,b). But such an effect does not seem to significantly influence the availability of soil P to the host plant (Merryweather and Fitter, 1996). A reduction in the extent of AM fungal colonization by benomyl has been reported in several crops (Kough et al., 1987; Jalali and Domsch, 1975; Ocampo and Hayman, 1980; Fitter, 1986). This reduction in AM fungal colonization closely reflects a decline in the host dry matter accumulation and AM fungal effectiveness measured as leaf P content (Habte, 1997). A reduction in alkaline phosphatase activity of both intraradical and extraradical AM fungal hyphae accompanies benomyl application (Kjøller and Rosendahl, 2000). Carr and Hinckley (1985) suggested that benomyl has a fungistatic rather than a fungicidal effect on AM fungi. However, the rapid decrease in succinate dehydrogenase (SDH) activity (Kough et al., 1987; Ames et al., 1982; Toth and Toth, 1982) in AM structures including arbuscules in response to the benomyl application indicates its fungicidal rather than a fungistatic effect. Benomyl affects the development and spread of intra radical mycelium within roots possibly by preventing nuclear division (Chioechio et al., 2000). But investigations also indicate least effects of benomyl on the colonization of corn roots by AM fungi (Samarbakhah et al., 2009) and pinnule P content or shoot yield at a high soil P (Habte, 1997). The lack of adverse effect of benomyl on non-target soil microorganisms has also been reported by De Bertoldi et al. (1977) and Vyas (1988).

Carbendazim, a benzimidazole fungicide, is a breakdown product of benomyl and both chemicals are known to severely inhibit the formation and function of AM (Dodd and Jeffries, 1989; Larsen et al., 1996; Merryweather and Fitter, 1996). These fungicides bind specifically to fungal tubulin, thereby inhibiting cellular processes involving microtubules, like mitosis and intracellular transport of particles (Davidse, 1986). Carbendazim has the most negative effect at concentrations of 0.001 μg/g and above leading to a sharp decline in hyphal P uptake. Concentrations above 0.1 μg/g almost completely inhibits hyphal P uptake. Similar observation has been reported for propiconazole but only at a higher concentration (1μg/g). Fungicide like fenpropimorph do not affect hyphal P uptake even when applied at concentrations of 125 or 46 μg/g (Schweiger and Jakobsen, 1998). In contrast, propiconazole and fenpropimorph inhibits the biosynthesis of ergosterol leading to the decreased capacity of the fungus to maintain normal membrane processes (Schmitz et al., 1992). Contrary to the above studies it has been shown that a mixture of carbendazim and propiconazole/ fenpropimorph stimulate hyphal P uptake by AM fungi, but was very significant with carbendazim. As fenpropimorph inhibits the formation and function of both intraradical and extraradical hyphae at 125 μg/g soil, propiconazole affect the intraradial hyphae only at a low concentration of 0.21 μg/g soil.

Bavistin and dithane M-45 has deleterious effect on AM fungal spore numbers and percentage of root length colonization (Aggarwal et al., 2005). Mycorrhizal colonization shows an inverse trend to increasing concentrations of bavistin with minimum mycorrhizal root colonization occurring at 1% (Aggarwal et al., 2005). Arbuscules tend to be fewer in roots growing in bavistin treated soil (Edgington and Barron, 1967). Further, the effect of bavistin persists in the soil for a long period, while the toxicity of dithane M-45 becomes neutralized within a short period of time (Sugavanam et al., 1994). A field study carried out by Udaiyan et al. (1999) to evaluate the effect of six fungicides on AM colonization and sporulation in different millets indicated that the effect could vary depending on the millet species involved. At the recommended application rates some fungicides reduced AM colonization, while others had no effect or a stimulatory effect on AM colonization and sporulation.

The time as well as the mode of fungicide application significantly influences the effect of fungicides on AM fungi. Foliar applications of trobilurin fungicides azoxystrobin and kresoxim-methyl to 5-week old *Zea mays* plants affect the development of mycorrhiza and the enzymatic activity of *Glomus coronatum*. Whereas, the same fungicides when applied as a soil drench during the same period decreased mycorrhizal activity after 7 days and completely inhibited it after 21 days (Diedhiou et al., 2004). The applications of these fungicides to 10-days old maize reduced the development of mycorrhiza by G. coronatum and enzyme activities. Seedling of legumes raised in fytolan (a non-systematic fungicide) drenched beds had normal growth, AM colonization (53.4-100%) and higher spore density (36-82.8 g soil^{-1}) (Udaiyan et al., 1996). The effect of fungicides tends to vary with AM fungal species and structures. Pentachloronitrobenzene, benomyl or captan when applied at 20 mg active ingredient Kg^{-1} soil did not affect the root length colonized by *Glomus etunicatum*, where as, these fungicides at a similar concentration reduced the root length colonized by *Gigaspora rosea* at 48 and 82 days after transplantation (Schreiner and Bethlenfalvay, 1997b). As Pentachloronitrobenzene reduce root length colonized by *G. mosseae* and sporulation at 48 and 82 days, fungicides like benomyl only affects early stages of root length colonization. Captan does not affect sporulation by *G. mosseae, G. etunicatum* and *Gi. rosea* while benomyl reduces sporulation by *Gi. rosea* (Schreiner and Bethlenfalvay 1997b). In addition the effect of fungicides also appears to be influenced by the nature of the substrate. For example benomyl inhibits root colonization by AM fungi in sand but not in soil (Dodd and Jeffries, 1989). This varied influence of benomyl could be attributed to the deeper penetration of benomyl in high sandy soils and its absorption in heavy texture soils.

Some fungicides like metalaxyl (methyoxyacetyl)-N-(2, 6-xylyl)-DL-alaninate), which do not exhibit any deleterious effect on AM fungi has been known to enhance AM colonization (Afek et al., 1991; Hwang et al., 1993). Saleh Al-Garni (2006) reported that lower concentrations of mancozeb stimulated mycorrhizal colonization and sporulation while higher concentrations were conductive to mycorrhizal activities. Likewise, metalaxyl (acylalanine group) and fosetyl-Al (organic phosphates group) tend to promote the AM colonization (Aziz et al., 1990; Hetrick and Wilson, 1991; Cardoso and Lambais, 1993). Stimulation of AM fungal colonization by fosetyl-Al can be attributed to an increase in the quality and quantity of root exudates (Jabaji-Hare and Kendrick, 1987), which favour AM fungal spore germination and directs germ tube growth towards the roots (Sieverding, 1991). Trimorph has both stimulatory and inhibitory effect on AM root colonization (Dodd and Jeffries, 1989).

PESTICIDES

Pesticides have profound effect on AM fungi and affect AM formation and host growth at different stages. The effect of pesticides on AM fungi varies with their chemical formulation, application rate and experimental conditions (Schreiner and Bethlenfalvay, 1996; Pattinson et al., 1997). Some pesticides like copper oxychloride have a deleterious effect on AM colonization and sporulation even at recommended levels (Sreenivasa and Bagyaraj, 1989).

The detrimental effect of copper oxychloride on AM fungal spore germination subsequently affects the initiation and extent of colonization spread within root (Nemec, 1980; Smith and Gianinazzi-Pearson, 1988; Dodd and Jefffries, 1989). As the growth and nutrient status of host tend to depend on microbial associations like AM symbiosis or nodulation, any effect of pesticides on the microsymbionts would ultimately affect host growth (Abd-Alla et al., 2000). Pesticides like brominal, gramoxone, selecron and sumi oil significantly reduces AM fungal colonization and spore number in legumes like *Vigna sinensis, Phaseolus vulgaris* and *Lupinus albus*. But on the other hand, pesticides stimulate sporulation in *Vigna unguiculata* after 60 days after planting (Abd-Alla et al., 2000). Also pesticides may affect root exudation from the macrosymbiont (Rathnayake et al., 1978; Schwab et al., 1982) or the production of cell wall degrading enzymes that are essential for root colonization by AM fungi.

The decrease in macronutrient content in plants in response to pesticide application may be related to the decline in AM fungal colonization (Abd-Alla et al., 2000; Mathew and Johri, 1989). An indirect influence of pesticides in legumes involves inhibition of N_2 fixation, as nodulation and N_2 fixation requires an optimum level of P in host tissues (Barea and Azcon-Aquilar, 1983). Stimulatory effect of folpet on cardoon seedling growth indicates their positive effect on AM fungal colonization (Marin et al., 2002). Two possible mechanisms have been proposed to account for the beneficial effects of pesticides on AM fungi. First, pesticides alter the host plant's physiology so the amount of soluble sugars allocated to root exudates is increased thereby increasing root colonization (Schwab et al., 1982; Jabaji-Hare and Kendrick, 1985). Second, stimulation of mycorrhizal formation through reduction in populations of organisms antagonistic to AM fungi (Hetrick and Wilson, 1991).

Like pesticides, fumigation with formaldehyde also reduces AM colonization and spore abundance in *Cajanus cajan, Dolichos biflorus, Vigna mungo* and *Vigna unguiculata* under field conditions (Udaiyan et al., 1995; Sreenivasa and Bagyaraj, 1989; Islam et al., 1980). This reduction can be due to a reduction in companion organisms (eg. mycorrhiza helper bacteria) that facilitate mycorrhizal formation (Aguilar and Barea, 1985). Though formaldehyde fumigation reduces AM colonization (18-25.3%) and spore density (3.1-10.6 g $soil^{-1}$), a gradual increase in AM colonization occurs at the later stages of plant growth (Udaiyan et al., 1995; 1996).

INSECTICIDES

Application rate appears to be a key factor in mediating effects of insecticides on AM fungi. Some insecticides have either neutral or stimulatory effect on AM fungal symbiosis, at low concentrations, while at high rates these insecticides may decrease root colonization (Trappe et al., 1984; Veeraswamy et al., 1993). Carbamate insecticide like carbofuran reduce mycorrhizal colonization in *Arachis hypogaea* both under field and greenhouse conditions (Backmann and Clark, 1977), while the same insecticide when applied at half the recommended level stimulated root colonization and sporulation by *Glomus fasciculatum* (Sreenivasa and Bagyaraj, 1989). High concentrations (5-10 kg ha^{-1}) of endosulfan, quinalphos and carbofuran has deleterious effects on AM fungal colonization in sorghum. Dimethoate has neutral effect on the growth of soybean colonized by *G. mosseae* and by the indigenous AM fungi and the spore germination of the former. However, as 0.5 mg/l of dimethoate increase germination of *Gi. rosea* spores, the same concentration tend to decrease germination of *Scutellospora castaneae* spores (Menendez et al., 1999).

Mycorrhizal formation by *G. mosseae* was reduced by isofenphos and oxamyl, whereas, the same insecticides increased root colonization and decreased plant root:shoot ratios in native AM fungal inoculations (Marin et al., 2002). Isofenphos even had a positive effect on *G. mosseae* and its colonization (Marin et al., 2002). The reduction in AM colonization may be due to the loss of photosynthetic tissue which impairs the ability of plants to support the carbon demand of the mycorrhiza (Gehring and Whitham, 1994; Gange and Bower, 1997).

HERBICIDES

Information is limited on the effect of herbicides on AM fungi compared to fungicides. Available information do not indicate any adverse effects of herbicides when applied at recommended levels as they mainly target weed plants. Phenylurea herbicides diuron and chlorotoluron do not adversely affect AM fungal sporulation or root colonization (Smith et al., 1981; Nemec and Tucker, 1983; Ocampo and Hayman, 1980; Dodd and Jeffries, 1989) and high concentrations of diuron actually increases AM fungal spore density in the soil (Smith et al., 1981). Though low concentrations of Ceridor (befenox, mecoprop) and Harrier (mecoprop, ioxynil, clopyralid) inhibits spore germination of *Glomus* spp., higher concentrations of these herbicides have a stimulatory effect (Dodd and Jeffries, 1989). Avenge (difenzoquat methyl sulphate) prevents spore germination, while Dicurane (chlorotpluron) has been shown to have neutral effect. Harrier significantly affects the colonization rate of *Glomus geosporum* (Dodd and Jeffries, 1989). Atrazine (2-Chloro-4-(ethylamine)-6-(isopropylamine)-8-triazine), the widely used herbicide in the world is highly phytotoxic. The presence of atrazine in soil at a concentration of upto 2.0 mg/kg decrease mycorrhizal colonization whereas, a higher concentration of 5.0 mg/kg actually increase colonization (Huang et al., 2006; 2007). Even though the exact mechanism for the increased AM fungal colonization at high concentrations of atrazine is not known, it has been speculated that AM fungi tend to develop tolerance with high atrazine application rates (Huang et al., 2007). There is no direct evidence of AM mediated degradation of aromatic pollutants as shown for ectomycorrhizal fungi (Meharg and Cairney, 2000; Meharg, 2001). Using a compartmented system Huang et al. (2009) showed that the extramatrical hyphae of *G. etunicatum* could degrade atrazine in the soil by stimulating soil enzymes and microbial activities. High rate of chlorate application decrease the colonization rate of AM fungi leading to reduced P uptake and plant biomass (Li et al., 2008). The effects of herbicides on AM fungi can also vary with the host plant as shown for the herbicide diclofop which inhibits root colonization by *Glomus deserticola* in wheat but not in rye grass (Rejon et al., 1997).

Carbamate herbicides like chlorpropham sulfallate and phenmedipham are photosynthetic inhibitors and show phytotoxic effects when applied as foliar spray or when applied to the roots of different food and forage plants (Chueca et al., 1981). The effect of these herbicides on AM formation may be indirect through their effect on host photosynthesis. However, the effect of carbamates on photosynthesis and AM fungi is only transitional and disappears within a short period of time (Santakumari and Das, 1979). AM fungi may also help in recovery of the hosts from the deleterious effect of phenmedipham and also cyanazine, an another photosynthetic inhibitor, when applied at moderate levels (Ocampo and Barea, 1985; Garcia-Romera et al., 1988). The effect of herbicides on soil AM fungal inocula may be indirect, as they eliminate several weed species that are mycorrhizal. However, this effect may be severe in cultivating systems involving non-mycorrhizal hosts.

Nematicides

Nematicides have greatest potential to affect mycorrhizal association as they are usually applied as soil drenches (Trappe et al., 1984). Nematicides like 1, 2-dibromo-3-chloropropane (DBCP) increases the production of AM fungal structures (arbuscules, vesicles and spores) in cotton roots grown in soils with a history of excellent cotton productivity, while no significant increase in arbuscule production was evident in soil with a poor history of cotton productivity (Bird et al., 1974). Treatment with methyl bromide reduces nematode populations (*Hoplolainus columbus, Pratylenchus brachyurus, Meloidogyne* sp.) thereby resulting in increased mycorrhization as evidenced by increased presence of AM fungal structures (Bird et al., 1974). Chemical control of the citrus nematode *Tylenchulus semipenetrans* with the nematicides aldicarb and fenamiphos was more effective in the presence of mycorrhiza (Abu-Elamayem et al., 1989). Fenamiphos and aldicarb though does not effect mycorrhizal formation, it reduces the number of mycorrhizal spores in the soil (Nemec, 1985; Abu-Elamayem et al., 1989).

Fertilizers

The proportion of nutrients in fertilizer influences the mycorrhizal response (Johnson and Pfleger, 1992). A balanced fertilization with recommended levels of fertilizers but with low P minimizes the deleterious effects of fertilizer application on AM fungi (Gryndler et al., 1990; Saif, 1986). The interactions between macronutrients like N, P and potassium (K) mediate the mycorrhizal response. Both N and K status of host plant influences the mycorrhizal response to P (Sylvia and Neal, 1990; Johnson and Pfleger, 1992). Application of the macronutrients individually reduces AM propagule density in the soil (Johnson and Pfleger, 1992; Strzemska, 1975; Baltruschat and Dehne, 1988). In contrast, addition P and N individually, in the form of super phosphate and urea at normal levels seems to have no marked effect on hyphal colonization in rice plants, whereas higher concentrations of these agrochemicals were detrimental to AM formation (Rajeshkannan et al., 2009). Stimulatory effect on AM colonization with the addition of P has also been shown (Bethlenfalvay et al., 1983; Hetrick and Bloom, 1986). The sensitiveness of AM fungi to fertilizers differs with the crop species and also fertilizers. Nitrogen fertilizer amendments reduce AM spore populations in wheat but not in *Vicia faba* (Hayman, 1975). *Glomus* taxa that were under *Sclerocystis* (Sieverding, 1991) disappear and mycorrhizal structures of *Acaulospora longula* and *Gigaspora margarita* are suppressed at high P application rates whereas, *G. intraradices* was unaffected by P application rates (Douds and Schenck, 1990; Thomson et al., 1986).

AM fungal species differ in their response to different concentrations of P. Some researchers have shown that the addition of P typically reduces the extent of AM fungal formation (Jasper et al., 1979; Abbott et al., 1984; Dhillion and Amornpan, 1992) and the extent of root length colonization (Ross, 1971; Daft and Nicolson, 1972; Mosse, 1973). The reason for the decline may be the reduction in the concentration of fungal metabolites such as soluble carbohydrates and free amino nitrogen compounds in the roots (Jasper et al., 1979; Same et al., 1983) or their exudation from the roots (Ratnayake et al., 1978; Graham et al., 1981). These factors may lead to the reduction in the growth of external hyphae which reduces the rate of spread of AM fungi through secondary colonization (Schwab et al., 1983) leading to a decline in the rate of formation of the entry points and colonization units (Sanders and Sheikh, 1983). Reductions in the rate of colonization by *Gigaspora calospora* and *G. fasciculatum* has been shown at high concentrations of P (Thomson et al., 1986). At low P concentration, the growth of the fungi within the root reduces, while at higher concentrations of P, the fungal growth outside the root reduces. The lethal effect of increasing concentrations of P was more pronounced in roots colonized with *Gi. calospora* than those colonized with *G. fasciculatum* due to differences in the concentration of carbohydrate in the roots (Thomson et al., 1986).

CONCLUSION

Mycorrhizal association offers considerable benefits in form of plant growth and productivity, both in natural and agro ecosystems. Studies investigating the effect of organic pollutants on AM fungi have produced ambiguous results. There is a considerable gap in our understanding on the underlying mechanisms of how organic pollutants actually affect AM fungi. Most of the observations on the effect of organic pollutants are made under simulated conditions and realistic field observations are scarce. The effects produced by different organic pollutants on AM fungi are not similar and the response of AM fungi also varies greatly. Restriction on the intentional use of organic chemicals that are deleterious to AM fungi is essential to fully exploit the benefits of AM association. This may be true for certain organic chemicals like herbicides which needs to be restricted in farming systems involving crops that are non-hosts to AM fungi. In addition the large variation in the reactivity for organic chemical among AM fungal conspecifics could be exploited to achieve the full range of mycorrhizal benefits. If AM fungi are to be utilized to remediate environment polluted with organic chemicals, our understanding of organic pollutants on AM fungi needs to be greatly improved.

REFERENCES

Abbott LK, Robson AD, de Boer G (1984) The effect of phosphorus on the formation of hyphae in the soil by the vesicular arbuscular mycorrhizal fungus *Glomus fasciculatus*. *New Phytol* 97: 191-196.

Abd-Alla MH, Omar SA, Karanxha S (2000) The impact of pesticides on arbuscular mycorrhizal and nitrogen-fixing symbioses in legumes. *Appl Soil Ecol* 14: 191-200.

Abu-Elamayem MM, El-Shoura MY, Rabie RS, Ibrahim IKA, Fawaz KS (1989) Effect of certain systemic nematicides on the interaction between *Tylenchulus semipenetrans* and *Acaulospora trappei* on sour orange and *Cleopatra mandarin*. *Nematol Medit* 17: 17-20.

Afek U, Menge JA, Johnson ELV (1991) Interaction among mycorrhizae, soil solarization, metalaxyl, and plants in the field. *Plant Dis* 75: 665-671.

Aggarwal A, Sharma D, Parkash V, Sharma S, Gupta A (2005) Effect of Bavistin and Dithane M-45 on the mycorrhizae and rhizosphere microbes of sunflower. *Helia* 28: 75-88.

Aguilar CA, Barea JM (1985) Effect of soil microorganisms on formation of vesicular-arbuscular mycorrhizae. *Trans Br Mycol Soc* 84: 536-537.

Alarcon A, Davies Jr FT, Autenrieth RL, Zuberer DA (2008) Arbuscular mycorrhiza and petroleum degrading microorganisms enhance phytoremediation of petroleum contaminated soil. *Int J Phytorem* 10: 251-263.

Alarcon A, Delgadillo-Martinez J, Frabco-Ramirez A, Davies Jr FT, Ferrera-Cerrato R (2006) Influence of two polycyclic aromatic hydrocarbons on spore germination and phytoremediation potential of *Gigaspora margarita- Echynochloa polystachya* symbiosis in benzo[a]pyrene polluted substrate. *Rev Int Contam Ambient* 22: 39-47.

Allen MF, MacMahon JA, Anderson DC (1984) Reestablishment of Endogonaceae on Mount St Helens: survival of residuals. *Mycologia* 76: 1031-1038.

Ames RN, Ingham ER, Reid CPP (1982) Ultraviolet induced auto fluorescence of arbuscular mycorrhizal root infections: an alternative to clearing and staining methods for assessing infections. *Can J Microbiol* 28: 351-355.

Audet P, Charest C (2006) Effects of AM colonization on wild tobacco plants grown in Zinc-contaminated soil. *Mycorrhiza* 16: 277-283.

Auge RM (2001) Water relations, drought and vesicular-arbuscular mycorrhizal symbiosis. *Mycorrhiza* 11: 3-42.

Aziz T, Yven JE, Habte M (1990) Response to mycorrhizal inoculation and fosetyl- Al treatment. *Commu Soil Sci Plant Anal* 21: 2309-2318.

Backman PA, Clark EM (1977) Effect of carbofuran and other pesticides on vesicular arbuscular mycorrhizae in peanuts. *Nematropica* 7: 14-18.

Bailey JE, Safir GR (1978) Effect of benomyl on soybean endomycorrhizae. *Phytopathol* 68: 1810-1812.

Baltruschat H, Dehne HW (1988) The occurrence of vesicular arbuscular mycorrhiza in agro-ecosystems. I. Influence of nitrogen and green manure in continuous mono culture and in crop rotation on the inoculum potential of winter wheat. *Plant Soil* 107: 279-284.

Barea JM, Azcon-Aguilar C (1983) Mycorrhizas and their significance in nodulating nitrogen fixing plants. *Adv Agron* 36: 1-54.

Bethlenfalvay GJ, Bayne HG, Pacovsky RS (1983) Parasitic and mutualistic associations between a mycorrhizal fungus and soybean: The effect of phosphorus on host plant-endophyte interactions. *Physiol Plant* 57: 543-548.

Binet P, Portal M, Leyval C (2000) Fate of polycyclic aromatic hydrocarbons (PAH) in the rhizosphere of rye grass. *Plant Soil* 227: 207-213.

Bird GW, Rich JR, Glover SU (1974) Increased endomycorrhizae of cotton roots in soil treated with nematicides. *Phytopathol* 64: 48-51.

Boatman N, Paget D, Hayman DS, Mosse B (1978) Effects of systemic fungicides on vesicular arbuscular mycorrhizal infection and plant phosphate uptake. *Trans Br Mycol Soc* 70: 443-450.

Bollen GJ (1979) Side effects of pesticides on microbial interaction. In: *Soil borne plant pathogens*, Schippers B, Gams W (eds.), Academic press, New York pp 451-481.

Brundrett M, Bougher N, Dell B, Grove T, Malajczuk N (1996) Working with mycorrhiza in forestry and agriculture. ACIAR monograph no. 32. *Australian Centre for International Agricultural Research,* Canberra, Australia.

Cabello MN (1997) Hydrocarbon pollution: its effect on native arbuscular mycorrhizal fungi (AMF). *FEMS Microbiol Ecol* 22: 233-236.

Cairney JWG, Meharg AA (1999) Influences of anthropogenic pollution on mycorrhizal fungal communities. *Environ Pollut* 106: 169-182.

Cardoso EJBN, Lambais MR (1993) Efeito de aldicarb e fosetil- Al no desenvolvimento e na colonizacao micorrizica de tabgerina cleópatra. *Rev Bras Cieňc Solo 17:* 179- 184.

Carr GR, Hinckley MA (1985) Germination and hyphal growth of *Glomus caledonium* on water agar containing benomyl. *Soil Biol Biochem* 17: 313-316.

Cerniglia CE (1992) Biodegradation of polycyclic aromatic hydrocarbons. *Biodegradation* 3: 351-368.

Chiocchio V, Venedikian N, Martinez AE, Menendez A, Ocampo JA, Godeas A (2000) Effect of the fungicide benomyl on spore germination and hyphal length of the arbuscular mycorrhizal fungus *Glomus mosseae*. *Int Microbiol* 3: 173-175.

Chueca A, Baron M, Lopez-Gorge (1981) Are bis-carbamates photosynthetic inhibitors by root application? In: Photosynthesis VI. *Photosynthesis and productivity, photosynthesis and environment,* Akoyunoglou G (eds), Balaban International Science Services, Philadelphia. pp 555-563.

Clark RB (1997) Arbuscular mycorrhizal adaptation, spore germination, root colonization, and host plant growth and mineral acquisition at low pH. *Plant Soil* 192: 15-22.

Corgie SC, Fons F, Beguiristain T, Leyval C (2006) Biodegradation on phenanthrene, spatial distribution of bacterial populations and dioxygenase expression in the mycorrhizosphere of *Lolium perenne* inoculated with *Glomus mosseae*. *Mycorrhiza* 16: 207-212.

Daft MJ, Nicholson TH (1972) Effect of *Endogone* mycorrhiza on plant growth, IV. Quantitative relationships between the growth of the host and the development of the endophyte in tomato and maize. *New Phytol* 71: 287-295.

Davidse LC (1986) Benzimidazole fungicides: mechanism of action and biological impact. *Annu Rev Phytopathol* 24: 43-65

Davies JrFT, Puryear JD, Newton RJ, Egilla JN, Saraiva-Grossi JA (2001) Mycorrhizal fungi enhance accumulation and tolerance of chromium in sunflower (*Helianthus annus). J Plant Physiol* 158: 777-786.

De Bertoldi M, Giovannetti M, Griselli M (1977) Effects of soil application of benomyl and captan on the growth of onion and the occurrence of endophytic mycorrhizas and rhizosphere microbes. *Ann Appl Biol* 86: 111-115.

Dhillion SS, Amornpan L (1992) The influence of inorganic nutrients fertilization on the growth nutrient composition and vesicular-arbuscular mycorrhizal colonization of pre-transplant stage rice (*Oryza sativa* L.) plants. *Biol Fertil Soils* 13: 85-91.

Diedhiou PM, Oerke EC, Dehne HW (2004) Effects of the strobilurin fungicides azoxystrobin and kresoxim-methyl on arbuscular mycorrhiza. *J Plant Dis Protect* 111: 545-556.

Dodd JC, Jeffries P (1989) Effect of fungicides on three vesicular-arbuscular mycorrhiza fungi associated with winter wheat (*Triticum aestivum* L.) *Biol Fertil Soils* 7: 120-128.

Douds DD, Schenck NC (1990) Relationship of colonization and sporulation by VA mycorrhizal fungi to plant nutrient and carbohydrate contents. *New Phytol* 116: 621-627.

Edgington LV, Barron GL (1967) Fungitoxic spectrum of oxathiin compounds. *Phytopathol* 57: 1256-1257.

Fitter AH (1986) Effect of Benlate on leaf phosphorus concentration in alpine grasslands: a test of mycorrhizal fungi. *New Phytol* 103: 767-776.

Fitter AH, Nichols R (1988) The use of benomyl to control infection by vesicular arbuscular mycorrhizal fungi. *New Phytol* 110: 201-206.

Franco-Ramirez A, Ferrera-Cerrato R, Varela-Fregoso L, Perez-Moreno J, Alarcon A (2007) Arbuscular mycorhizal fungi in chronically petroleum contaminated soils in Mexico and the effects of petroleum hydrocarbons on spore germination. *J Basic Microbiol* 41: 378-383.

Gange AC, Bower E (1997) Interactions between insects and mycorrhizal fungi. In: *Multitrophic interactions in terrestrial systems,* Gange AC, Brown VK (eds), Blackwell Science, Oxford. pp 115-131.

Gao Y, Cheng Z, Ling W, Huang J (2010) Arbuscular mycorrhizal fungal hyphae contribute to the uptake of polycyclic aromatic hydrocarbons by plant roots. *Bioresource Technol* 101: 6895-6901.

Garcia-Romera I, Ocampo JA (1988) Effect of the herbicide MCPA on VA mycorrhizal infection and growth of *Pisum sativum. Z Pflanzenemaehr Bodenkd* 151: 225-228.

Gasper ML, Cabello MN, Cazau MC (2002) Effect of phenanthrene and *Rhodotorula glutinis* on arbuscular mycorrhizal fungus colonization of maize roots. *Mycorrhiza* 12: 55-59.

Gehring CA, Whitham TG (1994) Interactions between above ground herbivores and the mycorrhizal mutualists of plants. *Tree* 9: 251-255.

Geng CN, Li PJ, Chen SH, Xiao BY, Zhang HR, Han GY, Zhang CG (2002) Effects of different arbuscular mycorrhizal fungi on oil tolerance of *Trifolium subterraneum* L. *Chinese J Appl Environ Biol* 8: 648-652.

Graham JH, Leonard RT, Menge JA (1981) Membrane mediated decrease in root exudation responsible for phosphorus inhibition of vesicular arbuscular mycorrhiza formation. *Plant Physiol* 68: 548-552.

Gryndler M, Lestina J, Moravec V, Prikyl Z, Lipavsky (1990) Colonization of maize roots by VAM-fungi under conditions of long term fertilization of varying intensity. *Agric Ecosyst Environ* 29: 183-186.

Gupta BK, Borse SS (1997) Relative toxicity of some insecticides as contact poison against third instar larvae of *Hyblea purea* (*Lepidoptera: Hyblaeidae*). *Indian Forester* 123: 427-429.

Habte M (1997) Use of Benlate to obtain soil free of arbuscular mycorrhizal fungal activity for greenhouse investigations. *Arid Soil Res Rehab* 11: 151-161.

Haugen LM, Smith SE (1992) The effect of high temperature and fallow period on infection of mung bean and cashew roots by the vesicular–arbuscular mycorrhizal fungus *Glomus intraradices*. *Plant Soil* 145: 71–80.

Hayman DS (1975) the occurrence of mycorrhiza in crops as affected by soil fertility. In: *Endomycorrhizas,* Mosse B et al. (eds), Academic Press, New York. pp 495-509.

Hetrick BAD, Bloom J (1986) The influence of host plant on production and colonization ability of vesicular arbuscular mycorrhizal spores. *Mycologia* 78: 32-36.

Hetrick BAD, Wilson GWT (1991) Effect of mycorrhizal fungus species and metalaxyl application on microbial suppression of mycorrhizal symbiosis. *Mycologia* 83: 97-102.

Hodge A, Campbell CD, Fitter AH (2001) An arbuscular mycorrhizal fungus accelerates decomposition and acquires nitrogen directly from organic material. *Nature* 413: 297-299.

Huang H, Zhang S, Chen BD, Wu N, Shan XQ, Christy P (2006) Uptake of atrazine and cadmium from soil by Maize (*Zea mays* L.) in association with the arbuscular mycorrhizal fungus *Glomus etunicatum*. *J Agric Food Chem* 54: 9377-9382.

Huang H, Zhang S, Shan XQ, Chen BD, Zhu YG, Bell JNB (2007) Effect of arbuscular mycorrhizal fungus (*Glomus caledonium*) on the accumulation and metabolism of atrazine in maize (*Zea mays* L.) and atrazine dissipation in soil. *Environ Pollut* 146: 452-457.

Huang H, Zhang S, Wu N, Luo L, Christie P (2009) Influence of *Glomus etunicatum*/*Zea mays* mycorrhiza on atrazine degradation, soil phosphatase and dehydrogenase activities, and soil microbial community structure. *Soil Biol Biochem* 41: 726-734.

Hwang SF, Chakravarty P, Prévost D (1993) Effects of *Rhizobia*, metalaxyl and VA mycorrhizal fungi on growth, nitrogen fixation and development of *Phythium* root rot of Sainfoin. *Plant Dis* 77: 1093-1098.

Islam RA, Ayanaba A, Sanders FE (1980) Response of cowpea (*Vigna unguiculata*) to inoculation with VA mycorrhizal fungi and to rock phosphate fertilization in some unsterilized Nigerian soil. *Plant Soil* 54: 107-109.

Jabaji-Hare S, Kendrick B (1987) Response of an endomycorrhizal fungus on *Allium porrum* L. to different concentration of the systemic fungicides, metalaxyl (Ridomil) and fosetyl-Al (Aliette). *Soil Biol Biochem* 19: 95-99.

Jalali BL, Domsch KH (1975) Effect of systemic fungi-toxicants on the development of endotrophic mycorrhiza. In: *Endomycorrhizas,* Sanders FE, Mosse B, Tinker PB (eds.), Academic Press, New York. pp 619-626.

Jasper DA, Robson AD, Abbott LK (1979) Phosphorus and the formation of vesicular arbuscular mycorrhizas. *Soil Biol Biochem* 11: 501-505.

Jasper DA, Robson AD, Abbott LK (1987) The effect of surface mining on the infectivity of vesicular-arbuscular mycorrhizal fungi. *Aus J Botany* 35: 641-652.

Johnson NC, Pfleger FL (1992) Vesicular arbuscular mycorhizae and cultural stresses. In*: VA mycorrhizae in sustainable agriculture,* Bethlenfalvay G et al. (eds.), ASA Publication, Am Soc Agronomy, Madison. pp 71-99.

Joner EJ, Corgie SC, Amellal N, Leyval C (2002) Nutritional constraints to degradation of polycyclic aromatic hydrocarbons in a simulated rhizosphere. *Soil Biol Biochem* 34: 859-864.

Joner EJ, Johansen A, Loibner AP, Dela Cruz MA, Szolar OHJ, Portal JM, Leyval C (2001) Rhizosphere effects on microbial community structure and dissipation and toxicity of polycyclic aromatic hydrocarbons (PAHs) in spiked soil. *Environ Sci Technol* 35: 2773-2777.

Joner EJ, Leyval C (2001) Influence of arbuscular mycorrhiza on clover and rye grass grown together in a soil spiked with polycyclic aromatic hydrocarbons. *Mycorrhiza* 10: 155-159.

Jurgensen MF, Larsen MJ, Harvey AE (1977) Effects of timber harvesting on soil biology. *Proc Annu Meet Soc Am Fores*. pp 244-250.

Kehri HK, Chandra S (1990) Mycorrhizal association in crops under sewage farming. *J Indian Bot Soc* 69: 267-269.

Kirk JL, Moutoglis P, Klironomos J, Lee H, Trevors JT (2005) Toxicity of diesel fuel to germination, growth and colonization of *Glomus intraradices* in soil and in vitro transformed carrot root cultures. *Plant Soil* 270: 23-30.

Kjøller R, Rosendahl S (2000) Effects of fungicides on arbuscular mycorrhizal fungi: differential responses in alkaline phosphatase activity of external and internal hyphae. *Biol Fertil Soils* 31: 361-365.

Kough JL, Gianinazzi-Pearson V, Gianinazzi S (1987) Depressed metabolic activity of vesicular arbuscular mycorrhizal fungi after fungicide applications. *New Phytol* 106: 707-715.

Larsen J, Thingstrup I, Jakobsen I, Rosendahl S (1996) Benomyl inhibits phosphorus transport but not fungal alkaline phosphatase activity in a *Glomus*- cucumber symbiosis. *New Phytol* 132: 127-134.

Leyval C, Binet P (1998) Effect of polyaromatic hydrocarbons (PAHs) in soil on arbuscular mycorrhizal plants. *J Environ Qual* 27: 402-407.

Leyval C, Joner EJ, del Val C, Haselwandter K (2002) Potential of arbuscular mycorrhizal fungi for bioremediation. In*: Mycorrhizal technology in Agriculture*, Gianinazzi S, Schuepp H, Bare JM, Haselwandter K (eds), Birkhäuser Verlag, Switzerland. pp 175-186.

Li H, Zhang X, Lin C, Wu Q (2008) Toxic effects of chlorate on three plant species inoculated with arbuscular mycorrhizal fungi. *Ecotoxicol Environ Saf* 71: 700-705.

Liu SL, Luo YM, Ho ZH, Wu LH, Ding KQ, Christe P (2004) Degradation of benzo[a]pyrene in soil with arbuscular mycorrhizal alfalfa. *Environ Geochem Hlth* 26: 285-293.

Manjunath A, Bagyaraj DJ (1984) Effects of fungicides on mycorrhizal colonization and growth of onion. *Plant Soil* 80: 147-150.

Marin M, Ybarra M, Fe A, Garcia-Ferriz L (2002) Effect of arbuscular mycorrhizal fungi and pesticides on *Cynara cardunculus* growth. *Agric Food Sci Finland* 11: 245-251.

Mathew J, Johri BN (1989) Effect of indigenous and introduced VAM fungi on growth of mungbean. *Mycol Res* 92: 491-493.

Meharg AA (2001) The potential for utilizing mycorrhizal associations in soil bioremediation. In: *Fungi in Bioremediation,* Gadd GM (ed), Surrey, British Mycological Society, UK. pp 445-455.

Meharg AA, Cairney JWG (2000) Ectomycorrhizas- extending the capacities of rhizosphere remediation? *Soil Biol Biochem* 32: 1475-1484.

Menendez A, Martinez A, Chiocchio V, Venedikian N, Ocampo JA, Godeas A (1999) Influence of the insecticide dimethoate on arbuscular mycorrhizal colonization and growth in soybean plants. *Int Microbiol* 2: 43-45.

Merryweather J, Fitter A (1996) Phosphorus nutrition of an obligately mycorrhizal plant treated with the fungicide benomyl in the field. *New Phytol* 132: 307-311.

Miller RL, Jackson LE (1998) Survey of vesicular arbuscular mycorrhizae in lettuce production in relation to management and soil factors. *J Agric Sci* 130: 173-182.

Mosse B (1973) Plant growth responses to vesicular-arbuscular mycorrhiza. IV. In soil given additional phosphate. *New Phytol* 72: 127-136.

Muthukumar T, Vediyappan S (2010) Comparison of arbuscular mycorrhizal and dark septate endophyte fungal associations in soils irrigated with pulp and paper mill effluent and well-water. *Eur J Soil Biol* 46: 157-167

Nemec S (1980) Effect of 11 fungicides on endomycorrhizal development in sour orange. *Can J Bot* 58: 522-526.

Nemec S (1985) Influence of selected pesticides on *Glomus* species and their infection in citrus. *Plant Soil* 84: 133-137.

Nemec S, Tucker D (1983) Effects of herbicides on endomycorrhizal fungi in Florida citrus (*Citrus* spp.) soils. *Weed Sci* 31: 427-431.

Ocampo JA, Barea JM (1985) Effect of carbamate herbicides on VA mycorrhizal infection and plant growth. *Plant Soil* 85: 375-383.

Ocampo JA, Hayman DS (1980) Effects of pesticides on mycorrhiza in field grown barley, maize and potatoes. *Trans Br Mycol Soc* 74: 413-416.

Olexa TJ, Gentry TJ, Hartel PG, Wolf DC, Fuhrmann JJ, Reynolds CM (2000) Mycorrhizal colonization and microbial community structure in the rhizosphere of annual ryegrass grown in pyrene-amended soils. *Int J Phytorem* 2: 213–231.

Ortega-Larrocea MP, Siebe C, Becard G, Mendez I, Webster R (2001) Impact of a century of wastewater irrigation on the abundance of arbuscular mycorrhizal spores in the soil of the Mezquital Valley of Mexico. *Appl Soil Ecol* 16: 149-157.

Pandey PC, Singh A, Karnatak DC, Bhartari BK (1996) *Melamspora larci-poplina* on poplar in India and its control in nursery. *Indian Forester* 122: 1062-1067.

Parke JL, Linderman RG, Trappe JM (1983) Effects of forest litter on mycorrhiza development and growth of Douglas-fir and western redcedar seedlings. *Can J Forest Res* 13: 666-671.

Parvathi K, Venkateshvarlu K, Rao AS (1985) Toxicity of soil applied fungicides to the vesicular arbuscular mycorrhizal fungus *Glomus mosseae* in groundnut. *Can J Bot* 63: 1673-1675.

Pattinson GS, Warton DI, Misman R, McGee PA (1997) The fungicides Terrazole and Terraclor and the nematicide Fenamiphos have little effect on root colonization by *Glomus mosseae* and growth of cotton seedlings. *Mycorrhiza* 7: 155-159.

Rajeshkannan V, Sumathi CS, Manian S (2009) Arbuscular mycorrhizal fungi colonization in upland rice as influenced by agrochemical application. *Rice Sci* 16: 307-313.

Ratnayake M, Leonard RT, Menge JA (1978) Root exudation in realtion to supply of phosphorus and its possible relevance to mycorrhizal formation. *New Phytol* 81: 543-552.

Rejon A, Garcia-Romera I, Ocampo JA, Bethlenfalvay GJ (1997) Mycorrhizal fungi influence competition in a wheat-rye grass association treated with the herbicide diclofop. *Appl Soil Ecol* 7: 51-57.

Rilling MC, Steinberg PD (2002) Glomalin production of an arbuscular mycorrhizal fungus: a mechanism of habitat modification. *Soil Biol Biochem* 34: 1371-1374.

Rivera-Beacerril F, van Tuinen D, Martin-Laurent F, Metwally A, Dietz KJ, Gianinazzi S, Gianinazzi-pearson V (2005) Molecular changes in *Pisum sativum* L. roots during arbuscular mycorrhiza buffering of cadmium stress. *Mycorrhiza* 16: 51-60.

Ross JP (1971) Effect of phosphate fertilization on yield of mycorrhizal and non-mycorrhizal soybean. *Phytopathol* 61: 1400-1403.

Safir GR, Siqueira JO, Burton TM (1990) Vesicular-arbuscular mycorrhizae in a wastewater irrigated oldfield ecosystem in Michigan. *Plant Soil* 121: 187-196.

Saif SR (1986) Vesicular arbuscular mycorrhizae in tropical forage species as influenced by season, soil texture, fertilizers, host species and ecotypes. *Angew Bot* 60: 125-139.

Saleh Al-Garni SM (2006) Influence of Malathion and Mancozeb on Mycorrhizal colonization and growth of *Zea mays* and *Vicia faba*. *World J Agric Sci* 2: 303-310.

Samarbakhah S, Rejali F, Ardakani MR, Pak Nejad F, Miransari M (2009) The combined effects of fungicides and arbuscular mycorrhiza on Corn (*Zea mays* L.) growth and yield under field conditions. *J Biol Sci* 9: 372-376.

Same BI, Robson AD, Abbott LK (1983) Phosphorus,soluble carbohydrates and endomycorrhizal infection. *Soil Biol Biochem* 15: 593-597.

Sanders FE, Sheikh NA (1983) The development of vesicular-arbuscular mycorrhizal infection in plant root systems. *Plant Soil* 71: 223-256.

Santakumari M, Das VSR (1979) Photosynthetic carbon fixation and changes in the levels of carbohydrates under the influence of diuron or atrazine in some crop and weed species. *Indian J Plant Physiol* 22: 263-268.

Schmitz O, Danneberg G, Hundeshagen B, Klinger A, Bothe H (1992) Quantification of vesicular- arbuscular mycorrhiza by biochemical parameters. *J Plant Physiol* 139: 106-114.

Schreiner RP, Bethlenfalvay GJ (1996) Mycorrhizae biocides and biocontrol. 4. Response of a mixed culture of arbuscular mycorrhizal fungi and hosts plant to three fungicides. *Biol Fertil Soils* 23: 189-195.

Schreiner RP, Bethlenfalvay GJ (1997a) Mycorrhizae, biocides, and biocontrol. 3. Effects of three different fungicides on developmental stages of three AM fungi. *Biol Fertil Soils* 24: 18-26.

Schreiner RP, Bethlenfalvay GJ (1997b) Plant and soil response to single and mixed species of arbuscular mycorrhizal fungi under fungicide stress. *Appl Soil Ecol* 7: 93-102.

Schwab SM, Johnson ELV, Menge JA (1982) Influence of simazine on formation of vesicular arbuscular mycorrhizae in *Chenopodium quinona* Wild. *Plant Soil* 64: 283-287.

Schwab SM, Menge JA, Leonard RT (1983) Comparison of stages of vesicular-arbuscular mycorrhiza formation in sudan grass grown at two levels of phosphorus nutrition. *Am J Bot* 70: 1225-1232.

Schweiger PF, Jakobsen I (1998) Dose response relationships between four pesticides and phosphorus uptake by hyphae of arbuscular mycorrhizas. *Soil Biol Biochem* 30: 1415-1422.

Sieverding E (1989) Ecology of VAM fungi in tropical agrosystems. *Agric Ecosyst Environ* 29: 369-390.

Sieverding E (1991) *Vesicular arbuscular mycorrhizal management in tropical agrosystems.* Technical cooperation, GTZ Press, Germany, Eschborn. pp 371.

Smith SE, Gianinazzi-Pearson V (1988) Physiological interactions between symbionts in vesicular-arbuscular mycorrhizal plants. *Annu Rev Physiol Plant Mol Biol* 39: 221-244.

Smith SE, Read DJ (1997) *Mycorrhizal symbiosis.* Academic Press Inc., San Diego, CA, USA.

Smith TF, Noack AJ, Cosh SM (1981) The effect of some herbicides on vesicular arbuscular endophyte abundance in soil and on infection of host roots. *Pestic Sci* 12: 91-97.

Sreenivasa MN, Bagyaraj DJ (1989) Use of pesticides for mass production of vesicular-arbuscular mycorrhizal inoculum. *Plant Soil* 119: 127-132.

Srinivasan D, Muthukumar T, Udaiyan K (1995) Occurrence of vesicular arbuscular mycorrhizae in a waste water irrigated field in Bharathiar University, Coimbatore. *Pollut Res* 14: 157-164.

Strzemska J (1975) Mycorrhiza in farm crops grown in monoculture. In: *Endomycorrhizas,* Sanders FE, Mosse B, Tinker PB (eds), Academic Press, New York. pp 527-536.

Sugavanam V, Udaiyan K, Manian S (1994) Effect of fungicides on vesicular mycorrhizal infection and nodulation in groundnut (*Arachis hypogaea* L.). *Agric Ecosyst Environ* 48: 285-293.

Sylvia DM, Neal LH (1990) Nitrogen affects the phosphorus response of VA mycorrhiza. *New Phytol* 115: 303-310.

Tang M, Chen H, Huang JC, Tian ZQ (2009) AM fungi effects on the growth and physiology of *Zea mays* seedlings under diesel stress. *Soil Biol Biochem* 41: 936-940.

Tarafdar JC, Marschner H (1994) Phosphatase activity in the rhizosphere and hyphosphere of VA mycorrhizal wheat supplied with inorganic and organic phosphorus. *Soil Biol Biochem* 26: 387-395.

Thomson BD, Robson AD, Abbott LK (1986) Effects of phosphorus on the formation of mycorrhizas by *Gigaspora calospora* and *Glomus fasiculatum* in relation to root carbohydrates. *New Phytol* 103: 751-765.

Tisserant B, Gianinazzi-Pearson V, Gianinazzi S, Gollotte A (1993) In planta histochemical staining of fungal alkaline phosphatase activity for analysis of efficient arbuscular mycorrhizal infections. *Mycol Res* 97: 245-250.

Tommerup IC (1984) Persistence of infectivity by germinated spores of vesicular arbuscular mycorrhizal fungi in soil. *Trans Br Mycol Soc* 82: 275-282.

Tommerup IC, Briggs GG (1981) Influence of agricultural chemicals on germination of vesicular arbuscular endophyte spores. *Trans Br Mycol Soc* 76: 326-328.

Toth R, Toth D (1982) Quantifying vesicular arbuscular mycorrhizas using a morphometric technique. *Mycologia* 74: 182-187.

Trappe JM, Molina R, Castellano M (1984) Reaction of mycorrhizal fungi and mycorrhiza formation to pesticides. *Annu Rev Phytopathol* 22: 231-259.

Udaiyan K, Greep S, Muthukumar T, Chitra A (1999) Effect of fumigation and pesticide drenches on VAM status and growth in cereals. *J Environ Biol* 20: 167-175.

Udaiyan K, Manian S, Muthukumar T, Greep S (1995) Biostatic effect of fumigation and pesticide drenches on an endomycorrhizal- *Rhizobium* legume tripartite association under field conditions. *Biol Fertil Soils* 20: 275-283.

Udaiyan K, Sugavanam V, Manian S (1996) Effects of formaldehyde fumigation and Fytolan drench on VAM fungi and nodulation in some leguminous forest tree seedlings in India. *Pertanika J Trop Agric Sci* 19: 33-41.

Vani K, Kumar GS, Selvaraj T (2005) Studies of VA-Mycorrhizal fungi associated with some weed plants in sugar mill effluent soils. *J Environ Biol* 26: 675-680.

Veeraswamy T, Padmavathi T, Venkateswarlu K (1993) Effect of selected insecticides on growth and mycorrhizal development in sorghum. *Agric Ecosyst Environ* 43: 337-343.

Verdin A, Lounes-Hadj Sahraoui A, Fontaine J, Grandmougin-Ferjani A, Durand R (2006) Effects of anthracene on development of an arbuscular mycorrhizal fungus and contribution of the symbiotic association to pollutant dissipation. *Mycorrhiza* 16: 397-405.

Volante A, Lingua G, Cesaro P, Cresta A, Puppo M, Ariati L, Berta G (2005) Influence of three species of arbuscular mycorrhizal fungi on the persistence of aromatic hydrocarbons in contaminated substrates. *Mycorrhiza* 16: 43-50.

Vyas SC (1988) Nontarget effects of agricultural fungicides. CRC Press, Boca Raton, Florida.

Wallenda T, Kottke I (1998) Nitrogen deposition and ectomycorrhizas. *New Phytol* 139: 169-187.

Wang B, Qiu YL (2006) Phylogenetic distribution and evolution of mycorrhizas in land plants. *Mycorrhiza* 16: 299–363.

Whipps JM (2004) Prospects and limitations for mycorrhizas in biocontrol of root pathogens. *Can J Bot* 82: 1198-1227.

Wilson GWT, Hartnett DC (1998) Interspecific variation in plant responses to mycorrhizal colonization in tallgrass prairie. *Am J Bot* 85: 1732-1738.

Wu N, Huang H, Zhang S, Zhu YG, Christie P, Zhang Y (2009) Phenanthrene uptake by *Medicago sativa* L. under the influence of an arbuscular mycorrhizal fungus. *Environ Pollut* 157: 1613-1618.

In: Mycorrhizal Fungi
Editor: Susanne M. Fulton

ISBN 978-1-61122-659-1
© 2011 Nova Science Publishers, Inc.

Chapter 6

MYCORRHIZOSPHERE INTERACTIONS MEDIATED THROUGH RHIZODEPOSITIONS AND ARBUSCULAR MYCORRHIZAL HYPHODEPOSITION AND THEIR APPLICATION IN SUSTAINABLE AGRICULTURE

Walid Ellouze[1,2,3]***, Chantal Hamel***[1,2]***,***
Sadok Bouzid[3] ***and Marc St-Arnaud***[1]

[1]Institut de recherche en biologie végétale, Université de Montréal, Jardin botanique de Montréal, 4101 East Sherbrooke Street, Montreal, QC, H1X 2B2, Canada
[2]Semiarid Prairie Agricultural Research Centre, Agriculture and Agri-Food Canada, 1 Airport Rd., Swift Current SK, S9H 3X2, Canada
[3]Departement de Biologie, Faculté des Sciences Mathématiques, Physiques et Naturelles, Universite Tunis El Manar, Campus Universitaire, Tunis 1060, Tunisia

ABSTRACT

Arbuscular mycorrhizal (AM) fungi are ubiquitous in soils around the world where they form symbiotic associations with the majority of plant species. The beneficial effects of AM fungi are well known. They include improved plant nutrient uptake, enhanced N_2-fixation in legumes, increased plant tolerance to biotic and abiotic stress and improved soil structural quality, and are seen as the cornerstone of sustainability in agro-ecosystems. In soil, AM fungi form extensive mycelial networks connected to plant roots. These AM mycelial networks, which increase the surface for nutrient and water absorption, also offer large surfaces for interactions with other microorganisms. These fungi often interact positively with biofertilizers and biocontrol agents. The presence of AM fungi in plant roots and soils was repeatedly shown to reduce the incidence of diseases caused by pathogenic nematodes and fungi and to improve plant health. AM fungi exert profound effects on soil microorganisms through competition for nutrients, and modification of root exudation, plant physiology and signaling. Better understanding of the interactions between AM fungi and other microorganisms is necessary for the development of sustainable management of soil fertility and crop production. The purpose of this review is to outline the current knowledge on the mechanisms by which the symbiotic association with mycorrhizal fungi alters the existing mycorrhizosphere microbial community and discusses the possible applications these mechanisms may have in sustainable agriculture.

1. INTRODUCTION

The arbuscular mycorrhizal symbiosis is a widespread mutualistic association between plants and fungi from the phylum Glomeromycota (Redecker and Raab, 2006). The origin of AM associations corresponds with the appearance of land plants, 450 million years ago (Pirozynski and Dalpé, 1989; Simon et al., 1993). This association is fundamental in the plant kingdom; strictly speaking, most plants do not have roots, they have mycorrhizae (Wilhelm, 1966). AM fungi play a crucial role in agro- and natural ecosystems by acting as biofertilizers, bioremediator, soil conditioner, bioprotectors and biocontrol agents and have wide applications in agriculture, horticulture, forestry, and biomass production.

The beneficial effects of AM fungi in ecological agriculture have been recently reviewed by Ellouze et al., (2008) and Fraser et al., (2009). The mycorrhizal symbiosis occupies a central position in rhizosphere development and many types of interactions involving this symbiosis and important microbial groups have been reported (Barea et al., 2004). The functional and morphological complexity of the rhizosphere is increased by mycorrhizal fungi, which in turn support a mycorrhizosphere (Linderman, 1988; Nehl and Knox, 2006).

Various microorganisms have been tested in interaction studies with AM fungi, including saprophytes (McAllister et al., 1995), pathogens (Lioussanne et al., 2008; Meyer and Linderman, 1986), biocontrol agents (Wyss et al., 1992) and other plant symbionts (Barea et al., 1987). The purpose of this review is to outline the current knowledge on the mechanisms by which the symbiotic association with mycorrhizal fungi alters the existing mycorrhizosphere microbial community and their possible application in sustainable agriculture.

2. THE MYCORRHIZOSPHERE

The rhizosphere is colonized by bacteria, fungi and other microorganisms and thus constitutes a living interface between roots and the soil. It is a dynamic environment determined by interactions between the soil, plants and roots-associated microflora. The rhizosphere is a unique habitat where only a subset of the soil microflora is able to survive and multiply (Foster 1986; Marilley and Aragno 1999; Marschner and Timonen 2005; Manoharachary and Mukerji 2006). In this environment, plant beneficial microorganisms are of peculiar importance: they interact and determine plant success (Aragno, 2003).

The mycorrhizosphere refers to the soil zone near and influenced by mycorrhizal roots, hyphae, spores, and fruit bodies. Arbuscular mycorrhizal (AM) fungi are typically abundant in soil, and may account for some 25% of the soil microbial biomass (Hamel, 2007; Hamel et al., 1991; Olsson et al., 1999) and up to 80% of the fungal biomass (Bååth et al., 2004; Kabir et al., 1997) in certain agricultural soils. A gram of soil can contain up to 30 m of AM fungal extraradical hyphae (Smith and Read, 1997). Most crop plant species form mycorrhizae, thus the concept of mycorrhizosphere applies to most crops (Timonen and Marschner, 2006).

Arbuscular mycorrhizal fungi have a large influence on soil microorganisms as they increase the size of C and energy input to soil by stimulating plant productivity and modifying the physical and chemical environment of the soil (Johansson et al., 2004). They influence mycorrhizosphere communities quantitatively and qualitatively with impact on plant health (Azcón et al., 1991; Barea, 2000; Linderman, 1988; Meyer and Linderman, 1986; St-Arnaud and Elsen, 2005; St-Arnaud and Vujanovic, 2007; St-Arnaud et al., 1995). Plants provide reduced carbon compounds to their mycorrhizal fungal partners to grow mycorrhizal mycelial networks and improve their capacity for soil nutrient and water extraction. In turn, better plant nutrition reduces root exudation and, thus, the food supply for rhizospheric microorganisms. The supply of photosynthates, as substrates to soil microbiota, is a key factor in mycorrhizosphere formation (Barea, 2000). Mycorrhizal fungi induced shifts in soil microbial communities were also attributed to the regulation of the symbiosis. For example, mycorrhizal colonization of roots triggers plant defense mechanisms against potential pathogens attack (Cordier et al., 1998; Pozo et al., 2009; Pozo et al., 2002; Vigo et al., 2000).

Arbuscular mycorrhizal fungi exert profound effects on other mycorrhizosphere microorganisms either directly or indirectly via their impact on host, but also through direct effects. Arbuscular mycorrhizal fungi are competitive soil microorganisms because they tap directly on the ultimate source of carbon and energy for soil organisms, plant photosynthesis, and have first access to carbon of plant origin. Direct effects include provision of energy-rich carbon compounds derived from host assimilates, which are transported to the mycorrhizosphere via fungal hyphae, changes in soil pH, competition for nutrients, and exudation of inhibitory or stimulatory compounds.

3. Mycorrhizosphere Interactions through Rhizodepositions

Rhizosphere soil is strongly influenced by the root, through rhizodeposition (Marx et al., 2007) and respiration, as well as a result of ions and minerals uptake (Aragno, 2003). Roots release 1 to 25% of plants' net photosynthesis as soluble and insoluble compounds into the rhizosphere (Merbach et al. 1999; Marschner and Timonen 2006). Rhizodeposition products are composed of exudates, lysates, mucilage, secretions and dead cell material, and include gaseous compounds (Lynch and Whipps, 1990). Root exudates constitute the major part of the rhizodeposition and are mainly composed of soluble low-molecular weight substances such as carbohydrate monomers, amino acids, organic acids, phytosiderophores, flavonoids, plant hormones and vitamins (Farrar et al., 2003; Lynch and Whipps, 1990; Marschner, 1995). Mycorrhiza establishment is known to change mineral nutrient composition, hormonal balance, C allocation patterns, and other aspects of plant physiology (Barea, 2000; Marschner and Timonen, 2006) including root exudates composition (Bansal and Mukerji, 1994; Marschner et al., 1997), which influences the rhizosphere environment. Mycorrhizal fungi may reduce the amount of root exudates in the rhizosphere by taking up carbohydrates directly from root cells before they reach the surrounding soil (Timonen and Marschner,

2006). Mycorrhizal plants transfer more assimilates to their roots than non-mycorrhizal plants (Eissenstat et al., 1993). Larger C investments in AM roots are explained by the need for fungal growth, respiration and synthesis of the wide range of compounds that may stimulate beneficial microorganisms or antagonize phytopathogenes (Finlay and Söderström, 1992; Kope and Fortin, 1989; Kucey and Paul, 1982). The variation in the amount and composition of rhizodeposition induced by AM symbiosis formation can result in dramatic changes in the activity and composition of soil microbial communities, as microbial species differ in metabolism and have different ability to use different carbon sources (Marschner and Timonen, 2006; Neumann, 2005)

There are conflicting reports in the literature regarding the impact of AM fungi on rhizodeposition. While some researchers have reported reduced amounts of carbohydrates (Bansal and Mukerji, 1994; Mada and Bagyaraj, 1993; Ocampo and Azcon, 1985), altered composition of amino acids (Laheurte et al., 1990; Mada and Bagyaraj, 1993) and stimulated release of phenolics, gibberellins and nitrogen (Mada and Bagyaraj, 1993) in roots colonized by AM fungi (Bansal and Mukerji, 1994; Marschner et al., 1997), others found no evidence of AM fungal influence on rhizodeposition (Azaizeh et al., 1995). Lioussanne et al. (2008) reported higher concentrations of proline and isocitrate in root exudates in presence of *G. intraradices* after 24 weeks of growth, while after 16 weeks, proline concentration in exudate was not affected and the isocitrate concentration was reduced. Furthermore, the presence of *G. intraradices* had no influence on the amounts of amino acids, organic acids and measured sugars released into the mycorrhizosphere in this study. Other studies hypothesize that mycorrhizal fungi regulate rhizosphere events such as secondary AM infection and *Striga* germination in reducing the production of strigolactones, which is a group of sesquiterpenes lactones known so far as rhizospheric signaling molecules (Akiyama and Hayashi, 2006; Lendzemo et al., 2009; Lendzemo et al., 2007; Matusova et al., 2005). A variety of factors could explain these contrasting results, including differences in the plant and fungus species involved, the experimental system used, and environmental conditions (Bending et al., 2006).

3.1. Interaction of AM Fungi with Rhizobacteria

Interactions between functional groups of the soil microflora affect both plant and soil development (Andrade et al., 1997; Schreiner et al., 1997). Bacteria and fungi influence each other (Garbaye, 1994; Paulitz and Linderman, 1991) and, thus, their influence on the plant-soil system depends on the outcome of their interaction (Bethlenfalvay et al., 1997; Meyer and Linderman, 1986). In particular, mycorrhizal infection influences the composition of the rhizosphere bacterial community (Bansal and Mukerji, 1994; Buwalda and Goh, 1982; Christensen and Jakobsen, 1993; Dixon et al., 1989; Graham et al., 1981; Jones et al., 2004; Mansfeld-Giese et al., 2002; Marschner et al., 1997; Shachar-Hill et al., 1995) and the density of bacterial population within the mycorrhizosphere (Bending et al., 2006). Negative as well as positive effects of AM fungi on bacteria have been reported.

Beneficial fungi and bacteria are a major focus in rhizosphere research. The large group of plant growth-promoting rhizobacteria (PGPRs) includes bacteria with different lifestyles (Perotto and Bonfante, 1997). PGPRs are either symbiotic, for example the *Rhizobium* spp., or free-living (Glick, 1995; Kloepper et al., 1980). PGPRs are often associated with mycorrhizal fungal spores and hyphae and may even be found within these structures (Bonfante and Anca, 2009; St-Arnaud and Elsen, 2005).

3.1.1. Interactions of AM Fungi with Free-living PGPRs

Several studies have confirmed synergism between AM fungi and free-living PGPRs such as *Burkholderia cepacia* Palleroni & Holmes (Ravnskov et al., 2002) and *Pseudomonas fluorescens* Migula (Edwards et al., 1998). Root exudates collected from tomato roots colonized by *Glomus fasciculatum* were more attractive to *Azotobacter chrococcum* and *Pseudomonas fluorescens* than exudates collected from non-colonized roots (Sood, 2003). Meyer and Linderman (1986) saw more facultative anaerobic bacteria and less fluorescent pseudomonas in the rhizosphere of mycorrhizal plants, but no effect of AM fungi on the total number of bacteria. Secilia and Bagyaraj (1987) reported an increase in total bacteria, nitrogen fixers and Gram negative bacteria in the rhizosphere of mycorrhizal plants. Later studies reported an increase in the total number of plant associated bacteria in presence of the AM fungus *G. intraradices* (Ravnskov and Jakobsen, 1999). More recent studies show that the AM fungi might affect plant and soil microbial activity by stimulating the production of root exudates, phytoalexins, and phenolic compounds (Dalpé and Monreal, 2004; Norman and Hooker, 2000).

In a study with split-root pepper plants (Marschner et al., 1997), ^{14}C exudation was decreased on the root side colonized by *Glomus intraradices* but not on the non-mycorrhizal side. However, the population density of the introduced *P. fluorescens* 2-79RL in the rhizosphere was decreased on both sides of the root system. This result was confirmed by a subsequent study on the effect of *G. mosseae* colonization on bacterial community structure in a split-root maize (Marschner and Baumann, 2003). Both studies showed that the effect of AM fungi colonization on rhizospheric bacterial community structure is systemic and involves more than mere changes in the abundance of root exudation.

The mycorrhizosphere may be favorable or unfavorable to microbial proliferation. Free-living N_2-fixing bacteria were more numerous in the mycorrhizosphere of *Panicum maximum* (Secilia and Bagyaraj, 1987). Similarly, the number of autotrophic NH_4^+-oxidizing bacteria were higher in the mycorrhizosphere of *G. mosseae* and *G. fasciculatum* growing with *Zea mays* than in non-mycorrhizosphere soil, with the reverse situation for the numbers of denitrifying and NH_4^+-producing organisms (Amora-Lazcano et al., 1998; Bending et al., 2006).

3.1.2. Interactions of AM Fungi with Symbiotic N_2-fixing Bacteria

AM fungi and symbiotic N_2-fixing bacteria are particularly important. The N_2-fixing bacteria belong to the genera *Rhizobium*, *Sinorhizobium*, *Bradyrhizobium*, *Mesorhizobium* and *Azorhizobium*, collectively termed rhizobia. It has been shown that dual inoculation with AM fungi and rhizobia increases plant growth and N_2-fixation to a greater extent than inoculation with rhizobia alone (Ibijbijen et al., 1992; Vejsadova et al., 1992; Vejsadova et al., 1993). Measurements of the $^{15}N/^{14}N$ ratio in plant shoots indicated enhancement of the N_2-fixation rates in nodulated mycorrhizal plants, relative to that achieved by the same rhizobium strain in non-mycorrhizal plants (Jeffries et al., 2003; Toro et al., 1998). Mycorrhizal fungi may also enhance rhizobial cell densities in root nodules, which tend to decrease under drought stress (Tate, 1995). The AM symbiosis can influence nodule function and rhizobia population by changing the rates of water movement into, through and out of host plants in addition to reducing oxidative damage. The interactions between mycorrhizal fungi and rhizobium are complex and may change during the development of the nodules (Marschner and Timonen, 2006).

Establishment of an AM fungus within a root can alter rhizosphere populations (Ames et al., 1984; Meyer and Linderman, 1986) and this may affect the distribution or development of nodules throughout the root system (Ames and Bethlenfalvay, 1987). Bethlenfalvay et al. (1985) demonstrated a competitive interaction between an AM fungus and *Bradyrhizobium japonicum,* where prior establishment of the fungus inhibited subsequent nodule development in soybean. Several studies indicate that rhizobia and AM fungi can compete for colonization sites on legume roots and thus reduce symbiotic efficiency (Chalk et al., 2006). Moreover, AM symbiosis formation initially reduces the rate of nodule formation, presumably as a result of competition for plant carbon, but at later stages normally enhances plant nodulation and N_2-fixation by enhancing water uptake and more importantly, plant P nutrition (Barea and Azcón-Aguilar, 1983). De Varennes and Goss (2007) suggests that the positive interaction of rhizobia, AM fungi and legumes is modulated by the rate of early AM colonization, and soil disturbance created by tillage impairs this interaction by delaying the colonization of roots by the fungal partner in a field situation.

3.2. Interactions of AM Fungi with Mycorrhizosphere Fungal Community

3.2.1. Interactions of AM Fungi with Saprotrophic Microorganisms Community

Arbuscular mycorrhizal fungi have a competitive advantage over saprophytic soil microorganisms because they have first access to carbon of plant origin. Olsson et al. (1998) reported that AM fungal mycelium can reduce the size of saprophytic fungal communities in calcareous dune sand. Green et al. (1999) observed that the presence of external mycelium of *G. intraradices* suppressed *T. harzianum* population development and β-glucuronidase (GUS) activity, and suggest that nutrient competition is a likely means of interaction.

Interactions with saprophytic populations could influence decomposition processes (Finlay and Söderström, 1992). Hodge et al. (2001) reported that *Glomus hoi*, an AM fungus, enhances organic matter decomposition. In absence of AM effect on the abundance of microbial PLFA biomarkers, they concluded to the direct involvement of *G. hoi* in N mineralization from organic residues, although hyphospheric effects could have been masked by a large soil volume in the organic residue compartment used. Arbuscular mycorrhizae could stimulate organic matter decomposition through their effect on soil microorganisms. Qualitative as well as quantitative changes in microbial community structure may also be responsible for faster organic matter decomposition in presence of arbuscular mycorrhizae in a process that links mineralization to plant nutrient demand (Atul-Nayyar et al., 2009).

The AM fungi can modify the composition the soil microbial community through different mechanisms and, thus, change soil functionality. Extraradical hyphae of AM fungi may also bring available C to microorganisms of the hyphosphere allowing them to mineralize recalcitrant soil organic matter, as described in the model of Schimel and Weintraub (2003). More research is needed to clarify the impact of AM fungi on saprotrophic microorganisms.

3.2.2. Interactions of AM Fungi with the Pathogenic Fungal Community

Root colonization with AM fungi changes root exudate composition which, in turn, exhibits a different bioactive effect on pathogens within the soil (Jones et al., 2004; Lioussanne et al., 2008; Meyer and Linderman, 1986; Norman and Hooker, 2000; Sharma et al., 1992). The antagonistic effects of mycorrhizal fungi against plant pathogenic fungi can be due to the inability of pathogens to sense the presence of root when exudation levels is reduced by AM fungi (Timonen and Marschner, 2006), but there might be more to this AM effect, as exudates composition are modified by AM fungi. Exudates released from strawberry roots colonized by *Glomus etunicatum* and *Glomus monosporum* reduced sporulation and zoospore production by the pathogen *Phytophthora fragariae*, in vitro (Norman and Hooker, 2000). Recently, it was shown that proline concentration in exudates liberated by AM tomato roots was higher than in exudates from nonmycorrhizal roots (Lioussanne et al., 2008) suggesting that AM colonization can modify the composition of tomato root exudates. Increased proline concentration in tomato plants infected by *Phytophthora nicotianae*, was thought to prevent pathogen-root encounter by reducing the accumulation of zoospores in the vicinity of roots (Lioussanne et al., 2008). These results support earlier observations of reduced *Fusarium* populations in the soil surrounding mycorrhizal tomato roots and suggest the involvement of AM fungi in the control of soil-borne diseases (Caron, 1989)

A positive effect of tomato root exudates from plants colonized by the AM fungus *Glomus mosseae* on microconidia germination of *Fusarium oxysporum* f. sp. *lycopersici* (*Fol*) was reported by Scheffknecht et al. (2006). The higher the level of AM colonization, the higher the rate of microconidial germination, indicating that alterations of the exudation pattern depended on the degree of AM root colonization. These quantitative and/or qualitative alterations in root exudation of the *Fol* host tomato through mycorrhization are not specific to the *Fol*-tomato interaction. Similar changes in the bioactivity of root exudates of 12 non-host species from eight plant families showed that mycorrhizae-induced changes in root exudates were unrelated to plant susceptibility to *Fol* (Scheffknecht et al., 2007).

Information on the effect of AM root colonization on pathogens comes from greenhouse experiments using treated soil or in vitro systems using transformed roots. Artificial conditions allow the isolation of the organisms under study from other biotic and abiotic effects present in field situation but may be misleading. Field studies should always be used to validate results obtained under artificial conditions (Filion et al., 1999; St-Arnaud and Elsen, 2005). Observations from field surveys (e.g. Hamel et al., 2005), interpreted under the light of controlled conditions study results, would greatly help us understand the AM-pathogenic fungi interactions that are actually occurring in real situation.

4. Mycorrhizosphere Interactions Mediated through Arbuscular Mycorrhizal Hypho-deposition

Research has been focused on the biology of AM symbiotic plant root systems. However, attention has begun to shift to the biology of AM extraradical mycelia and their contribution to the spatial and temporal heterogeneity of microbial diversity and function in the soil. This mycelium may be an important site for microbial interactions. In fact, recent evidence suggest that rhizosphere microbial community could be much more sensitive to the presence of AM mycelia than to the modification in root exudation mediated by the AM symbiosis (Lioussanne et al., 2010).

The extraradical mycelium of AM fungi constitutes a large surface area interacting with the surrounding soil environment and representing an interesting source of C for the soil microflora (George et al., 1995). The AM extraradical mycelium exudes various compounds into the soil (Wright and Upadhyaya, 1996; Wright et al., 1996) influencing the chemical composition and pH of the soil environment (Bago and Azcón-Aguilar, 1997). The analysis of exudates from a *Glomus* revealed the presence of low-molecular-weight sugars and organic acids, and also unidentified high-molecular-weight compounds (Toljander et al., 2007).

The hyphosphere effect is created by the release of carbonaceous exudates by AM hyphae (Toljander et al., 2007), but also by the rapid (5 to 6 days) turnover of these hyphae (Staddon et al., 2003). The C compounds released have different bioactivity on different soil microorganisms of the hyphosphere and as a result, mycelial exudates can not only increase microbial growth and vitality, but can also change the composition of the microbial community (Duponnois et al., 2008; Toljander et al., 2007). For example, Filion et al. (1999) found a stimulation of *Pseudomonas chlororaphis* growth and *Trichoderma harzianum* germination in the presence of *G. intraradices* mycelia. In contrast, conidial germination of *F. oxysporum* f. sp. *chrysanthemi* was reduced, while the growth rate of *Clavibacter michiganensis* subsp. *michiganensis* was not affected. Scheublin et al. (2010) found specificity in bacterial attachment to AM hyphae and suggested that hyphal exudates shape hyphosphere bacterial communities through specific signaling that makes AM hyphae attractive to certain types of bacteria and not to others.

The hyphosphere effects can be nutritional or not. For example, Toljander et al. (2006) reported preferential attachment of some soil bacterial isolates to living AM fungal hyphae, but also to dead hyphae. Preferential attachment to dead hyphae suggests the existence of physical mechanisms for bacterial attachment to AM hyphae.

The bi-compartment *in vitro* system, where the AM fungi is grown on genetically transformed carrot (*Daucus carota*) roots in one compartment and where only AM mycelium is allowed to grow in the other (St-Arnaud et al., 1996), has been an important tool to study the direct effects of AM fungi on soil microorganisms. It revealed the AM mycelium as an important microbial community structuring agent in soil.

5. Mycorrhizospheric Bioactive Molecules and their Application in Sustainable Agriculture

Public concerns about climate change, food safety, and water quality, raise the need for a profound change in the way we produce crops. Biotechnologies harnessing biodiversity to enhance the contribution of natural plant symbiotic relationships to crop nutrition appear an option to improve the efficiency of nutrient use by crop plants and reduce the environmental impacts of fertilizers. Recent advances in research on chemical signaling in microorganisms open new promising avenues in biotechnology development.

The soil rooting zone is a place where communication between plants and soil microorganisms through signaling molecules is a key determinant of the soil microbial community composition, with impact on the success of the plant involved. The development and function of the AM symbiosis is governed by the cross-talk between plants and AM fungi (Brachmann, 2006). The identification of plant signal molecules promoting AM fungal infectivity and fungal signal molecules stimulating root receptivity could lead to the development of biotechnology tools for the management of the AM symbiosis in crop production (Gianinazzi and Vosátkatka, 2004).

Knowledge on the regulation of the AM symbiosis is accumulating and the stimulating activity of some compounds naturally produced by plants such as flavonoids, sesquiterpenes, ethylene and polyamines has been reported (Brachmann, 2006; Horii et al., 2009). Alginate oligosaccharide (Ishii et al., 2000) and nucleoside derivative (Kuwada et al., 2006) were also shown to enhance AM fungi growth. Stimulation of AM hyphal growth (Ishii et al., 1997) and root colonization (Cruz et al., 2004) were found in response to eupalitin, a flavonoid of Bahiagrass roots, and a bioactive peptide was recently discovered in the root extract of this plant (Horii et al., 2009). Tryptophan dimer promotes the growth and induces a chemotaxic response in germinating AM fungal spores, attracting growing hyphae. Short molecular peptides, such as tryptophan dimer (Trp–Trp) and Leu–Pro, remarkably stimulated spore formation of *G. clarum*, *G. etunicatum* and *Gigaspora albida* in absence of roots or root exudates (Ishii and Horii, 2009). When applied to *Citrus iyo*, a 25% MeOH eluate of an extract of brown alga, *Laminaria japonica*, raised AM colonization level and spore production (Kuwada et al., 2000).

Research on bioactive molecules may lead to improved AM inoculants production and may lower their price. Alternatively, cultivars with improved signaling capacity could be developed through marker-assisted selection by plant breeding programs. Care should be taken, however, with the manipulation of plant signals, as some signal molecules may result in the attraction of undesirable organisms (Steinkellner et al., 2007) and reduce crops' ability to manage their rhizosphere.

The AM fungi also release diffusible signal molecules offering potential for the development of green biotechnologies. These AM fungal signals could be formulated and applied to crops to enhance the contribution of indigenous AM fungi to crop production. They could be introduced into the formulation of AM inoculants to promote their infectivity (Gianinazzi and Vosátkatka, 2004; Hamel and Strullu, 2006; Mabood et al., 2006).

6. Mycorrhizosphere Inoculation with Multiple Beneficial Microorganisms in Sustainable Agriculture

Considering the beneficial effects of many microorganisms in promoting plant growth and health, co-inoculation of crop plants with beneficial microorganisms appears as an attractive complement or alternative to conventional agrochemicals (Avis et al., 2008). However, the interaction between AM fungi and other beneficial microorganisms may vary from mutualistic to antagonistic. A good understanding of the microbe–microbe and plant-microbe interactions taking place in the mycorrhizosphere is the first step toward the development of effective co-inoculation technologies. Several research works have compared the effects of plant inoculation with more than one beneficial microorganism to single inoculation (Table 1). The effects of co-inoculation on plants vary from additive or synergistic, to negative or neutral according to the microbial combination used (Avis et al., 2008; Gianinazzi and Vosátkatka, 2004). The development of technologies for plant inoculation with multiple microorganisms is currently attracting much interest in the inoculant industry.

The potential value of inoculation with microbial consortia in crop production was demonstrated mostly in controlled condition studies, but it is important to test the effect of co-inoculated microorganisms in field trial. One possible advantage of inoculating beneficial microorganisms with the AM fungi is that these fungi may offer an opportunity to improve the ability of beneficial microorganisms to establish and persist in soil after inoculation. AM fungi are very competitive in soil as they have first access to plant resources. They may favor the persistence of microorganisms introduced through inoculation if these ones are adapted to life in the hyphosphere. Therefore, it is important to better understand the interactions occurring in the mycorrhizosphere and the role of roots and AM extraradical mycelia exudates in the dynamics of microbial populations.

7. Conclusion

The mycorrhizosphere is a complex environment where AM fungi play a central role in increasing the nutrition and stress tolerance of host plants by modifying their metabolism and improving their reaction against pathogens (Smith and Read, 1997). The influence of AM extraradical mycelia on soil microbial growth and community composition has large ecological significance.

More efforts should be made to study the mycorrhizosphere under field conditions, because observations made in situ have much more relevance than those made in controlled condition studies. The new molecular and biotechnological methods to measure the abundance of AM fungi in soil, which are now brewing in specialized laboratories, will certainly open new possibilities in mycorrhizosphere ecology and contribute solutions for a more sustainable world. Important and rapid progress in the field of AM-soil microbial interaction is expected.

Acknowledgment

This study is supported by a scholarship to Mr. Walid Ellouze from the Islamic Development Bank Merit Scholarship Programme for High Technology.

Table 1. Reports of effect of inoculation of plants with multiple beneficial microorganisms

Host plant	AM fungus	Other microorganisms	Outcome	Reference
Tomato (*Solanum lycopersicon* L.) in pot and field grown	*G. intraradices*	*Pseudomonas fluorescens, Trichoderma harzianum*	All the biocontrol agents were effective at controlling Fusarium wilt and stimulating yield; their combination was most effective.	Srivastava et al., 2010
Tomato (*Solanum lycopersicon* L.) in pot	*G. intraradices*	*Clonostachys rosea*	Synergistic microbial effect in dual inoculation despite a mutual inhibition of isolates.	Ravnskov et al., 2006
Chickpea (*Cicer arietinum* L.) in pot	*G. mosseae*	*P. fluorescens, Azotoacter chroococcum, Azospirillum brasilense, Rhizobium sp.*	*P. fluorescens* and *G. mosseae* improved growth better than any combination treatment.	Siddiqui and Mahmood, 2001
Chickpea (*Cicer arietinum* L.) in pot	*G. intraradices*	*Pseudomonas striata, Rhizobium sp.*	Combined inoculation best reduced galling caused by the root-rot disease complex (*Meloidogyne incognita* and *Macrophomina phaseolina*) and improved plant growth and nutrition.	Sayeed Akhtar and Siddiqui, 2008
Chickpea (*Cicer arietinum* L.) in pot	*G. fasciculatum*	*Rhizobium sp., Pseudomonas striata, Penicillium variable*	The AM fungus in combination with the Rhizobium and the Pseudomonas improved nutrition and yield, but decreased nutrition and yield resulted from co-inoculation with the AM fungi, the Rhizobium and P. variable.	Zaidi et al., 2003
Clover (*Trofolium repens* L.) in pot	*G. mosseae*	*actinomycetes*	Synergistic effect of the microorganisms on growth and nutrition	Franco-Correa et al., 2010
Cucumber (*Cucumis sativus* L.) in pot	*G. intraradices*	*Paenibacillus macerans* and *Paenibacillus polymixa*	The bacteria had suppressive effects on the AM fungus, with co-inoculation causing growth depression.	Larsen et al., 2009

REFERENCES

Akiyama K., Hayashi H. (2006) Strigolactones: Chemical Signals for Fungal Symbionts and Parasitic Weeds in Plant Roots. *Ann Bot* 97:925-931. DOI: 10.1093/aob/mcl063.

Ames R.N., Bethlenfalvay G.J. (1987) Localized Increase in Nodule Activity but No Competitive Interaction of Cowpea Rhizobia Due to Pre-Establishment of Vesicular-Arbuscular Mycorrhiza. *New Phytologist* 106:207-215.

Ames R.N., Reid C.P.P., Inghame R. (1984) Rhizosphere bacterial population responses to root colonization by a vesicular-arbuscular mycorrhizal fungus. *New Phytologist* 96:555-563.

Amora-Lazcano E., Vázquez M.M., Azcón R. (1998) Response of nitrogen-transforming microorganisms to arbuscular mycorrhizal fungi. *Biology and Fertility of Soils* 27:65-70.

Andrade G., Mihara K.L., Linderman R.G., Bethlenfalvay G.J. (1997) Bacteria from rhizosphere and hyphosphere soils of different arbuscular-mycorrhizal fungi. *Plant and Soil* 192:71-79.

Aragno M. (2003) Beneficial soil microorganisms: a key for sustainable agriculture. Seventh International Conference on Kyusei Nature Farming *Proceedings of the conference held at Christchurch,* New Zealand, 15 18 January 2002.

Atul-Nayyar A., Hamel C., Hanson K., Germida J. (2009) The arbuscular mycorrhizal symbiosis links N mineralization to plant demand. *Mycorrhiza* 19:239-246.

Avis T.J., Gravel V., Antoun H., Tweddell R.J. (2008) Multifaceted beneficial effects of rhizosphere microorganisms on plant health and productivity. *Soil Biology and Biochemistry* 40:1733-1740. DOI: 10.1016/j.soilbio.2008.02.013.

Azaizeh H.A., Marschner H., Römheld V., Wittenmayer L. (1995) Effects of a vesicular-arbuscular mycorrhizal fungus and other soil microorganisms on growth, mineral nutrient acquisition and root exudation of soil-grown maize plants. *Mycorrhiza* 5:321-327.

Azcón R., Rubio R., Barea J.M. (1991) Selective Interactions between Different Species of Mycorrhizal Fungi and *Rhizobium meliloti* Strains, and Their Effects on Growth, N_2-Fixation (^{15}N) and Nutrition of *Medicago sativa* L. *New Phytologist* 117:399-404. .

Bååth E., Nilsson L.O., Goransson H., Wallander H. (2004) Can the extent of degradation of soil fungal mycelium during soil incubation be used to estimate ectomycorrhizal biomass in soil? *Soil Biology and Biochemistry* 36:2105-2109.

Bago B., Azcón-Aguilar C. (1997) Changes in the rhizospheric pH induced by arbuscular mycorrhiza formation in onion (*Allium cepa* L.). *Zeitschrift für Pflanzenernährung und Bodenkunde* 160:333-339.

Bansal M., Mukerji K.G. (1994) Positive correlation between VAM-induced changes in root exudation and mycorrhizosphere mycoflora. *Mycorrhiza* 5:39-44.

Barea J.M. (2000) Rhizosphere and mycorrhiza of field crops, in: J. P. Toutant, et al. (Eds.), *Biological resource management: connecting science and policy* (OECD), Springer-Verlag, Berlin. pp. 110–125

Barea J.M., Azcón-Aguilar C. (1983) Mycorrhizas and their significance in nodulating nitrogen-fixing plants. *Advance in Agronomy* 36:1-54.

Barea J.M., Azcón-Aguilar C., Azcón R. (1987) Vesicular arbuscular mycorrhiza improve both symbiotic N_2-fixation and N-uptake from soil as assessed with a N^{15} technique under field conditions. *New Phytologist* 106:717-725. DOI: 10.1111/j.1469-8137.1987.tb00172.x.

Barea J.M., Azcón R., Azcón-Aguilar C. (2004) Mycorrhizal fungi and plant growth promoting rhizobacteria, in: A. Varma, et al. (Eds.), *Plant surface microbiology,* Springer-Verlag, Heidelberg, Germany.

Bending G.D., Aspray T.J., Whipps J.M. (2006) Significance of Microbial Interactions in the Mycorrhizosphere, *Advances in Applied Microbiology,* Academic Press. pp. 97-132.

Bethlenfalvay G.J., Brown M.S., Stafford A.E. (1985) Glycine-Glomus-Rhizobium Symbiosis: II. Antagonistic Effects between Mycorrhizal Colonization and Nodulation. *Plant Physiol.* 79:1054-1058.

Bethlenfalvay G.J., Andrade G., Azcón-Aguilar C. (1997) Plant and soil responses to mycorrhizal fungi and rhizobacteria in nodulated or nitrate-fertilized peas (*Pisum sativum* L.). *Biology and Fertility of Soils* 24:164-168.

Bonfante P., Anca I.-A. (2009) Plants, Mycorrhizal Fungi, and Bacteria: A Network of Interactions. *Annual Review of Microbiology* 63:363-383.

Brachmann A. (2006) Plant-fungal symbiosis en gros and en detail. *New Phytologist* 171:242-246.

Buwalda J.G., Goh K.M. (1982) Host-fungus competition for carbon as a cause of growth depressions in vesicular-arbuscular mycorrhizal ryegrass. *Soil Biology and Biochemistry* 14:103-106.

Caron M. (1989) Potential use of mycorrhizae in control of soilborne diseases. *Canadian Journal of Plant Pathology 1*1:177–179.

Chalk P.M., Souza R.d.F., Urquiaga S., Alves B.J.R., Boddey R.M. (2006) The role of arbuscular mycorrhiza in legume symbiotic performance. *Soil Biology and Biochemistry* 38:2944-2951.

Christensen H., Jakobsen I. (1993) Reduction of bacterial growth by a vesicular-arbuscular mycorrhizal fungus in the rhizosphere of cucumber (*Cucumis sativus* L.). *Biology and Fertility of Soils* 15:253-258.

Cordier C., Pozo M.J., Barea J.M., Gianinazzi S., Gianinazzi-Pearson V. (1998) Cell defense responses associated with localized and systemic resistance to *Phytophthora parasitica* induced in tomato by an arbuscular mycorrhizal fungus. *Molecular Plant-Microbe Interactions* 11:1017-1028.

Cruz A.F., Ishii T., Matsumoto I., Kadoya K. (2004) Relationship between arbuscular mycorrhizal fungal development and eupalitin content in bahiagrass roots grown in a satsuma mandarin orchard. *Journal of the Japanese Society for Horticultural Science* 73:529-533.

Dalpé Y., Monreal M. (2004) Arbuscular mycorrhiza inoculum to support sustainable cropping systems. Online. *Crop Management.* DOI: 10.1094/CM-2004-0301-09-RV.

De Varennes A., Goss M.J. (2007) The tripartite symbiosis between legumes, rhizobia and indigenous mycorrhizal fungi is more efficient in undisturbed soil. *Soil Biology and Biochemistry 39*:2603-2607.

Dixon R.K., Garrett H.E., Cox G.S. (1989) Boron fertilization, vesicular-arbuscular mycorrhizal colonization and growth of *Citrus jambhiri* Lush. *J. Plant Nutr.* 12:687–700.

Duponnois R., Galiana A., Prin Y. (2008) The Mycorrhizosphere Effect: A Multitrophic Interaction Complex Improves Mycorrhizal Symbiosis and Plant Growth, Mycorrhizae: *Sustainable Agriculture and Forestry.* pp. 227-240.

Edwards S.G., Young J.P.W., Fitter A.H. (1998) Interactions between *Pseudomonas fluorescens* biocontrol agents and *Glomus mosseae*, an arbuscular mycorrhizal fungus, within the rhizosphere. *FEMS Microbiology Letters* 166:297-303.

Eissenstat D.M., Graham J.H., Syvertsen J.P., Drouillard D.L. (1993) Carbon Economy of Sour Orange in Relation to Mycorrhizal Colonization and Phosphorus Status. *Ann Bot* 71:1-10. DOI: 10.1006/anbo.1993.1001.

Ellouze W., Hanson K., Nayyar A., Perez J.C., Hamel C. (2008) Intertwined Existence: The Life of Plant Symbiotic Fungi in Agricultural Soils, in: A. Varma (Ed.), *Mycorrhiza : State of the Art, Genetics and Molecular Biology, Eco-Function, Biotechnology, Eco-Physiology, Structure and Systematics*, Springer-Verlag Berlin Heidelberg. pp. 507-528.

Farrar J., Hawes M., Jones D., Lindow S. (2003) How roots control the flux of carbon to the rhizosphere. *Ecology* 84:827-837.

Filion M., St-Arnaud M., Fortin J.A. (1999) Direct interaction between the arbuscular mycorrhizal fungus *Glomus intraradices* and different rhizosphere microorganisms. *New Phytologist* 141:525-533. DOI: 10.1046/j.1469-8137.1999.00366.x.

Finlay R., Söderström B. (1992) Mycorrhiza and carbon flow to the soil, in: M. Allen (Ed.), *Mycorrhizal Functioning*, Chapman & Hall, New York. pp. 134–160.

Franco-Correa M., Quintana A., Duque C., Suarez C., Rodríguez M.X., Barea J.-M. (2010) Evaluation of actinomycete strains for key traits related with plant growth promotion and mycorrhiza helping activities. *Applied Soil Ecology* 45:209-217. DOI: 10.1016/j.apsoil.2010.04.007.

Fraser T., Nayyar A., Ellouze W., Perez J.C., Hanson K., Germida J., Bouzid S., Hamel C. (2009) Arbuscular mycorrhiza: where nature and industry meet, in: D. Khasa, et al. (Eds.), *Advances in Mycorrhizal Science and Technology,* NRC Research Press and CABI Publishing, Canada. pp. 71-86.

Garbaye J. (1994) Tansley Review No. 76 Helper bacteria: a new dimension to the mycorrhizal symbiosis. *New Phytologist* 128:197-210. DOI: 10.1111/j.1469-8137.1994.tb04003.x.

George E., Marschner H., Jakobsen I. (1995) Role of arbuscular mycorrhizal fungi in uptake of phosphorus and nitrogen from soil. *Crit. Rev. Biotechnol.* 15:257–270.

Gianinazzi S., Vosátkatka M. (2004) Inoculum of arbuscular mycorrhizal fungi for production systems: science meets business. *Canadian Journal of Botany* 82:1264-1271.

Glick B.R. (1995) The enhancement of plant growth by free-living bacteria. *Canadian Journal of Microbiology* 41:109–117.

Graham J.H., Leonard R.T., Menge J.A. (1981) Membrane-Mediated Decrease in Root Exudation Responsible for Phosphorus Inhibition of Vesicular-Arbuscular Mycorrhiza Formation. *Plant Physiol.* 68:548-552.

Green H., Larsen J., Olsson P.A., Jensen D.F., Jakobsen I. (1999) Suppression of the Biocontrol Agent *Trichoderma harzianum* by Mycelium of the Arbuscular Mycorrhizal Fungus *Glomus intraradices* in Root-Free Soil. *Appl. Environ. Microbiol.* 65:1428-1434.

Hamel C. (2007) Extraradical arbuscular mycorrhizal mycelia: shadowy figures in the soil, in: C. Hamel and C. Plenchette (Eds.), *Arbuscular mycorrhizae in crop production.* Haworth's Food Products Press, Binghampton; New York. pp. 1-36.

Hamel C., Strullu D.G. (2006) Arbuscular mycorrhizal fungi in field crop production: Potential and new direction *Canadian Journal of Plant Science* 86:941-950

Hamel C., Neeser C., Barrantes-Cartín U., Smith D.L. (1991) Endomycorrhizal fungal species mediate ^{15}N-transfer from soybean to corn in non-fumigated soil. *Plant and Soil* 138:41-47.

Hamel C., Vujanovic V., Nakano-Hylander A., Jeannotte R., St-Arnaud M. (2005) Factors Associated with Fusarium Crown and Root Rot of Asparagus Outbreaks in Quebec. *Phytopathology 95*:867-873. DOI: 10.1094/PHYTO-95-0867.

Hodge A., Campbell C.D., Fitter A.H. (2001) An arbuscular mycorrhizal fungus accelerates decomposition and acquires nitrogen directly from organic material. Nature 413:297-299.

Horii S., Matsumura A., Kuramoto M., Ishii T. (2009) Tryptophan dimer produced by water-stressed bahia grass is an attractant for *Gigaspora margarita* and *Glomus caledonium. World Journal of Microbiology and Biotechnology* 25:1207-1215.

Ibijbijen J., Urquiaga S., Ismaili M., Alves B.J.R., Boddey R.M. (1992) Effect of arbuscular mycorrhizal fungi on growth, mineral nutrition and nitrogen fixation of three varieties of common bean. *New Phytologist* 134:353–360.

Ishii T., Horii S. (2009) Importance of short molecular peptides to axenic culture of arbuscular mycorrhizal (AM) fungi, *The 6th International Conference on Mycorrhiza,* Belo Horizonte-Minas Gerais, Brazil. pp. 24.

Ishii T., Narutaki A., Sawada K., Aikawa J., Matsumoto I., Kadoya K. (1997) Growth stimulatory substances for vesicular-arbuscular mycorrhizal fungi in Bahia grass (*Paspalum notatum* Flugge.) roots. *Plant and Soil* 196:301-304.

Ishii T., Kitabayashi H., Aikawa J., Matumoto I., Kadoya K., Kirino S. (2000) Effects of alginate oligosaccharide and polyamines on hyphal growth of vesicular-arbuscular mycorrhizal fungi and infectivity on citrus roots. *Proc Int Soc Citriculture* 9:1030–1032.

Jeffries P., Gianinazzi S., Perotto S., Turnau K., Barea J.-M. (2003) The contribution of arbuscular mycorrhizal fungi in sustainable maintenance of plant health and soil fertility. *Biology and Fertility of Soils* 37:1-16.

Johansson J.F., Paul L.R., Finlay R.D. (2004) Microbial interactions in the mycorrhizosphere and their significance for sustainable agriculture. FEMS Microbiology *Ecology* 48:1-13.

Jones D.L., Hodge A., Kuzyakov Y. (2004) Plant and mycorrhizal regulation of rhizodeposition. *New Phytologist* 163:459-480. DOI: 10.1111/j.1469-8137.2004.01130.x.

Kabir Z., O'Halloran I.P., Fyles J.W., Hamel C. (1997) Seasonal changes of arbuscular mycorrhizal fungi as affected by tillage practices and fertilization : Hyphal density and mycorrhizal root colonization. *Plant and Soil* 192:285-293.

Kloepper J.W., Leong J., Teintze M., Schroth M.N. (1980) Enhanced plant growth by siderophores produced by plant growth-promoting rhizobacteria. *Nature* 286 286.

Kope H.H., Fortin J.A. (1989) Inhibition of Phytopathogenic Funig Invitro by Cell Free Culture Media of Ectomycorrhizal Fungi. *New Phytologist* 113:57-63.

Kucey R.M.N., Paul E.A. (1982) Carbon flow, photosynthesis, and N_2-fixation in mycorrhizal and nodulated faba beans (*Vicia faba* L.). *Soil Biology and Biochemistry* 14:407-412.

Kuwada K., Utamura M., Matsushita I., Ishii T. (2000) Effect of tangle stock ground extracts on in vitro hyphal growth of vesicular-arbuscular mycorrhizal fungi and their in vivo infections of citrus roots. *Proc Int Citrus Cong* 9:1034–1037.

Kuwada K., Kuramoto M., Utamura M., Matsushita I., Ishii T. (2006) Isolation and structural elucidation of a growth stimulant for arbuscular mycorrhizal fungus from *Laminaria japonica* Areschoug. *Journal of Applied Phycology* 18:795-800.

Laheurte F., Ieyval C., Bertheien J. (1990) Root exudates of maize, pine and beech seedlings influenced by mycorrhizal and bacterial inoculation. *Symbiosis* 9:111–116.

Larsen J., Cornejo P., Barea J.M. (2009) Interactions between the arbuscular mycorrhizal fungus *Glomus intraradices* and the plant growth promoting rhizobacteria *Paenibacillus polymyxa* and *P. macerans* in the mycorrhizosphere of *Cucumis sativus*. *Soil Biology and Biochemistry* 41:286-292. DOI: 10.1016/j.soilbio.2008.10.029.

Lendzemo V., Kuyper T., Vierheilig H. (2009) Striga seed-germination activity of root exudates and compounds present in stems of Striga host and nonhost (trap crop) plants is reduced due to root colonization by arbuscular mycorrhizal fungi. *Mycorrhiza* 19:287-294.

Lendzemo V.W., Kuyper T.W., Matusova R., Bouwmeester H.J., Van Ast A. (2007) Colonization by Arbuscular Mycorrhizal Fungi of Sorghum Leads to Reduced Germination and Subsequent Attachment and Emergence of *Striga hermonthica*. *Plant Signal Behavior* 2:58-62.

Linderman R.G. (1988) Mycorrhizal interactions with the rhizosphere microflora: The mycorrhizosphere effect. *Phytopatholology 78*:366–371.

Lioussanne L., Jolicoeur M., St-Arnaud M. (2008) Mycorrhizal colonization with *Glomus intraradices* and development stage of transformed tomato roots significantly modify the chemotactic response of zoospores of the pathogen *Phytophthora nicotianae*. *Soil Biology and Biochemistry* 40:2217-2224.

Lioussanne L., Perreault F., Jolicoeur M., St-Arnaud M. (2010) The bacterial community of tomato rhizosphere is modified by inoculation with arbuscular mycorrhizal fungi but unaffected by soil enrichment with mycorrhizal root exudates or inoculation with *Phytophthora nicotianae*. *Soil Biology and Biochemistry* 42:473-483. DOI: 10.1016/j.soilbio.2009.11.034.

Lynch J.M., Whipps J.M. (1990) Substrate flow in the rhizosphere. *Plant and Soil* 129:1-10.

Mabood F., Gray E.J., Lee K.D., Smith D.L. (2006) Exploiting inter-organismal chemical communication for improved inoculants. *Canadian Journal of Plant Science* 86:951–966.

Mada R.J., Bagyaraj D.J. (1993) Root exudation from *Leucaena leucocephala* in relation to mycorrhizal colonization. *World Journal of Microbiology and Biotechnology* 9:342-344.

Mansfeld-Giese K., Larsen J., Bodker L. (2002) Bacterial populations associated with mycelium of the arbuscular mycorrhizal fungus *Glomus intraradices*. *FEMS Microbiology Ecology* 41:133-140. DOI: 10.1111/j.1574-6941.2002.tb00974.x.

Marschner H. (1995) Mineral nutrition of higher plants Academic Press, Inc, San Diego.

Marschner P., Baumann K. (2003) Changes in bacterial community structure induced by mycorrhizal colonisation in split-root maize. *Plant and Soil* 251:279-289.

Marschner P., Timonen S. (2006) Bacterial Community Composition and Activity in Rhizosphere of Roots Colonized by Arbuscular Mycorrhizal Fungi, in: K. G. Mukerji, et al. (Eds.), *Microbial Activity in the Rhizoshere,* Springer Berlin Heidelberg. pp. 139-154.

Marschner P., Crowley D.E., Higashi R.M. (1997) Root exudation and physiological status of a root-colonizing fluorescent pseudomonad in mycorrhizal and non-mycorrhizal pepper (*Capsicum annuum* L.). *Plant and Soil* 189:11-20.

Marx M., Buegger F., Gattinger A., Zsolnay A., Munch J.C. (2007) Determination of the fate of ^{13}C labelled maize and wheat exudates in an agricultural soil during a short-term incubation. *European Journal of Soil Science* 58:1175-1185. DOI: 10.1111/j.1365-2389.2007.00911.x.

Matusova R., Rani K., Verstappen F.W.A., Franssen M.C.R., Beale M.H., Bouwmeester H.J. (2005) The Strigolactone Germination Stimulants of the Plant-Parasitic *Striga* and *Orobanche* spp. Are Derived from the Carotenoid Pathway. *Plant Physiol.* 139:920-934. DOI: 10.1104/pp.105.061382.

McAllister C.B., Garcia-Romera I., Martin J., Godeas A., Ocampo J.A. (1995) Interactions between *Aspergillus niger* van Tiegh. and *Glomus mosseae.* (Nicol. and Gerd.) *Gerd. and Trappe. New Phytol.* 129:309- 316.

Meyer J.R., Linderman R.G. (1986) Selective influence on populations of rhizosphere or rhizoplane bacteria and actinomycetes by mycorrhizas formed by *Glomus fasciculatum. Soil Biology and Biochemistry* 18:191-196.

Nehl D., Knox O. (2006) Significance of Bacteria in the Rhizosphere, in: K. G. Mukerji, et al. (Eds.), *Microbial Activity in the Rhizoshere*, Springer Berlin Heidelberg. pp. 89-119.

Neumann G. (2005) Rhizodeposition: Composition and Quantitative Aspects, Encyclopedia of Plant and Crop Science Taylor & Francis pp. 1-4.

Norman J.R., Hooker J.E. (2000) Sporulation of *Phytophthora fragariae* shows greater stimulation by exudates of non-mycorrhizal than by mycorrhizal strawberry roots. *Mycol. Res.* 104:1069-1073.

Ocampo J.A., Azcon R. (1985) Relationship between the concentration of sugars in the roots and VA mycorrhizal infection. *Plant and Soil* 86:95–100.

Olsson P.A., Francis R., Read D.J., Söderström B. (1998) Growth of arbuscular mycorrhizal mycelium in calcareous dune sand and its interaction with other soil microorganisms as estimated by measurement of specific fatty acids. *Plant and Soil* 201:9-16.

Olsson P.A., Thingstrup I., Jakobsen I., Bååth F. (1999) Estimation of the biomass of arbuscular mycorrhizal fungi in a linseed field. S*oil Biology & Biochemistry* 31:1879-1887.

Paulitz T.C., Linderman R.G. (1991) Mycorrhizal interactions with soil organisms, in: D. K. Arora, et al. (Eds.), *Soil and Plant: Handbook of Applied Mycology.* , Marcel Dekker Inc., New York, USA. pp. 77–129.

Perotto S., Bonfante P. (1997) Bacterial associations with mycorrhizal fungi: Close and distant friends in the rhizosphere. *Trends in Microbiology* 5:496-501.

Pirozynski K.A., Dalpé Y. (1989) Geological history of the Glomaceae, with particular reference to mycorrhizal symbiosis. *Symbiosis* 7:1–36.

Pozo M.J., Verhage A., García-Andrade J., García J.M., Azcón-Aguilar C. (2009) Priming Plant Defence Against Pathogens by Arbuscular Mycorrhizal Fungi, in: C. Azcón-Aguilar, et al. (Eds.), *Mycorrhizas - Functional Processes and Ecological Impact,* Springer, Berlin Heidelberg. pp. 1-13.

Pozo M.J., Cordier C., Dumas-Gaudot E., Gianinazzi S., Barea J.M., Azcon-Aguilar C. (2002) Localized versus systemic effect of arbuscular mycorrhizal fungi on defence responses to *Phytophthora* infection in tomato plants. *J. Exp. Bot.* 53:525-534. DOI: 10.1093/jexbot/53.368.525.

Ravnskov S., Jakobsen I. (1999) Effects of *Pseudomonas fluorescens* DF57 on growth and P uptake of two arbuscular mycorrhizal fungi in symbiosis with cucumber. *Mycorrhiza* 8:329-334.

Ravnskov S., Larsen J., Jakobsen I. (2002) Phosphorus uptake of an arbuscular mycorrhizal fungus is not affected by the biocontrol bacterium *Burkholderia cepacia*. Soil Biol. *Biochem.* 34:1875-1881.

Ravnskov S., Jensen B., Knudsen I.M.B., Bødker L., Funck Jensen D., Karlinski L., Larsen J. (2006) Soil inoculation with the biocontrol agent *Clonostachys rosea* and the mycorrhizal fungus *Glomus intraradices* results in mutual inhibition, plant growth promotion and alteration of soil microbial communities. *Soil Biology and Biochemistry* 38:3453-3462. DOI: 10.1016/j.soilbio.2006.06.003.

Redecker D., Raab P. (2006) Phylogeny of the Glomeromycota (arbuscular mycorrhizal fungi): recent developments and new gene markers. *Mycologia* 98:885-895. DOI: 10.3852/mycologia.98.6.885.

Sayeed Akhtar M., Siddiqui Z.A. (2008) Biocontrol of a root-rot disease complex of chickpea by *Glomus intraradices, Rhizobium sp.* and *Pseudomonas straita. Crop Protection* 27:410-417. DOI: 10.1016/j.cropro.2007.07.009.

Scheffknecht S., Mammerler R., Steinkellner S., Vierheilig H. (2006) Root exudates of mycorrhizal tomato plants exhibit a different effect on microconidia germination of *Fusarium oxysporum* f. sp. *lycopersici* than root exudates from non-mycorrhizal tomato plants. *Mycorrhiza* 16:365-370.

Scheffknecht S., St-Arnaud M., Khaosaad T., Steinkellner S., Vierheilig H. (2007) An altered root exudation pattern through mycorrhization affecting microconidia germination of the highly specialized tomato pathogen *Fusarium oxysporum* f. sp. *lycopersici* (Fol) is not tomato specific but also occurs in *Fol* nonhost plants *Canadian Journal of Botany* 85:347-352.

Scheublin T.R., Sanders I.R., Keel C., van der Meer J.R. (2010) Characterisation of microbial communities colonising the hyphal surfaces of arbuscular mycorrhizal fungi. *The ISME Journal* 4:752–763.

Schimel J.P., Weintraub M.N. (2003) The implications of exoenzyme activity on microbial carbon and nitrogen limitation in soil: a theoretical model. *Soil Biol Biochem 35*:549-563.

Schreiner R.P., Mihara K.L., McDaniel H., Bethlenfalvay G.J. (1997) Mycorrhizal fungi influence plant and soil functions and interactions. *Plant and Soil* 188:199-209.

Secilia J., Bagyaraj D.J. (1987) Bacteria and actinomycetes associated with pot cultures of vesicular-arbuscular mycorrhizas. *Can. J. Microbiol.* 33:1069–1073.

Shachar-Hill Y., Pfeffer P.E., Douds D., Osman S.F., Doner L.W., Ratcliffe R.G. (1995) Partitioning of Intermediary Carbon Metabolism in Vesicular-Arbuscular Mycorrhizal Leek. *Plant Physiol.* 108:7-15.

Sharma A.K., Johri B.N., Gianinazzi S. (1992) Vesicular-arbuscular mycorrhizae in relation to plant disease. *World Journal of Microbiology and Biotechnology* 8:559-563.

Siddiqui Z.A., Mahmood I. (2001) Effects of rhizobacteria and root symbionts on the reproduction of *Meloidogyne javanica* and growth of chickpea. *Bioresource Technology* 79:41-45. DOI: 10.1016/s0960-8524(01)00036-0.

Simon L., Bousquet J., Levesque R.C., Lalonde M. (1993) Origin and diversification of endomycorrhizal fungi and coincidence with vascular land plants. *Nature* 363:67–69.

Smith S.E., Read D.J. (1997) *Mycorrhizal Symbiosis*. 2nd ed. Academic Press, London.

Sood S.G. (2003) Chemotactic response of plant-growth-promoting bacteria towards roots of vesicular–arbuscular mycorrhizal tomato plants. *FEMS Microbiology Ecology* 45:219–227.

Srivastava R., Khalid A., Singh U.S., Sharma A.K. (2010) Evaluation of arbuscular mycorrhizal fungus, fluorescent *Pseudomonas* and *Trichoderma harzianum* formulation against *Fusarium oxysporum* f. sp. *lycopersici* for the management of tomato wilt. Biological Control 53:24-31. DOI: 10.1016/j.biocontrol.2009.11.012.

St-Arnaud M., Elsen A. (2005) Interaction of Arbuscular Mycorrhizal Fungi with Soil-Borne Pathogens and Non-Pathogenic Rhizosphere Micro-Organisms, *In Vitro Culture of Mycorrhizas*. pp. 217-231.

St-Arnaud M., Vujanovic V. (2007) Effect of the arbuscular mycorrhizal symbiosis on plant diseases and pests, in: C. Hamel and C. Plenchette (Eds.), *Mycorrhizae in crop production, Haworth Press,* Binghampton, New York. pp. 67-122.

St-Arnaud M., Hamel C., Caron M., Fortin J.A. (1995) Endomycorhizes VA et sensibilité aux maladies: synthèse de la littérature et mécanismes d'interaction potentiels, in: J. A. Fortin, et al. (Eds.), *La symbiose mycorhizienne - État des connaissances,* Orbis Publishing, Frelighsburg, Quebec. pp. 51-87.

St-Arnaud M., Hamel C., Vimard B., Caron M., Fortin J.A. (1996) Enhanced hyphal growth and spore production of the arbuscular mycorrhizal fungus *Glomus intraradices* in an *in vitro* system in the absence of host roots. *Mycological Research* 100:328-332. DOI: 10.1016/s0953-7562(96)80164-x.

Staddon P.L., Ramsey C.B., Ostle N., Ineson P., Fitter A.H. (2003) Rapid Turnover of Hyphae of Mycorrhizal Fungi Determined by AMS Microanalysis of ^{14}C. *Science* 300:1138-1140. DOI: 10.1126/science.1084269.

Steinkellner S., Lendzemo V., Langer I., Schweiger P., Khaosaad T., Toussaint J.-P., Vierheilig H. (2007) Flavonoids and Strigolactones in Root Exudates as Signals in Symbiotic and Pathogenic Plant-Fungus Interactions. *Molecules* 12:1290-1306.

Tate R.L. (1995) *Soil microbiology (symbiotic nitrogen fixation),* John Wiley & Sons, Inc., New York, N.Y. pp. 307-333.

Timonen S., Marschner P. (2006) Mycorrhizosphere Concept, in: K. G. Mukerji, et al. (Eds.), *Microbial Activity in the Rhizoshere*, Springer Berlin Heidelberg. pp. 155-172.

Toljander J.F., Artursson V., Paul L.R., Jansson J.K., Finlay R.D. (2006) Attachment of different soil bacteria to arbuscular mycorrhizal fungal extraradical hyphae is determined by hyphal vitality and fungal species. *FEMS Microbiology Letters* 254:34-40. DOI: 10.1111/j.1574-6968.2005.00003.x.

Toljander J.F., Lindahl B.D., Paul L.R., Elfstrand M., Finlay R.D. (2007) Influence of arbuscular mycorrhizal mycelial exudates on soil bacterial growth and community structure. *FEMS Microbiology Ecology* 61:295-304. DOI: 10.1111/j.1574-6941.2007.00337.x.

Toro M., Azcón R., Barea J.M. (1998) The use of isotopic dilution techniques to evaluate the interactive effects of Rhizobium genotype, mycorrhizal fungi, phosphate-solubilizing rhizobacteria and rock phosphate on nitrogen and phosphorus acquisition by *Medicago sativa*. *New Phytol.* 138:265–273.

Vejsadova H., Siblikova D., Hrselova H., Vancura V. (1992) Effect of the AM fungus *Glomus* sp. on the growth and yield of soybean inoculated with *Bradyrhizobium japonicum*. *Plant and Soil:*121–125.

Vejsadova H., Siblikova D., Gryndler M., Simon T., Miksik I. (1993) Influence of inoculation with *Bradyrhizobium japonicum* and *Glomus claroideum* on seed yield of soybean under greenhouse and field conditions. *Journal of Plant Nutrition* 16:619–629.

Vigo C., Norman J.R., Hooker J.E. (2000) Biocontrol of the pathogen *Phytophthora parasitica* by arbuscular mycorrhizal fungi is a consequence of effects on infection loci. *Plant Pathology* 49:509-514.

Wilhelm S. (1966) Chemical treatments and inoculum potential of soil. *Ann. Rev. Phytopath.* 4:53-78.

Wright S.F., Upadhyaya A. (1996) Extraction of an abundant and unusual protein from soil and comparison with hyphal protein of arbuscular mycorrhizal fungi. *Soil Science* 161:575-586.

Wright S.F., Franke-Snyder M., Morton J.B., Upadhyaya A. (1996) Tirne-course study and partial characterization of a protein on hyphae of arbuscular mycorrhizal fungi during active coloilizatioil of roots. *Plant and Soil* 181:193-203.

Wyss P., Boiler T.H., Wiemken A. (1992) Testing the effect of biological control agents on the formation of vesicular arbuscular mycorrhiza. *Plant Soil* 147.

Zaidi A., Khan M.S., Amil M. (2003) Interactive effect of rhizotrophic microorganisms on yield and nutrient uptake of chickpea (*Cicer arietinum* L.). *European Journal of Agronomy* 19:15-21. DOI: 10.1016/s1161-0301(02)00015-1.

In: Mycorrhizal Fungi
Editor: Susanne M. Fulton

ISBN 978-1-61122-659-1
© 2011 Nova Science Publishers, Inc.

Chapter 7

APPLICATION EFFICIENCY OF ARBUSCULAR MYCORRHIZAS IN SUSTAINABLE AGRICULTURE IN CHINA: A REVIEW

Wen Ke Liu
Institute of Environment and Sustainable Development in Agriculture, Key Lab. for Agro-Environment & Climate Change, Ministry of Agriculture, Chinese Academy of Agricultural Sciences, Beijing 100081, China

ABSTRACT

Sustainable agriculture, conservation-oriented production, is the optimal target of all countries worldwide with the increased concerns on food security and resources & environmental protection. More and more biological measures and recycle technologies were attempted to apply in crop production to reduce the chemicals' input, such as fertilizer and pesticides. Many studies have been carried out to determine the efficacy of arbuscular mycorrhizal fungi (AMF) in promoting plant fitness, productivity and produce quality over the past two decades in China except for the biodiversity and distribution investigations. Besides field crops, many industrial crops (fibre crops, sugar crops, tea plants, offcinals), horticultural crops (vegetables and fruits) as well as micropropagated seedlings, were proved to benefit from the arbuscular mycorrhizal symbiosis with respect to stress resistance, yield and quality enhancement. In this paper, the data were reviewed the beneficial functions of AMF in agricultural production, highlighting the environment-friendly and non-pollution advantages of AMF application in agriculture, and the application strategies were also discussed. In particular, we hope that our review will make some of the recent work published in the Chinese literature more accessible to a wider international audience.

Keywords: arbuscular mycorrhizal fungi (AMF); crops; quality; fitness; plant micropropagation

INTRODUCTION

Sustainable agriculture is characterized by environment-friendly and resources-saving with appropriate chemical input (pesticide and fertilizer) as much as possible. Sustainable agriculture, providing greater yield and better quality produce, is a long-term target of agricultural development in China. In sustainable agricultural production, chemicals and the alternatives, e.g. biological resources and technologies (such as beneficial microorganisms, organic fertilizer, biological control), should be integrated properly. Therefore, they are urgently needed to be developed to meet need of modern agriculture. Arbuscular mycorrhizal fungi (AMF) are multi-functional microorganisms with high biodiversity that can form symbioses with more than 90% of plant species by colonizing root tissues and soil [1,2]. They could promote plant fitness by improving the uptake of water and nutrients, resistance to pathogens etc. [1]. Therefore, many researchers had suggested that AMF potentially play important roles in sustainable plant production systems, by which plant fitness was improved with reduced or without input of synthetic chemicals, particularly in agriculture and horticulture [3]. In ecosystems, AMF play an important role in determining plant diversity, ecosystem variability and productivity [4,5]. Furthermore, AM could improve soil stability by increasing aggregate formation and stability [6]. In view of this, AMF can be regarded as soil amendments, biofertilizers, bioprotectors and bioregulators [7] based on their above- and below-ground functions.

AMF exist in almost all kinds of soil ecosystems, and have high affinity with local crops during long-term evolution. Therefore, it is believed that there are efficient isolates in indigenous communities at specific geographic sites for local mycorrhizal plant species with positive inoculation efficiency. AMF are thus regarded as the most promising beneficial microorganisms that can be produced and applied commercially in the practice of agricultural production, revegetation and soil restoration. Biodiversity, especially functional diversity between or within species of AMF made great application potential after selection [8]. Functional complementary within species were also revealed [9,10]. Smith et al. (2000) [11] showed that both Scuttellospo calospora and Glomus caledonium could increase P uptake of host plants. However, the former only explored P in the vicinity of roots, but the latter could uptake P both in distance and nearby roots. The functional complementary between fungi are the reason of increased P uptake. Therefore, selection and inoculation with functional complementary isolates is a efficient pathway to enhance mycorrhizal efficiency.

China is a country with large proportion population engaging agricultural production. Some emerging issues, e.g. inventory of resource decrease of non-renewable factors such as fertilizer (phosphorus and potassium etc.) and land annually, soil degradation and more and more serious the environment pollution force China to adopt sustainable ways in agricultural production. Under the circumstances, more and more attentions are paid to develop biological avenues to be a substitution for chemicals to guarantee food security. Among them, AMF are the most important microorganisms that are widely attempted in crop production in China in laboratory and field conditions. Up to now, the inoculation efficiency of almost all common crop species were examined in terms of the contribution in nutrient uptake, stress resistance, functional component content, yield and quality. Except mycorrhizal efficiency of crops, germplasm resources of AMF associated with the specific were widely investigated. After being fully surveyed, isolated, identified and mass-propagated, more and more AMF or isolates were screened for application in crop production at specific climate zone with particular crop species. It is roughly estimated that about 200 scientific papers had been published in the past ten more years on this topic. Gai et al. (2006) [12] reviewed the biodiversity and distribution of AMF in China conducted in the past twenty years, partially including some literatures covering the germplasm resources of AMF living in rhizosphere soil of some cultivated crops. In China, many studies had been conducted to examine the symbiotic efficiency with different crops. Because of the diversity in soil adaptability [32] and functions, selecting AMF at great length is necessary. However, no review that collected the previous data on inoculation efficiency of AMF on crop fitness, productivity and quality had been published. In this paper, all useful results issued updated are summarized, and constraints, strategy and current problems are highlighted in order to provide a reference to the colleagues worldwide.

1. Research Status of Mycorrhization on Crop Fitness and Productivity

1.1. Field Crops

1.1.1. Cereal Crops

In China, cereal crops mainly include wheat, corn, rice, sorghum and millet, which contribute 68.8 Percent planting area and 456 million tons yield in 2007. Cereal crops constitute the majority of species of grain crops, which provide the major source of starch, mineral elements, amino acids, and carbohydrates and energy source. AMF are efficient in improving growth and nutrient uptake for cereal crops, such as maize [14,15], wheat and soybean [16] , sorghum [17] and rice [18,19] barley [20]. However, the majority of studies were conducted with pot culture method, and harvested at seedling stage to investigate inoculation efficiency, including growth enhancement and nutrient uptake. Only few studies were carried out at field conditions to reveal yield and quality effects of arbuscular mycorrhizas. Hu et al. (2001) [18] showed that inoculating AMF enhanced rice growth of dry rice-nursery and yield. Chen et al.(2000) [14] showed that inoculation with G.m, G.v and S.s could increase the maize yield by 17, 22 and 11 compared with the control in sterilizing soil, and 10, 20, and 5 in non-sterilizing soil respectively.

Actually, application of AMF on cereal crops is scarce in China for two apparent reasons existing currently. One is poor maneuverability for field AM inoculation due to inoculum requirement in large quantity; the other is low commercial value of cereal crops. Therefore, cereal crops have been frequently used in most cases as host plants to test the physiochemical properties of mycorrhizal fungi in studies.

1.1.2. Legume Crops

In China, soybean, common bean, pea, mungbean, kidney bean and adzuki bean are dominant species planted at large scales. In 2007, the acreage of legume crops accounted for 7.68 percent of the total crop area in 2007. Major documents about mycorrhizal effects on legume crops were focused on growth promotion and nutrient uptake. One paper studied the dual-inoculation of arbuscule mycorrhiza and Rhizobia on growth of Kidney Bean [21].The inoculation also increased the root weight and diameter of the kidney beans, the weight and volume of the pods of the kidney beans, and the contents of the total nitrogen and quick-acting phosphorus elements in the soil. The output of the kidney bean was thus significantly improved, reaching to a maximum of 30% increase. The results demonstrated that inoculation of arbuscule mycorrhiza and rhizobia can promote growth of bean family plants and reduce utilization of fertilizer [21]. Most research was aimed to investigate the growth, P uptake and water utilization.

1.1.3. Tuber Crops

Tuber crops are important constituents in Chinese food, which mainly include sweet potato, potato, yambean, cassava and yam. All species of tuber crops are mycorrhizal plants. The infective abilities of three mycorrhizal species of G. intraradices, Glomus manhotis and Glomus etunicatum, to the roots of cassava and their effects on growth of cassava were studied by Huang et al. (2005) [22]. Gai et al. (2004) [23] showed that AMF, BEGl67, BEG14l and WUM26 increased P uptake of sweetpotato and vegetative growth, and efficiencies of BEG167和BEG14l are higher than WUM26. Bai et al. (2006) [24] found that the yield of micropropagated potato inoculated with BEG167 was 25.2% higher than that of controls on unsterilized soil conditions. However. inoculation with G.v reduced the yield of microtuber in both sterilized and unsterilized soil. When mixed inoculums of BEG167 and G.v (v/v=1:1) were applied under filed conditions, the yield of microtuber was increased by 21%. It is concluded that inoculating with AM fungal inoculum in both seedling plate an d field culture condition improved the yield of micro propagated potato.

A greenhouse study was conducted to study the efficiency of 14 isolates of arbuscular mycorrhizal (AM) fungi isolated from a local agricultural soil on the productivity of sweetpotato (*Ipomoea batatas*) [25]. The different AM fungi enhanced the biomass and nutritional status of sweet potato seedlings to different extents. The genus Glomus was more effective than Acaulospora or Scutellospora. Efficiency also varied among isolates of Glomus irrespective of individual host plant or location of origin. Intraspecific differences were sometimes greater than interspecific differences. Benefits deriving from fungal isolates were positively correlated with the root-colonization rate and the abundance of extraradical propagules of the AM fungi. Taking plant yield parameters, nutritional status of the plants, and fungal attributes into consideration, GEGM (Glomus etunicatum together with Glomus mosseae) and GE6 (Glomus etunicatum) were the most effective AM symbionts for sweet potato under the experimental conditions.

Liu et al. (2006) [26] showed that AMF inoculation can increase mycorrhizal fungi colonization, growth and P uptake of sweet potato six weeks after transplantation and also improve the yield and tuber quality of sweet potato on harvest. Some studies suggested that the different species of AM fungi should be used for various crops when applied in field [27]. It is suggested that the different species of AM fungi should be used for various crops when applied in field, and the mycorrhizal fungi stain from the native soil performs better than that from exotic soil and mixed strains perform better than the single one. Therefore, selection is necessary for a specific crop species at specific site.

1.2. Industrial Crops

1.2.1. Fibre Crops

Inoculation with AM fungi increases the biomass of cotton plant under both artificial and nature salinized soils significantly. The effects of AM fungi on cotton growth are also promoted with the increase of salt levels. This may indicate that the mycorrhizal plant association is beneficial for plant growth in salinity environment. Phosphorus concentrations in non-mycorrhizal plant show a up trend with saline level increasing and mycorrhizal plants have higher phosphorus concentration at all salt levels than that of non-mycorrhizal plants[28]. Effects of AMF on growth and yield of cotton in fertilized and non-fertilized fields with phosphorus are studied. The results show that the AMF is able to increase the photosynthetic rate, improve water status, promote cotton growth and increase its yield, especially the yield before frost season. Glomus versiforme is the most effective species in increasing cotton yield [29]. Bu et al. (2009) [30] found that G. mosseae and Glomus etunicatum inoculation promoted the growth of cotton plants and increased the content of chlorophyll in leaves. In addition, the content of total phenol in cotton roots was increased, inducing the synthesis and activity enhancement of POD enzyme in cotton roots, and the disease incidence and disease index of *verticillium wilt* of cotton plants were decreased significantly.

1.2.2. Oil Crops

Oil crops includes rapeseed plants, canola plants, peanut, sesame and sunflower. An investigation was carried out to test the effect of AM and peanut/wheat intercropping on iron accumulation and mobilization in root apoplast and it's role in improving iron nutrition of peanut. Four treatments were set up, namely, peanut monoeropping without AM, peanut monocropping with AM, peanut/wheat intercropping without AM, peanut/wheat intercropping with AM. Plants were grown in sandy soil in pots separated by 30mm nylon nets. The results demonstrated that inoculated with AM improved iron nutrition of peanut under monocropping and intercropping condition. The content of root apoplastic iron pool of young leaves and iron content of shoot in peanut plant inoculated with AM were all greater than the control. There was the same effect on root apoplastic iron-chlorophyll content of young leaves and iron content of shoot as peanut inoculated with AM. There was also a positive interaction between peanut inoculated with AM and peanut/wheat intercropping on iron accumulation in root apoplast and improvement of iron nutrition of peanut plant [31]. Zhang et al. (2001) [32] found that some tested AMF enhanced peanut and sweet potato growth and yield.

1.2.3. Sugar Crops

Sugarcane plants were generally less responsive. Apart from a 30% DW increase with VAM when 2.7 mg P/kg was added, DW of +VAM plants was equivalent to, or worse than in the case of 222 mg P/kg, DW of VAM plants[33]. The results indicated that inoculations with AMF significantly increased reductase activity, promoted the absorption of N, P and K, chlorophyll content and biological yield of sugarcane on exposed to water stress [34]. Another study showed that to enhance the drought-resistance of sugarcane affected by water stresses [35].

1.2.4. Tobacco

As early as 1979, Rich and Schench had conducted a investigation into the diversity of AMF in tobacco farmland. With respect to AMF resources in tobacco field or rhizosphere, until 1990s in China, Liu et al. (1997) [36] and Zhang et al. (1997) [37] surveyed the AMF communities around tobacco fields in Shandong and Shanxi provinces. They found that *Glomus mosseae* and *Glomus claroideum* are the dominant AMF species. However, Rich and Schench (1979) [39] suggested that *Gigaspora margarita* and *Acaulospora trappei* are the dominant species in the tobacco farmland of Florida. In summary, the dominant AMF species changed with the geographic condition, so it is better to screen efficient AMF isolates for tobacco production in local tobacco plant rhizosphere. Liu and Huang (1999) [40] studied the effects of inoculation AMF at different P application rate on tobacco yield and quality. The results showed that tobacco yield and the nicotine could be significant increased. The role of AMF in enhancement of nutrient uptake has been evidenced by many studies [41,42,43].

The effects of AM fungus *Glomus rnosseae* inoculation on the potassium absorption of flue-cured tobacco were studied with un-sterilized soil in a pot experiment. Potassium application and AM fungi inoculation had a significant influence on potassium accumulation and distribution, and the difference of potassium contents among different growth stages and different leaf positions was significant. Different combinations of AM fungi inoculation and potassium application had an obvious influence on potassium content in tobacco leaf during the middle and later growth stage. AM fungi inoculation could enhance the potassium content of top, middle and beneath position leaf, and hence, the content of potassium in leaf was increased, but that in stem was relatively lower [42]. Later, He et al. (2003) [41] also found that mycorrhizal infection promoted the total dry weights of tobacco leaves (especially middle position). *G. mosseae* significantly increased distribution ratio of K in leaves, while decreased that in shoot, also accumulation amounts of K in leaves and K contents in leaves during the maturity period were increased. Inoculation of the suitable AM species or strains at sowing stage could be an efficient way to improve the qualities of tobacco seedlings in intensive culture of tobacco seedlings [44].

1.2.5. Tea Plants

Due to commercial part of tea plants are leaves, successive tea plucking will lead to continuous recovery leaves of tea plants, which acquires more and more mineral nutrients from soils. However, most top quality tea plants are grown in the mountain area with barren soils lacking of adequate nutrient supply, particularly nitrogen and phosphorus. In addition, tea garden often far away from water source, drought stress occurs frequently. As it is well acknowledged that AMF are efficient in improvement of abiotic stresses, thus AMF application on tea plants are paid attentions in China.

Vegetative tea seedlings were inoculated with AM fungi *Acaulospora lavis, Glomus manihot* and *Glomus caledonium* to study the effects of AM on the growth of vegetative tea trees and the quality of tea in greenhouse. The results indicated that AM promoted the growth of vegetative tea seedlings, not only plant height but also aboveground and underground biomass. Moreover, AM increased element uptake of tea tree, especially to P, Ca and Mg. More important, the concentration of water-soluble substances, amino acids, caffeine and polyphenols in tea leaves were increased by AM, indicating that the quality of tea was improved [45]. Some surveys on AMF resource around rhizosphere of tea plants had been conducted [46,47]. The morphology, species frequency, spore density and natural infection rate of the VAM fungi in tea gardens were studied in the southern parts of Henan province. The results showed that plant height, dry weight and the uptake of phosphorus and potassium were increased [48,49].

1.2.6. Offcinals Medicinal Plants

Collection and use of Traditional Chinese medicine is originated in China thousands of years ago. Today, it is more and more necessary to cultivate medicinal plants other than to collect wild plants for the sharp cutoff in quantity. Moreover, used as materials for pharmaceutical factory, the entire cultivation process should be ensured to be not polluted by contaminants. Agro-chemicals (such as fertilizer, pesticide and so on) should be applied as little as possible. Under these circumstances, how to supply sufficient nutrients and diseases-resistance are issue should be resolved. Therefore, AMF are useful to help medicinal plants to establish, grow, and health besides enhancement in yield and quality. About interactions between arbuscular mycorrhizal fungi and medicinal plants are focused on rhizosphere community convey

In China, the occurrence of arbuscular mycorrhzial fungi in rhizosphere of some medicinal plants has been investigated. Ren et al, (2006) [50]found that infection rate of *Panax notoginseng* ranged from 12% to 30%, but infection strength is weak. In addition, they pointed out AMF may be a potent resource of biology which can stimulate the growth. Gao and He (2007) [51] found that 44.5% medicinal plants are Arum type, and 55.5% are Paris type among nine species. Soil available N was positively correlated with spore density of *Salvia miltiorrhiza* and Dioscorea opposite, and hyphal and total percent colonization of Adenophora borealis, while negatively correlated with arbuscular colonization of *Aster tataricus*; Soil available P exhibited a positive correlation with arbuscular colonization of *Salvia miltiorrhiza* and vesicular colonization of Dioscorea opposite, while a negative correlation with spore density of *Platycodon grandiflorus* and vesicular colonization of *Atractylodes macrocephala*; Spore density of Bupleurum chinense had a positive affection with soil organic matter; Soil pH was negatively correlated with hyphal and total colonization of A. denophora borealis and spore density of Bupleurum chinense, while positively correlated with spore density of A. denophora borealis. An investigation was made of mycorrhizae in 38 medicinal plant species, belonging to 21 families and 37 genera-distributed in Chongqing. Mycorrhizal symbiosis with AMF was found on the roots of 30 species. Typical arbuscular mycorrhizal structures were observed-including arbusculars-vesicles-spores-BAS (branched absorbing structure)-appressorium and hyphae-indicating a great diversity of the arbuscular mycorrhizal structures of the main species of medicinal plants in Chongqing [52].

Up to date, many studies had been conducted to examine the functions of AMF in improve the fitness of medicinal plants, including nutritional status and biomass. Inoculation of AM fungi could promote the total nitrogen content of *Atractylodes macrocephala*, the dry weight of plant, also the use efficiency of nitrogen fertilizer [53]. Zhang et al. (2003) [54] showed that *Glomus mosseae*, *Glomus intraradices* and *Glomus versiforme* inoculation increased growth of Dateplum Persimmon seedlings and the chilling resistance, and Zhao et al. (2007) [55] found that drought resistance of *Forsythia suspense* was improved by AMF.

Natural colonization rate is 13.1%, and increase to 35.5% after inoculation. Mycorrhizal inoculation significantly improved the survival rate and yield of three-year ginseng [56]. Infected VAM can make the plant higher and increase the tillering number. With the right amount of compound fertilizer, the results were more significant. Compared with control on an average the plant heights, each plant's tillering number, the weight of one hundred seeds and yield separately increased 6.74cm, 1.58, 0.6g and 58.3g/m. Infected with VAM. output and quality of Coix lachryma-jobi L increased, and its disease and harm resistance strengthened[56]. Six strains were inoculated into the tissue culture seedlings of *Dendrobium candidum*. After 60 days, the average fresh weights of seedlings had been higher than those of the control. The average fresh weight of seedlings. inoculated with GS222, GDB162, GDB254, MLXl02, had been extremely significantly increased, and inoculated with CLN103, CLX104 had been significantly increased. The mineral elements contents of the seedlings had also been higher than those of the control (except Zn). And the fungal strains had already been misdated from the mycorrhiza[57].

1.3. Vegetables

Many vegetables, including leaf vegetables and fruit vegetables, are verified to be beneficial to AMF inoculation, such as lettuce, watermelon, eggplant, cucumber, tomato and spring onion and so on. Inoculation AMF could decrease chemicals input in greenhouse or in the open, which is beneficial for high-quality vegetable production.

Zheng and Feng (2006) [58]found that AM fungi improved the growth of lettuce plants at a lower soil salt (607μS/cm) level, but reduced the growth at a higher level (1866 μS/cm). Effects of AMF on growth, yield and quality of *Phaseolus vulgaris* in field have investigated from 1996 to 1997. Results show that growth, yield and quality were improved [59]. Guo et al. (2007) [60] found that the enzyme produced pyruvic acid (EPY) and plant S concentrations were usually higher in mycorrhizal spring onion (*Allium fistulosum*) than controls irrespective of S supply, and shoot total S concentrations and EPY were strongly correlated. Arbuscular mycorrhizal colonization appeared to make a substantial contribution to the sulfur status of spring onion, and may thus have had a strong influence on its flavor characteristics. Previous experiments showed that mycorrhizal fungi tended to affect the pungency of Allium cepa [61] and Allium fistulosum [62].The EPY production tended to be mediated by mycorrhizal colonization, but not by N or S supply level.

The influence of 4 species of AMF, *Glornus mosseae*, *Glomus versiforme*, *Gigaspora roseo* and *Sclerocystis sinousa* on the quality of watermelon, cucumber, taro and bean is investigated under field conditions. The results show that the contents of vitamin C, amino acids, raw proteins, sugar and minerals in these vegetables are increased significantly by inoculation with the tested AM fungi. The treatments with *G. mosseae* show that the contents of VC and phosphorus are 25% and 63% increased in bean. Raw proteins and total amino acids 19% and 24% in taro, soluble sugar, phosphorus and raw proteins 20%, 26% and 40% in cucumber, soluble solid and Vc 25% and 32% in watermelon, respectively. It is suggested that the inoculation with AM fungi should be an important measure in creasing production and quality of green vegetables.

Significant differences were observed in some treatments with respect to mycorrhizal colonization rates, mineral contents, growth, yield and soluble sugar contents of watermelon. Peat, G. rnosseae and G. versiforme treatments resulted in the most beneficial effects on growth and yield of watermelon. So peat, and media (peat: sand: vermiculite 3:1:2) should be used for vegetables grown in soilless media and amployed more widely in future [64]. Compared with control yield, soluble solids content and vitamin C content of cucumber respectively increased by 26.9%, 16.0% and 2.7% by inoculating AMF treatment, while nitrate content decreased by 31.3%. The differences of cucumber plant height, stem thickness, leaf area, yield and fruit nitrate content between inoculating AMF treatment and control arrived markedly significant levels[65].

1.4. Production of Micropropagated Seedlings

High quality seedlings are necessary for agricultural production and ecological restoration. However, weak adaptability to biotic and abiotic stresses of micropropagated seedlings leads to large loss in transplantation. It is well known that inoculating a selected AMF at transplantation stage will helpful for improvement the stress resistance and survival ratio. Nevertheless, inoculation at transplantation will result in competition between AMF and host plants, which may make the transplantation shock more serious. Liu and Yang (2008) [66] suggested that the integration of PM system and mycorrhization in vitro (here called mycorrhization-PM cultivation system) is highly feasible. Furthermore, combining mycorrhization with PM system in vitro was shown to be an efficient method for alleviating the injuries of plantlets in vitro and ex vitro caused by environmental stresses and for mass producing AMF inoculants through applying different cultivation strategies. Cultivation strategies for the mass production of mycorrhizal plants and AMF inoculants using mycorrhization-PM cultivation system were discussed. Taking into consideration the economy and operability, most previous studies focused on developing methods for AMF application in transplant systems due to the advantages of low inoculant dosage and early symbiosis formation ahead of field growth, features that are helpful for the post-transplantation establishment of plantlets.

More and more isolates of AMF were found and selected in China for crop production, and methods to propagate AMF inocula were also developed to mass production [66,12]. Utilization of inoculum of Glomus manihotis produced in the on-farm system stimulated a 20% increase in yield of cassava (*Manihot esculenta*) in Colombia[67]. Mycorrhization treatment of plantlets is currently considered an effective method for enhancing post-transplantation establishment. In the optimization of crop production it has been shown that yield quality and quantity can be improved by a light-dependent application Green pepper [68]. For achievable application of AMF in crop production, methods must be developed to combine them with the roots of crops as early as possible successfully. Many crops, especially for industrial crops, culture of seedling was necessary in order to assure cost-effective production. Generally, together with seedling preparation, pre-inoculation of AMF were conducted before transplantation into field growth. Mycorrhization was considered to be more efficient at more the early period. Two kinds of methods are used in AMF production, i.e. commercially available or produced via on-farm systems in compost and vermiculite. Inoculation AMF is a promising, sustainable technique to enhance plant growth Production of micropropagated plants is one of the most promising fields for application for AMF inoculation[7]. Liu and Yang (2008) [66] made a feasibility analysis of using a new method by integrating mycorrhization and photoautotrophic micropropagation.

AMF are ubiquitous soil microorganisms that can form symbiosis with the majority of crop species, mainly including field crops and industrial crops. AMF are potentially important tools in agricultural systems that reduce or eliminate chemical inputs in modern agriculture. Many AMF or isolates (their ecotypes) were tested to inoculate in pot cultivation or under field conditions in China or on international [69]. Potato 1 [69], leek [70]Yield of potato tubers on a fresh weight basis were increased 10% to 45% in two farming systems under field conditions in two years experiment. These results demonstrated the potential yield benefits of inoculation of crops with AMF produced on the farm. Particularly, recently studies showed that AMF could improve crop production in a high P soil. The responses of crops to mycorrhization were tested in terms of yield, quality and stress tolerance. However, research has shown that even in the absence of a grown or P uptake enhancement relative to non-mycorrhizal plants, the majority of P uptake can be attributed to the fungi [71].

2. Summary and Prospects

In sustainable, low-input cropping systems the natural roles of microorganisms in maintaining soil fertility and biocontrol of plant pathogens may be more important than in conventional agriculture where their significance has been marginalized by high inputs of agrochemicals [72]. Better understanding of the interactions between AMF and other microorganisms is necessary for the development of sustainable management of soil fertility and crop production. The People's Republic of China has a coastline of 18,000 km and a total land area of 9.6 million km^2. Its territory extends over 50° of latitude from north to south, embracing the equatorial belt, the tropics, the subtropics, the moderate temperate zone, and the cold temperate zone. From east to west, China extends over 62° of longitude with land covered by forests, grasslands, deserts, plains, hills, and mountains. Climate diversity and plant diversity make more diverse AMF occurrence in different soil ecosystems. Update, except for farmland, numerous studies have been conducted in all kinds of ecosystems, especially in desert, island, mountain, grassland, stream valley, oasis etc. to survey efficient isolates [12]. Diversity of AMF will provide efficient isolates that can improve production of the variety of crops no matter where they grow. Nowadays, some excellent isolates for specific crops had been selected. However, more works are needed to be done along the route, isolation, identification, selection and practical application in terms of various crop species and usage. Abundance of AMF provides a assurance in application in agriculture. Nowadays, many crop species have been examined in terms of inoculation efficiency with respect to growth, nutrient uptake, stress resistance, yield and quality. To sum up, the most results showed a positive effect on above index, which give us more confidence to future research on utilization AMF in agriculture.

In practice, two methods can be used to introduce AMF into crops grown under field conditions. The first method is to culture AMF inoculants on-farm level [73], which can be used in cereal crops, legume crops, oil crops, sugar crops and tea plants. That is, propagation AMF in farmland by growing suitable crop species and cropping system. The other method is to inoculate AMF manually at the stage of transplantation within seedling preparation, which can be used in tuber crops, fibre crops, tobacco, offcinal medicine plants, horticultural plants and micropropagated seedlings. For the latter, micropropagation is vital part, which accounts for large proportion in commercial seedlings. Moreover, some emerging method modified from conventional tissue culture method had made it possible to mass produce mycorrhizal seedlings through association of mycorrhization and photoautotrophic micropropagation. Micropropagation is an important method to mass-produce seedlings, and mycorrhization was suggested it is feasibility to introduce AMF at in vitro stage by integrating mycorrhization with photoautotrophic micropropagation. However, for most cases, the efficacy of the latter method is the interactive result by association the exotic AMF with the indigenous AMF because almost all soils contain indigenous AMF. In order to achieve better efficacy, the compatible AMF (one or mixed isolates or ecotypes) with complementary functions with should be selected in advance.

Numerous studies have shown that individual species of AM fungi, and even fungal isolates in one species, differ in their ability to promote plant growth, and that promotion of plant growth can depend on the particular matching of plant and fungal species[74,75,76]. The interspecific and intraspecific variation makes it essential to screen for efficient AM fungi for particular host-plant species. It is traditional practice to prepare vegetative seedlings of sweet potato in seedling beds and then transfer stem cuttings by hand to the field, which facilitates direct low-cost inoculation with AM fungi to increase sweet potato yield. There has been increasing interest in AMF application in crop production after substantially proving their multi-benefits. Nevertheless, mycorrhization method is still a constraint for AMF application. There some negative factors that limit the sustainable development of agriculture currently in China. One is increasing area extension of leanness and pollution soils. Urgent requirement on non-pollution produce with highlighted quality demand. Third, it had to lower the input of chemicals such as fertilizer, pesticide and heavy metals. AMF is a promising driver to solve above issues in sustainable agriculture in China. Based on current research results, it is concluded that AMF are suitable to improve the production of some crops, especially for horticultural,

Two kinds of application strategies of AMF on for different crop species are practiced in terms of feasibility, economical efficiency, and public quality demand grade. In China, biodiversity of AMF associated with crops and their beneficial effects on growth and nutrient uptake and stresses resistance have been highly investigated in pot or field culture. However, few field experiments have been conducted to assess the effects of inoculation AMF on crop yield and quality. Some crops, e.g. tuber crops, vegetables, tobacco etc., can introduce AMF at the nursery stage conveniently, but some crops, e.g. maize, wheat, soybean etc., are difficult to inoculate, which may attribute to the laborious operations. For the latter kind crops, it can improve mycorrhizal efficiency by regulating AMF community on-farm level but manually inoculation exotic inocula. Vegetables, offcinals medicinal plants and tea are urgently and strictly required for non-pollution quality control for human healthy due to direct eat as accessory food. Moreover, mycorrhizal inoculation is feasible from at nursery stage or field conditions. Roles of AMF in agricultural production included vegetative growth, nutrient uptake and stress resistance improvement, by which substantial chemicals, such as fertilizer and pesticides are reduced. Production cost, quality and environmental benefits are obvious. Screening efficient AMF for crop production in field in sustainable agriculture is feasible when the method of mass-production of AMF inocula is realized. Mycorrhization are of plants will benefit fitness of plants, also helpful in ecological protection and agricultural productivity. Technologies to mass-produce AMF inoculum are urgently needed to be developed in 21st century to cope with climatic change and resource crisis. In China, tea and traditional Chinese medicinal materials are export produce that can achieve the international green food standards with the aid of AMF to decrease fertilizer and pesticides application.

REFERENCES

[1] Smith, S. E., Read D. J. *Mycorrhizal symbiosis,* 2nd ed. 1997. Academic Press, San Diego, CA.

[2] Newsham, K.K., Fitter A.H., Watkinson A.R. Multifunctionality and biodiversity in arbuscular mycorrhizas. *Trends in Ecology Evolution,*1995,10,407-411.

[3] Douds D.D., Nagahash Jr. G., Reider C., Hepperly P.R. Inoculation with arbuscular mycorrhizal fungi increases the yield of potatoes in a high P soil. *Biological Agriculture and Horticulture,* 2007, 25, 67-78.

[4] Van der Heijden MG.A., Klironomos J.N., Ursic M., Moutogolis, P. Streitwolf-Engel R., Boller T., Wiemken A., Sanders I.R...Mycorrhizal fungal diversity determines plant biodiversity, ecosystem variability and productivity. *Nature,* 1998b, 396, 69-72.

[5] Van der Heijden E.W., Kuyper T.W. Does origin of mycorrhizal fungus or mycorrhizal plant influence effectiveness of the mycorrhizal symbiosis. *Plant and Soil,* 2001, 230: 161-174.

[6] Wright S.F., Upadhyaya A. Extraction of an abundant and unusual protein from soil and comparison with hyphal protein of arbuscular mycorrhizal fungi. *Plant and Soil,* 1996, 161, 575-586.

[7] Lovato P.E., Trouvelot A., Gianninazzi-Pearson V., Gianinazzi S. Enhanced growth of wild cherry using micropropagated plants and mycorrhizal inoculation. *Agronomy Sustainable Development,* 2006, 26,209-213.

[8] Munkvold L., Kjøller R., Vestberg M., Rosendahland S., Jakobsen I. High functional diversity within species of arbuscular mycorrhizal fungi. *New Phytologist,* 2004,164: 357–364.

[9] Read D.J. The ties that bind. *Nature.* 1998,396,22–23.

[10] Koide R.T. Functional complementarity in the arbuscular mycorrhizal symbiosis. *New Phytologist,* 2000, 147: 233–235.

[11] Smith F., Jakobsen A.I., Smith S.E. Spatial differences in acquisition of soil phosphate between two arbuscular mycorrhizal fungi in symbiosis with Medicagotruncatula. *New Phytologist,* 2000, 147, 357- 366.

[12] Gai JP, Christie P, Feng G., Li. X.L. Twenty years of research on biodiversity and distribution of arbuscular mycorrhizal fungi in China: a review. *Mycorrhiza*, 2006, 16: 229-239.

[13] Liu WK. N, P contribution and soil adaptability of four arbuscular mycorrhizal fungi. *Acta Agric Scand B Soil and Plant Sci,* 2008, 58,285-288.

[14] Chen J.M., Jiang D.F., Liu S.T., Wang W.H., Li X.L. Ecological effects of arbuscular mycorrhizal fungi on the growth of maize. *Eco-agriculture Research*, 2000,.8, 24-26.

[15] Liu W.K., Feng G., Li X.L. The growth characteristics and symbiotic effectiveness of six arbuscular mycorrhizal fungi on three soils associated with maize. *Plant Nutrition and Fertilizer Scienee,* 2006, 12,530-536.

[16] Zheng H.L., Xing J., Hu J., Li X. Effects of AMF on the growth of wheat and soybean. *Journal of Inner Mongolia Agricultural University,* 2002,23, 104-106.

[17] Zhou W.L., Zhang Z.Q., Yao L.Q. *Effects of AM inoculation on microelements and secondary nutrients uptake of sorghum.* 1998, 71,72-74.

[18] Hu Z.J., Shen F.C., Qin W.G., Wang Y.L. Effects of arbuscular mycorrhizal fungi and bacterial fertilizer on rice growth. *Journal of Huazhong Agricultural University,* 2001,246-247.

[19] Zhu S.C., Zhang S., Tang J.J., Jiang Q.Q., Chen X. Regulation effects of arbuscular mycorrhizal fungi on the interactions between barnyard grass and upland rice under enhanced nitrogen supply. *Chin. J. Appl. Ecol.,* 2007, 18 (10) : 2337- 2342.

[20] Liu C.H., Chen B.D., Zhu Y.G., Zhang C. Effects of arbuscular mycorrhizal fungi on growth and phosphorus nutrition of highland barley. *Acta Pedologica Sinica*, 2006, 43, 1052-1055.

[21] Yu B., Li Z.F. Efects of dual-inoculation of arbuscule mycorrhiza and Rhizobia on growth of Kidney Bean. *Tianjin Agricultural Sciences*, 2003, 9, 21-24.

[22] Huang H.C., Tang G.D., Luo X.Y., Zhuang X.Y., Chen Y.J. The effects of three Glomus fungi on growth of *Manihot esculenta*. *Journal of South China Agricultural University,* 2005, 26, 43-47.

[23] Gai, J. P., Feng, G., Christie, P., Li, X. L. Screening of arbuscular mycorrhizal fungi for symbiotic efficiency with sweet potato. *Journal of Plant Nutrition,* 2006,29, 1085-1094.

[24] Bai D.S., Sun L.B., Wang H., Shao L.G., Feng G. Effect of Inoculation of arbuscular mycorrhizal fungi on yield of micropropagated potato. *China Potato,* 2006, 329-332.

[25] Gai J. P., Feng G., Christie P., Li X. L. Screening of arbuscular mycorrhizal fungi for symbiotic efficiency with sweet potato. *Journal of Plant Nutrition,* 2006, 29,1085–1094.

[26] Liu W.K., Feng G., Li X.L. Effects of arbuscular mycorrhizal fungi inoculation on the yield and quality of sweet potato. *Chinese Journal of Eco-Agriculture,* 2006,14, 105-107.

[27] Liu R.J., Li M., Shi Z.Y., Han Y.Z., Li X.L. Effect of arbuscular mycorrhizal fungi on yield of peanut and sweet potato. *Chinese Journal of Eco-Agriculture,* 2003,11,36-37.

[28] Feng G., Zhang F.S. Effect of arbuscular mycorrhizal fungi on salinity tolerance of cotton. *Chinese Journal of Eco-Agriculture,* 2003, 11, 21-24.

[29] Jiang D.F., Li M., Liu R.J., Liu P.L., Wu J.B. Effects of arbuscular mycorrhizal fungi on growth and yield of cotton. Eco-Agriculture Research, 1998,6, 24-26.

[30] Bu J., Cui W.D., Lou M., Long X.Q., Yang R. The effect of inoculation with AM inoculum on growth and *verticillium wilt* of cotton. *Xinjiang Agricultural Sciences,* 2009,46, 549-55.

[31] Song Y.N., Zheng W.W., Wang H. Effect of VAM and intercropping of wheat with peanut on formation of root apoplastic iron pool in peanut. *Scientia Agricultura Sinica,* 2001,34(5), 568-571.

[32] Zhang L. R. Cheng X.G. Xing J. Effect of vesicular arbuscular mycorhizas and peanut rhizobium on peanut. *Journal of Mianyang Agricultural College,* 1992, 9, 6-10.

[33] Kelly R.M., Edwards D.G. , Thompson J.P., Magarey R.C. Responses of sugarcane, maize, and soybean to phosphorus and vesicular-arbuscular mycorrhizal fungi. *Aust. J. Agric. Res.,* 2001, 52, 731–743.

[34] Wang Y.Z., Zhang M.Q., Ke Y.Q., Pan T.G., Effects of mycorrhizal fungus on sugarcane growth under water stress. *Journal of Fujian Agricultural University* (Natural Sciences Edition), 1994, 23,383-385.

[35] Wang Y.Z., Zhang M.Q., Ke Y.Q., Pan T.G., Liu Wei. Effect of mycorrhizal preinoculation on the activity of oxygen metabolism sugarcane leaves under water stress conditions, *EAR,* 1995,3, 10-15.

[36] Liu Y.R; Ding ZL; Fang CY. Isolation and selection of arbuscular mycorrhizal fungi from tobacco. *Journal of Shandong Agricultural University*, 1997, 28 (3) : 269 - 273.

[37] Zhang GY; Wang YJ. Isolation and selection of arbuscular mycorrhizal fungi from The Loess Plateau. *Journal of Shanxi University(Natural science edition),* 1997 , 20 (3) : 326 - 32.

[38] Rich J R, Schenck NC. Survey of Florida flue-cured tobacco fieleds for root - knot nematodes and vesicular-arbuscular mycorrhizal fungi *J. Plant Disease Rep,* 1979, 63 (11) : 952 - 955.

[39] Effects of VA mycorrhizal fungi on the production and quality of tobacco leaf under different phosphate applied amount. *J Yunnan Univ,* 1999, 2l(3) : 239-242.

[40] Liu J., Huang XY, Li TF. Effects of inoculation of mycorrhizal fungi, phosphrus-soluble bacteria on growth of tobacco seedling. *Tobacco Science and Technology-Cultivation and Process,* 2002, (2),43-44.

[41] He X>L, Sun W., Zhao L.L. Effects of AM fungi on potassium accumulation and distribution in tobacco plant under deficient K levels. *Chinese Journal of Applied and Environmental Biology,* 2003, 9 (1):81-84.

[42] Li D.W., Wang D.M., He X.L. Effects of AM fungi on potassium uptake of tobacco. levels. *Chin JAppl Ecol., 2003,*14 (10):1719-1722.

[43] Zhang X; Cao ZH. Effects of potassium status of three tobacco planting soils on potassium nutrition and contents of other elements in flue-cured tobacco. *Acta Pedologica Sinica*, 1994, 31,50-60.

[44] Jiang L., Li Z.M.,Huang J.G., Yuan L.,Wu H.T. Influences of arbuscular mycorrhizal fungi on growth and selected physiological indices of tobacco seedlings. *Plant NutritJon and Fertilizer Science,* 2008,14(1) : 156-161.

[45] Wang S.G., Lin X. G., Dong Y.H., Luo Z.C., Shi Y.Q. Influence of AM on the growth of vegetative tea seedling and the quality of tea. *Chinese Bulletin of Botany,* 2002, 19, 462-468.

[46] Lu D.S., Wu X.Q. Species of VAM fungi around tea roots in the southern area of Henan province. *Journal of Nanjing Forestry University (Natural Sciences Edition),* 2005, 29, 33-36.

[47] Zhang MQ, Wang YS, Wang KN, Xing LJ. VA mycorrhizal fungi of the south and east coasts of China, seven new records of Acaulospora. *Mycosystema,* 1998, 17:15–18.

[48] He S.J., Fang D.H., Wang S.S., Wu G.Q. Wang D.J., Bie Z.L. Effect is of VA mycorrhizal fungus on phosphorus and potassium uptake in tea seedlings. *Acta Agriculturae Nucleate Sinica,* 1997, 11, 45-48.

[49] He S.L., Wang S.S., Lei Z.Z. Effects of VA mycorhizal on the mineral nutrition for Camellia sinensis and their mechanism. *Journal of Southwest Agricultural University,* 1994, 16, 492-496.

[50] Ren J.H., Liu R.X., Li Y.L. Study on arbuscular mycorrhizae of Panax notoginseng. *Bulletin of Microbiology*, 2007, 34, 224-227.

[51] Gao A.X., He X.L. Ecological study on AM fungi around roots of medicinal plants in the middle area of Hebei Province. *Agricultural Research in the Arid Areas,* 2007, 25, 196-202.

[52] Ma Y.F., Yang X.H., Li P.M., Tong R.J. Investigation of the diversity of arbuscular mycorrhizal structure of medicinal plants in Chongqing. *Journal of Southwest Agricultural University (Natural Science),* 2005, 27,406-409.

[53] Lu Y.Q., Cui Y., He X.L. Effects of AM fungi on biomass and nitrogen content of *Atractylodes macrocephala* under different nitrogen levels. *Agricultural Science of Henan,* 2008, 4, 94-96.

[54] Zhang G.P., Qi G.H., Guo Q. Effects of arbuscular mycorrhizal on growth and chilling resistance of *Dateplum Persimmon* seedlings. *Hebei Fruits,* 2003, 6-7.

[55] Zhao P.J., An F., Tang M. *Effects of arbuscular mycorrhiza fungi on drought resistance of Forsythia suspense.* 2007,27, 396-399.

[56] Li X.C. 2003. Effects of infecting vesiclar-arbuscular mycorhiza on growth and development of Coix Lachryma-jobi L. *J. Shanxi Agric. Univ*, 23, 351-353.

[57] Jin H., Kang Z.H., Chen H., Han S.F. Effects of mycorrhizal fungi on the growth and mineral nutrition absorption of *Dendrobium candidum*. *Journal of Fujian College of Forestry,* 2007, 27,80-83.

[58] Zheng Y. Y., Feng G. Effect of salt stress and AM on growth of lettuce. *Acta Pedologica Sinica*, 2006, 43, 966-971.

[59] Li M., Jiang D.F., Meng X.X., Liu R.J. Effects of arbuscular mycorrhizal fungi on growth, yield and quality of *Phaseolus Mgaris* L. in field. *Eco-Agriculture Research,* 1999, 7, 43-46.

[60] Guo T, Zhang JL, Christie P, Li XL. Pungency of spring onion as affected by inoculation with arbuscular mycorrhizal fungi and sulfur supply. *Journal of Plant Nutrition,* 2007,.30:1023-1034.

[61] Guo T, Zhang JL, Christie P, Li XL. Effects of arbuscular mycorrhizal fungi and ammonium: nitrate ratios on growth and pungency of onion seedlings. *Journal of Plant Nutrition,* 2006a,.29:1047-1059.

[62] Guo T,Zhang JL,Christie P, Li XL..Influence of nitrogen and sulfur fertilizers and inoculation with arbuscular mycorrhizal fungi on yield and pungency of spring onion. *Journal of Plant Nutrition,* 2006b, 29, 1767-1778.

[63] Li M., Liu R. J. Effects of arbuscular mycorrhizal fungi on vegetable quality. *Chinese Journal of Eco-Agriculture,* 2002,10, 62-64.

[64] Shi Z.Y., Diao Z.K., Xu Q., Li M., Liu R.J. Effects of media and arbuscular mycorrhizal fungi on growth and yield of watermelon. *Journal of Laiyang Agricultural College (Natural Science),* 2006, 23(1) : 1-6.

[65] Ren Z.Y., He C.X., Li S.H., Zhang Z.B. Effects of arbuscular mycorrhizal fungi on growth, yield and quality of cucumber in organic culture. *Acta Agriculturae Boreali-Sinica,* 2008, 23,1 35-138.

[66] Liu W.K., Yang Q.C. Integration of mycorrhization and photoautotrophic micropropagation in vitro: feasibility analysis for mass production of mycorrhizal

transplants and inoculants of arbuscular mycorrhizal fungi. *Plant Cell, Tissue and Organ Culture*, 2008, 95:131–139. 27.

[67] Sieverding E. Vesicular-Arbuscular Mycorrhiza Management in Tropical Agrosystems. 1991. *Deutsche Gesellschaft für Technische Zusammansabeit* (GT2) GmbH; Eschbon, Germany

[68] Douds D.D., Reider C. Inoculation with mycorrhizal fungi increases the yield of green peppers in a high P soil. *Biological and Horticulture,* 2003, 21, 91-102.

[69] Douds Jr., D.D., Nagahashi, G., Reider, C., Hepperly, P.R. Inoculation with arbuscular mycorrhizal fungi increases the yield of potatoes in a high P soil. *Biol. Agric. Hortic.* , 2007, 25, 67–78.

[70] Sorensen J.N., Larsen J., Jakobsen I. Pre-inoculation with arbuscular mycorrhizal fungi increases early nutrient concentration and growth of field-grown leeks under high productivity conditions. *Plant and Soil,* 2008, 307,135-147.

[71] Smith S.E., Smith F.A., Jakobsen I. Functional diversity in arbuscular mycorrhizal (AM) symbioses: the contribution of the mycorrhizal P uptake pathway is not correlated with mycorrhizal responses in growth or total P uptake. *New Phytologist,* 2004, 162: 511–524.

[72] Johansson J. F., Paul L.R., Finlay R. D. Microbial interactions in the mycorrhizosphere and their significance for sustainable agriculture. *FEMS Microbiology Ecology,* 2004, 48 1–13.

[73] Douds, D. D., Jr., Nagahashi, G., Pfeffer, P. E., Kayser, W. M. and Reider, C.. On-farm production and utilization of arbuscular mycorrhizal fungus inoculum. *Can. J. Plant Sci.,* 2005, 85: 15–21.

[74] Johnson, N. C. Can fertilization of soil select less mutualistic mycorrhizae *Ecological Applications*, 1993, 3: 749–757.

[75] Johnson, N. C., J. H. Graham, and F. A. Smith. Functioning of mycorrhizal associations along the mutualism-parasitism continuum. *New Phytologist,* 1997, 135: 575–586.

[76] Jansa, J., A. Mozafar, T. Anken, R. Ruh, I. R. Sanders, and E. Frossard. Diversity and structure of AMF communities as affected by tillage in a temperate soil. *Mycorrhiza,* 2002, 12: 225–234.

In: Mycorrhizal Fungi
Editor: Susanne M. Fulton
ISBN 978-1-61122-659-1
© 2011 Nova Science Publishers, Inc.

Chapter 8

Increased Soluble Carbon from Arbuscular Mycorrhizal Fungal Colonization Alters Carbon and Nitrogen Mineralization

Hida R. Manns*
School of Environmental Sciences, University of Guelph, Canada

Abstract

The fungal symbiont, Arbuscular mycorrhiza (AM) may benefit greenhouse gas reduction at the same time as increasing plant nutrients, moisture availability and soil carbon. The lack of increased mineralization observed in response to higher carbon availability in AM colonized soils is addressed through a literature review of studies from fields with a range of soil carbon, soil texture from sand to clay and over various time frames. Many studies pointed to highest carbon solubility in organic systems with carbon amendments and on undisturbed prairie, both of which host abundant AM colonization. Published data from several large field studies with wide variation in carbon levels, provided evidence of significantly decreased nitrogen mineralization in high soil carbon conditions, and more so in sand than clay soils. Multiple explanations were proposed in literature, including increased nitrogen immobilization with abundant soluble carbon, changes in carbon storage form and turnover time and the delayed effect of AM colonization on soil water repellency and aggregation. Our review is contrary to ideology found in some computer modeling programs, where nitrate levels and nitrogen mineralization were increased in high soil carbon conditions, and decreased with low soil carbon in response to nitrogen level. Arbuscular mycorrhizal fungal colonization may provide a unique system to retain carbon and nitrogen in the environment in association with plant growth. Further research in this area could provide necessary detail to understand soil carbon feedback from plant growth that may alter carbon and nitrogen mineralization with increased atmospheric CO_2 concentrations.

* Contact Address: 7700 Langstaff Rd. Kendal, Ont. L0A 1E0, Canada.v Email: hmanns@trentu.ca; hmanns@uoguelph.ca; Telephone: 905 983-8149.

INTRODUCTION

Arbuscular mycorrhizal (AM) fungi are in a symbiotic association with plant roots. This type of association doubles the spatial area for plant nutrient uptake and the range of soil moisture potential (Smith and Read, 1997; Carlile et al., 2001). Studies suggest that the increased soluble carbon observed in AM colonized root rhizospheres is not reflected in increased mineralization of carbon and nitrogen by microbial activity (Douds et al., 1988; Poudel et al., 2002; Raiesi and Ghollarata, 2006). Arbuscular mycorrhiza has evolved with the majority of plants (Smith and Read, 1997; Parniske, 2008), presumably from a positive effect on plant establishment, growth and decomposition. For example, trees have a highly developed symbiosis, where up to 45% of their carbon from photosynthesis is associated with microbial respiration in the soil (Godbold et al., 2006) indicating there is a major role of AM hyphae in soil processes.

Although AM fungal mycelia is a scale of magnitude less in pasture and cropping systems compared to forests (Smith and Read, 1997), significant benefits to the plant and soil are evident (Gosling et al., 2006; Piotrowski and Rillig, 2008). Arbuscular mycorrhizal root colonization has been linked to increases in nutrients (Smith and Read, 1997), plant water uptake (Marulanda et al., 2003), soil aggregation (Rillig et al., 2002) and soil organic carbon (Rillig et al., 2003). Photosynthetic carbon transferred to fungal hyphae changes structural form (Bago et al., 2000; Parniske, 2008) and increases the solubility of soil carbon (DeLuca and Keeney, 1993) and turnover time (Godbold et al., 2006; Kuzyakov, 2006). There is evidence from field and laboratory studies (De Luca and Keeney, 1993; Barrett and Burke, 2000; Smit and Velthof, 2010) that poses a great debate on the effect of AM associations on carbon quantity and solubility and subsequent effects on mineralization of carbon and nitrogen.

Modeling studies assume that carbon and nitrogen are highly coupled in soil (Li et al., 2005). Therefore, where carbon is increased and becomes subject to decomposition, there would be a corresponding loss in nitrogen as well. The chemical change of organic nitrogen into inorganic nitrogen forms of ammonium (NH_4^+) or nitrate (NO_3^-) (nitrogen mineralization) is assumed to be in the same proportion as the release of CO_2 from microbial respiration of carbon (carbon mineralization). Although results are variable with plant species and soil and climate conditions, studies on AM colonization comparisons show increased solubility of carbon and decreased soil respiration at the same time (Douds et al., 1988; Raiesi and Ghollarata, 2006) in contrast to expectations.

The net effect of soil microbial activity on carbon and nitrogen mineralization appears to be influenced by the initial soil carbon too. Where soil carbon is low the secretion of carbon from plant roots or hyphae stimulates microbial activity to mineralize nitrogen (Raynaud et al., 2006). In high carbon conditions, the immobilization of inorganic nitrogen into more stable organic nitrogen in soils can occur (Barrett and Burke, 2000). Although immobilization has often been referred to in a negative sense relating to fertilizer rendered unavailable for plant uptake, this may be an element of AM associations that links to maintenance of soil structure and prevention of nutrient loss.

This review will look first at the changes in carbon mineralization (Cmin) and nitrogen mineralization (Nmin) at field scale from changes in total soil carbon and carbon solubility (water-soluble) associated with AM colonization. The details of AM colonization influence on the form of carbon, turnover time of carbon and aggregate stability will be followed with laboratory studies that analyze the physiological details down to the molecular scale. A small amount of AM colonization may be disproportionately important for the maintenance of soil carbon and mineralization to atmospheric CO_2 and nitrous oxide.

CHANGES IN SOIL RESPIRATION FROM AM COLONIZATION

Arbuscular mycorrhizal colonization has been reported to increase the total carbon from photosynthesis by the plant (Paul and Kucey, 1981; Staddon et al., 2003; Raiesi and Ghollarata, 2006). Studies found positive effects of AM colonization on plant growth that accounted for 4 to 20% of carbon from photosynthesis (Rillig and Allen, 1999; Bago et al., 2000; Staddon et al., 2003). Paul and Kucey, (1981) reported that total carbon fixation increased 9% with AM colonization of fava beans (*Vicia faba* L.) followed with $^{14}CO_2$ labeling.

The allocation of photosynthetic carbon to roots, shoots and soil is also influenced by AM colonization (Paul and Kucey, 1981). Arbuscular mycorrhizal treatments increased both net root and shoot weight 20% with increased carbon from photosynthesis in fava beans. While shoot respiration increased 5% with AM colonization, root respiration decreased by the same percentage, possibly from the inclusion of hyphal activity (Paul and Kucey, 1981). The allocation of carbon to roots and shoots has great variation in the literature, as the environmental conditions and plant species also contribute to the distribution of carbon. Although most studies find increased plant biomass with AM colonization, shorter and heavier roots (Jakobsen and Rosendahl, 1990) and reduced plant height from an increased allocation of carbon to roots (Olsson and Johnson, 2005) are commonly found.

The influence of AM colonization on plant photosynthesis and soil carbon also varies with the time of the year, location and time of plant growth cycles. Changes in AM fungal activity have been shown to cycle with the season (Kabir et al., 1997) with a peak in September in temperate climates, after most crops have completed growth. As well, fungal colonization varies within the crop cycle. Mycorrhizal respiration comprised 25% of the mycorrhizosphere respiration following a crop of barley (*Hordeum vulgare* L., cultivar Naomi) when measured using 35µm mesh separation of roots and hyphae in soil cores in the field (Moyano et al., 2007). Basal respiration (that was used to evaluate existing organic matter) formed a large background volume, while mycorrhizal respiration peaked at 4-6 weeks and plant root respiration was maximum at 10-12 weeks (Figure 1) (Moyano et al., 2007).

Carbon mineralization is frequently measured as CO_2 released from microbial activity where fungi and bacteria take up glucose, amino acids, organic acids, and root/fungal cells and mucilage from soil solution (Jackson et al., 2008). The carbon that is easily measured by mixing soil with water is referred to as "water soluble carbon" (WSOC), while carbon that exists in the soil water in situ is termed "dissolved organic carbon" (DOC). The correspondence between microbial activity and solubility of carbon is not clearly defined. Soil respiration was not correlated to DOC in organic and conventional management comparisons (Lundquist et al., 1999a), nor was microbial biomass carbon related to WSOC in long term crop rotations (McGill et al., 1986) or $K_2SO_4^-$ soluble carbon with prairie soils (DeLuca and Keeney, 1993).

Sakamoto and Oba (1994) studied CO_2 evolved from a range of soils with varying carbon and nitrogen levels and in different management systems and related their results to fungal and bacterial biomass. Mineralization of carbon measured in terms of CO_2 released, increased in accordance with total soil carbon and fungal biomass. Fungal biomass and CO_2 were increased 2 x with low soil carbon (47.2 g kg^{-1}) compared to a small increase in soil with increased carbon (75.0 – 100.0 g kg^{-1}) (Sakamoto and Oba, 1994). With corn stalks + cattle manure addition and at high soil carbon levels, fungal biomass was more than double over other applications, but CO_2 decreased slightly rather than increasing in response to soil carbon (Sakamoto and Oba, 1994). The authors related their results to the greater efficiency of fungal decomposition which incorporates 30-40% of carbon into mycelium, compared to 5-10% for bacteria (Alexander, 1977 quoted in Sakamoto and Oba, 1994). The higher assimilation efficiency of fungi is believed to accommodate decomposition of high C:N substrates (Manzoni and Porporato, 2009) and may also relate to changes in soil respiration.

One of the major observations with organic agricultural methods is the high colonization of AM fungi in soils. Marinari et al. (2010) observed increased total carbon and soluble carbon along with AM colonization with organic management over conventional tillage. Lundquist et al. (1999b) also found increased DOC and soil respiration associated with organic matter inputs. While reduced tillage has been found to increase fungal activity in conventional agriculture (Hu et al., 1995), organic agriculture systems with continued tillage indicate the host plant species and soil organic matter availability in the soil are most important for continued support of AM fungi.

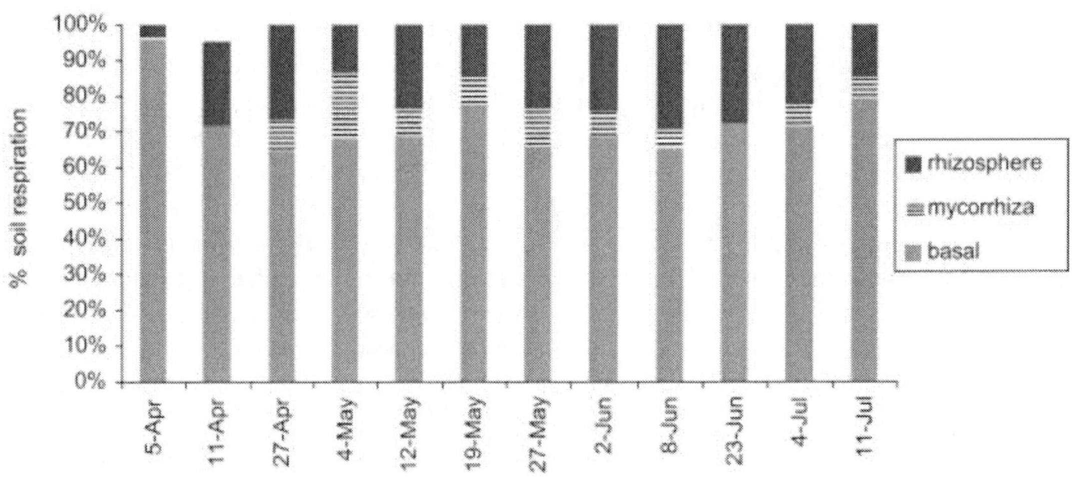

Figure 1. The distribution of total soil respiration (%) measured in µmol m^{-2} s^{-1} apportioned into background soil (basal in grey), plant roots (rhizosphere in dark) and AM colonization (mycorrhiza in white stripes) over the 3 month growing season of barley (*Hordeum vulgare* L., cultivar Naomi) (Moyano et al., 2008)[1].

[1] Reprinted from Soil Biology and Biochemistry Vol 39, Moyano et al., Response of mycorrhizal, rhizosphere and soil basal respiration to temperature and photosynthesis in a barley field, pages 834-853, 2008, with permission from Elsevier.

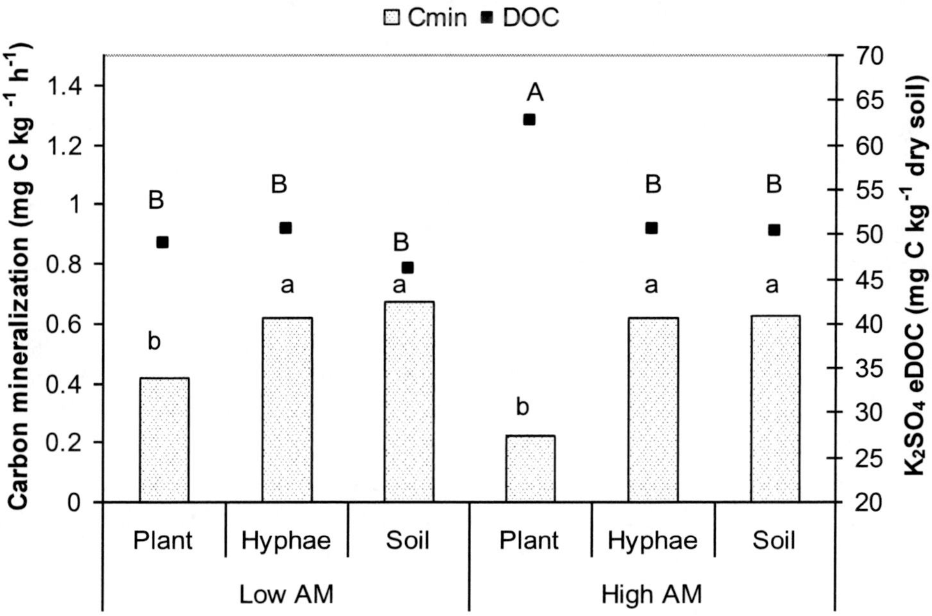

Figure 2. Differences in carbon mineralization (mg C kg^{-1} h^{-1}) and dissolved organic carbon (K_2SO_4 eDOC (mg C kg^{-1} dry soil)) after 20 weeks growth of flax in microcosms separating plant from hyphae (65 µm mesh) and soil (solid plate). Arbuscular mycorrhiza colonization re-established after Canola cropping to low (<10%) and high (10-20%) colonized root length. Differences between letters indicate statistical differences. n = 4 replications. (Manns unpublished data)

Manns (Under review) used cropping systems to test the influence of AM colonization on soil carbon solubility in soils. Flax (*Linum usitatisimum* cv A.C. Lightning) was grown in soils following Canola (*Brassica napus* L.) cropping (non-AM) compared to the organic rotations following soybean (*Glycine max* L.*)* (highly AM) less than 200 m away, from the Ian. N. Morrison Research Farm, Carman, Manitoba, Canada. The soils had initially the same soil carbon (38.0 and 34.9 mg kg^{-1} respectively), texture and phosphorous. The very high AM colonization with organic soil of around 50%, compared to low colonization in Canola soils up to 20% at 8 weeks growth of flax was highly correlated to higher carbon solubility and CO_2 released on aerobic incubation of soils (Manns, Under review). Comparing AM colonization within Canola soils, a similar result of increased soluble carbon was found with 20 vs 10% AM colonized root length at 20 weeks, while at the same time, carbon mineralization was significantly reduced by the flax plants (Figure 2) (Manns, Unpublished data).

Most studies indicate AM colonization increases the total volume and solubility of carbon, and in soils with low carbon content (< 5%), results in increased CO_2 release as well in association with active plant roots (Talbot et al., 2008). However, at higher carbon content in soil, or including time beyond the peak of plant growth, CO_2 emitted from soil mineralization can be reduced with AM colonization.

ARBUSCULAR MYCORRHIZAL EFFECTS ON NITROGEN MINERALIZATION

Nitrogen mineralization is more complex than carbon mineralization in soil building and decomposition. The release of carbon from organic complexes also releases organic nitrogen in the process. Organic nitrogen can then be transformed into NH_4^+ and subsequently into NO_3^- to be taken up by plants, fungi or bacteria. Alternatively, NH_4^+ and NO_3^- can also be re-assimilated into organic forms, complexed to soil particles (immobilization) or reduced further to nitrogen gas (Jackson et al., 2008). Organic carbon and nitrogen are generally in constant proportion in soil (approximately 10:1) (Thomsen et al., 2008) and likewise carbon and nitrogen mineralization from decomposition occur at a similar rate at the same time. However, studies indicate that the level of soil carbon, AM colonization and soil texture affect the carbon to nitrogen ratio (C:N) and particularly, the solubility of nitrogen.

The decomposability of organic matter has been associated with the rate of nitrogen loss (Burford and Bremner, 1975), as a close correlation was found between mineralization of soluble carbon and NO_3^- in soil incubation studies (Stanford et al., 1975). The acknowledged divergence in the literature on the influence of rhizosphere carbon on denitrification (McCarty and Bremner, 1992) could be explained if the same form of soluble carbon used in the conversion of NO_3^- to NO, N_2O and N_2, is also able to immobilize nitrate to organic nitrogen (Drury et al., 2003). Support for this theory has been found in grassland studies where labile organic matter correlated to rapid nitrogen immobilization, indicating a strong correlation of Cmin to nitrogen immobilization (Barrett and Burke, 2000).

There is evidence from several studies to support the activity of soluble carbon in nitrogen immobilization, where increased carbon from AM colonization appears to decrease nitrogen mineralization (Holland and Coleman, 1987; Poudel et al., 2002; Drury et al., 2003). Poudel et al. (2002) studied organic systems with tomato (*Lycopersicon esculentum* L.) in California, USA and found 50% lower nitrogen turnover despite increased nitrogen input and increased potential for nitrogen mineralization. There appeared lower nitrogen storage loss and greater nitrogen storage capacity that the authors suggested were possibly driven by the effect of crop rotations on AM (Poudel, et al., 2002).

Residue decomposition has been studied with regard to soil carbon and nitrogen retention. Holland and Coleman (1987) tested the role of fungi in decomposition, using placement of litterbags in the field, and found surface residue application associated with increased fungal hyphae length, soil carbon retention, and nitrogen immobilization. Residue incorporated into soil also decreased Nmin in a study on limits of soil moisture on mineralization through soil incubations (Drury et al., 2003). The decreased availability of nitrogen was attributed to a readily available carbon source from the preceding legume crop used to convert nitrate into nitrogen gas or organic nitrogen (Drury et al., 2003).

The comparison of grassland before and after cropping also provides a comparison of the effect of AM colonization on soil carbon and nitrogen. Smit and Velthof (2010) compared the nitrogen mineralization of 42 soils from 13 different sites in the Netherlands where pasture was established up to 30+ years prior to tillage, with wide variation in total soil carbon from 10.1 to 77.4 g kg^{-1}, and with soils classified according to clay or sand. While clay soils had greater total carbon, dissolved organic carbon (DOC) increased rapidly to form significant regressions to total carbon in both sand and clay soils (Figure 3). The increase of DOC in clay was 13.9x total carbon, that was much greater than the increase in sand of 2.25x. The nitrogen mineralization and NO_3^- levels increased significantly with chemical removal of sod, where there was decreased total carbon and soluble carbon along with a greater proportion of NO_3^- ($p < 0.001$) in clay than sand (means of 13.3 vs 3.6mg kg^{-1}) (Figure 4) (Smit and Velthof, 2010). Although AM levels were not included in their study, the reduction in AM without pasture may have contributed to reduced carbon availability and reduced immobilization of NO_3^- at the same time at all levels of total soil carbon at field scale.

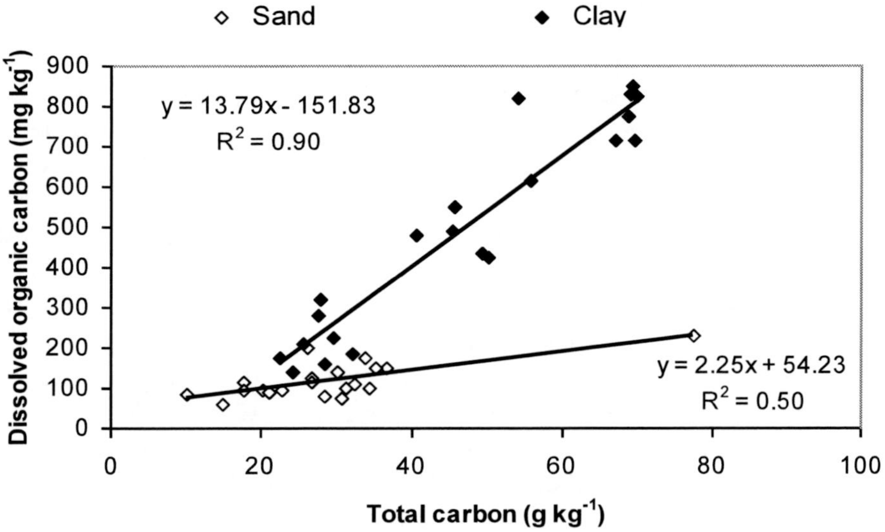

Figure 3. Comparison of dissolved organic carbon (DOC) to total carbon (g kg^{-1}) (Organic + Inorganic) in sand and clay soils from 42 soils from 13 sites in the Netherlands where pasture was tilled after variable lengths of time, up to 30 years. Linear regressions in sand and clay are significant $p< 0.001$). Data from Smit and Velthof, 2010, adapted with permission from the authors.

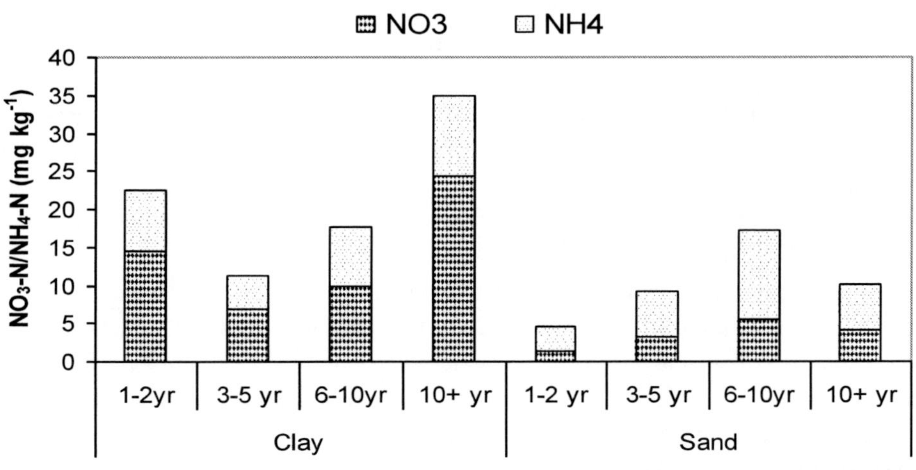

Figure 4. Changes in NO_3^-N and NH_4^+N (mg kg^{-1}) in sand and clay soils, and with time of pasture establishment in the Netherlands from 1 to 20+ years. Differences between NO_3^- in sand and clay are significant ($p = 0.001$). Data from Smit and Velthof, 2010, adapted with permission from the authors.

DeLuca and Keeney (1993) compared nitrogen and carbon changes from paired sites when Iowa prairie soils were cultivated. Changes in solubility were measured by extracting soil amino acids and carbohydrates by acid hydrolysis to measure soluble carbon and amino nitrogen. Cultivated soil had 40% less soluble carbon (extracted with Anthrone) than unbroken prairie soil within 1 year (DeLuca and Keeney, 1993). When the sites were compared according to tillage x soil type (sand or clay), there was significantly greater Nmin with tillage along with lower organic carbon. The Nmin was also greater in clay than sand (DeLuca and Keeney, 1993) similar to the results from Smit and Velthof (2010).

A study of widespread long-term management sites on the Canadian Prairie that paired native soil with tillage sites greater than 30 years, found Nmin was decreased with cultivation compared to native prairie except in very dry conditions or in soil with high C:N (Paul et al., 2004). With high soil organic carbon, the soil structure and microbial community composition may play a more important role in the retention and solubility of nutrients. Where there is predominantly sandy soil with a highly fungal soil microbial community (Degens et al., 1996), minimal change in Cmin and Nmin would be expected in soil with high carbon (> 5%) with field management.

The changes in mineralizable potential with AM colonization were compared between soil recolonizing from Canola cropping under conventional management and soils supporting organic crop rotation at Carman, Manitoba, Canada (Manns, under review). The extent of AM colonization corresponded positively to indices of carbon availability (WSOC and Cmin) and negatively to nitrogen availability (NO_3^- and Nmin). Although the potential for N_2O release increased with higher soluble carbon from AM colonization, the significant decreases in soil moisture and NO_3^- in the root zone decreased available NO_3^-. Nitrogen mineralization was highest when soil Cmin was low and remained constant with increasing Cmin (Figure 5) (Manns Under Review). Although Nmin normally refers to the percentage of inorganic to organic nitrogen, the ratio of NO_3^-:NH_4^+ was highly correlated to Nmin in the data of Smit and Velthof (2010) and was used to calculate Nmin in the preceding study as a measure of readily available soluble nitrogen in the soil.

The linkage of carbon to nitrogen mineralization is accepted in soil with low organic carbon, or without AM colonization, where increases in Cmin correspond linearly to increases in Nmin (Schimel and Bennett, 2004; Li et al., 2005; Manzoni and Porporato, 2009). A current synthesis on turnover of nitrogen forms in soil by Schimel and Bennett (2004) suggested enzymes are most important for rates of Nmin, especially in the root zone. However, there is growing evidence that where soluble carbon is elevated from high initial soil carbon or AM colonization, there may be increased immobilization of nitrogen rather than mineralization (Thomsen et al., 2008).

In forest ecosystems that are highly mycorrhizal, ectomycorrhizal mycelium was attributed to a net immobilization of 15% nitrogen (Staddon et al., 2003). Without ectomycorrhiza, there was net loss of nitrogen. The statement that only fungi can maintain high nitrogen and high carbon in soil at the same time (Staddon et al., 2003) in forest ecosystems, may also contribute to the explanation of changes in carbon and nitrogen mineralization in agricultural fields.

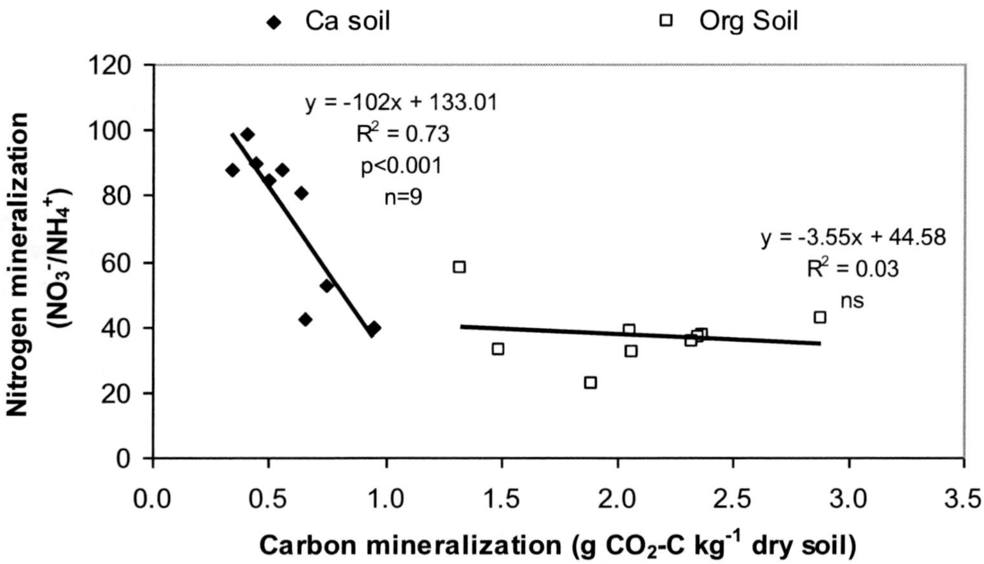

Figure 5. The tradeoff between nitrogen mineralization (ratio of NO_3^- to NH_4^{+}) and carbon mineralization (g CO_2-C kg^{-1} dry soil) in soil following conventional canola cropping (Ca) (significant at $p = 0.001$), or organic rotation of soybeans (Org) (ns). (Manns Under review).

The soil carbon to nitrogen ratio may be an indicator of the decomposability of soil organic matter and was found to correlate to Nmin. A data set of 36 arable soils with varying clay, sand and C:N was analyzed with CO_2-C evolution from incubation up to 238 days (Data by Springob and Kirchmann, 2002 in Thomsen et al., 2008). A decisive difference in the C:N ratio of soil was reported, with escalating CO_2-C release as C:N decreased below 10. With increasing C:N, there was a slight, but not significant negative trend in CO_2 from mineralization (Thomsen et al., 2008).

The results by Thomsen et al. (2008), confirm that a significant change in soil mineralization occurs between high and low soil carbon that determines mineralization or immobilization of nitrogen (Raynaud, 2006). In that way, organic carbon availability and solubility controls nitrogen retention in soil (Barrett and Burke, 2000). This is contrary to theory included in some modeling studies where nitrogen solubility is determined by the level of nitrogen (Manzoni and Porporato, 2009) implying there would be increased immobilization of nitrogen at low soil fertility and mineralization of nitrogen at high soil fertility. Soluble carbon and NO_3^- have been positively correlated with CO_2 emissions in studies, most often with C:N below 10 (Thomsen et al., 2008), and total organic carbon less than 3% (Lundquist et al., 1999b; Barrett and Burke, 2000; Li et al., 2005; Marinari et al., 2010). The immobilization of NO_3^- in high nutrient conditions may not be found in laboratory studies without AM colonization. Analysis at the molecular scale may suggest reasons for this counter-intuitive phenomenon.

Changes in the Form of Carbon with AM Colonization

Physiological differences between fungi and bacteria may affect the form of carbon, turnover time and resistance of soil to decomposition. Amino sugars produced in the soil, can be used to identify fungi and bacteria by the specific structures that form peaks in gas chromatography (Liang et al., 2007). Liang et al. (2007) interpreted glucosamine as an index of fungi (commonly 60% AM), galactosamine and muramic acid representative of bacteria, and mannosamine, a cumulative index of microbial activity in general. In coarse sand, bacteria-derived amino sugars contributed the most to soil organic matter during initial decomposition of leaves, but with fine sand and clay, there was higher glucosamine and higher microbial activity in general (Zhang et al., 1998). Liang et al. (2007) suggested the increase in amino sugars with the level of soil organic matter was representative of slower mineralization and faster synthesis of amino sugars relative to other organic constituents during early plant decomposition.

For a complete detailed review of the colonization and biological activity of AM fungus in plant roots see Parniske (2008). In AM fungal hyphae, glucosamine is directly incorporated into trehalose and glycogen, which are turned over in both intraradical and extraradical mycelium and spores (Bago et al., 2000). During the growth of extracellular AM hyphae, triacylglycerides and glycogen are immobilized for construction of cell walls (Bago et al., 2000) forming chitin (Swift et al., 1979) that is comparatively resistant to decomposition. A ^{13}C labeling study with in vitro dual-mycorrhizal culture in divided Petri plates concluded that sugar was taken up by AM fungus within the root in the form of hexose, and converted to lipids for transport to external mycelium (Pfeffer et al., 1999). Lipids are the main respiratory substrate for fungal hyphae, and are stored in spores as well (Bago et al., 2000). Lipids were reported to have a retention time of 10-24 minutes, compared to 45-65 minutes for phospholipids (Pfeffer et al., 1999).

The changes in forms of sugar in AM fungi along with rapid transit time in hyphae could contribute to the increased rate of carbon turnover, concomitant with a decreased rate of mineralization from fungal activity. Soil factors, such as carbon binding between sand and clay could also influence mineralization. Clay had the highest concentrations of hexose sugars (from root deposition), from the proportion of muramic acid that is stable only if bound to clay (Zhang et al., 1998). Hexose sugars are not stable forms of carbon in the soil and are turned over readily in the microbial food web. In contrast, glucosamine may include the recalcitrant glycoprotein glomalin that has been studied extensively in association with stable soil organic matter (Zhang et al., 1998).

Another study on Berseem clover (*Trifolium alexandrinum* L.) grown in greenhouse pots, with a single fungal species (*Glomus intraradices* Schenck and Smith) found AM colonization increased carbon input into soil and also reduced soil respiration, relative to plants grown equally with phosphorus addition (Raiesi and Ghollarata, 2006). The increase in carbon could be related to the changes in root exudates with AM plants found by Schwab et al. (1994). The AM species may lower exudation of amino acids and reducing sugars with colonization possibly from changes in plant allocation, or reduced membrane permeability (Schwab et al., 1994). Low phosphorous (Raiesi and Ghollarata, 2006) and possibly low nitrogen (Raynaud, 2006) could trigger increased root exudates, without AM colonization to help alleviate the deficiencies. Microbial biomass carbon was also reduced in the AM colonized Berseem clover (Raiesi and Ghollarata, 2006), adding to the quandary of how changes in carbon in the plant rhizosphere and hyphae affect total carbon storage in the soil.

CHANGES IN CARBON TURNOVER TIME IN SOIL

Literature concurs that AM root colonization increases the volume and speed of carbon that travels through plant roots to intraradical and extraradical hyphae (Bago et al., 2000; Staddon et al., 2003; Olsson and Johnson, 2005). Therefore, there is potential for greater soluble carbon in the soil. However, the transformation of sugars into glycogen and lipids in fungal hyphae slows the decomposition process relative to hexose sugars from plant roots. Jakobsen and Rosendahl (1990) measured ^{14}C in 4 week cucumber (*Cumumis sativus* L.) plants grown in PVC tubes with roots in a 60 μm mesh bag. The ratio of soluble to insoluble carbon was 2 x for roots compared to hyphae (Jakobsen and Rosendahl, 1990) indicating the increases in soluble carbon noticed in AM plant rhizospheres may be driven by plant root effects and not hyphae. The mycorrhizal roots were associated with much higher respiration than fungal hyphae alone in soil (Jakobsen and Rosendahl, 1990).

The crop plant may play an important role in the changes in CO_2 release depending on the proportion of activity of roots and hyphae. The separation of hyphae from root effects is difficult, but research indicates AM colonized roots are associated with increased soluble carbon and mineralization in soil while AM hyphae mainly result in reduced carbon solubility and microbial respiration (Jackobsen and Rosendahl, 1990). The plant root effect was noted in Hanson (2000) where CO_2 from soil respiration was primarily associated with root density, although AM colonization was not separately distinguished in the study. The proposed theory of Schimel and Bennett (2004) that Nmin and nitrogen immobilization interact at the root interface from enzyme activity is consistent with increased mineralization from AM colonized root activity followed with higher nitrogen immobilization at a distance where fungal hyphae are dominant. The AM colonized roots may increase soluble carbon, but this effect is not shared by extraradical fungal hyphae in studies where hyphal activity was measured in separate compartments (Manns Under review). This remains to be proven decisively.

Using pulse labeling, Kuzyakov (2006) found root exudates are turned over within hours and roots within days while fungal hyphae can take up to years. A difficulty in measuring turnover time is the large background level of organic matter, relative to a small carbon flux in a short time frame (Godbold et al., 2006, Moyano et al., 2007). Despite the difficulties, turnover time of mycorrhizal carbon in a forest study was estimated at greater than 30 days (Godbold et al., 2006).

Isotopic labeling with ^{13}C allows the changes in carbon within the plant and soil to be followed from CO_2 input by photosynthesis to respiration by plants and soil microorganisms in lab studies. This enables assessment of the carbon flux from plant roots, hyphae and soils over short periods of time. Increased mycorrhizas from addition of lime to soil were associated with increased carbon cycling in response to plant photosynthesis when grassland sites in Scotland were studied with $^{13}C\text{-}CO_2$ (Johnson et al., 2002; Staddon et al., 2003). Johnson et al. (2002) reported a 50% increase in photosynthate uptake of CO_2 with liming of soil, most of the effect being attributed to higher productivity of the plants. In the very high organic matter soil, plant biomass was 197 vs 264 g m^{-2} in non-mycorrhizal and mycorrhizal treatments respectively. In the lower soil carbon of Staddon et al. (2003), where plant biomass from regrowth of pasture was 18.7 g m^{-2}, the increase in labeled CO_2 in soil with liming was associated with a greater amount of $^{13}CO_2$ immediately respired. Liming generally increases soil respiration as well, possibly from increases in root exudations or bacterial transformations of carbon (Johnson et al., 2002). While AM colonization is constantly associated with increased soil carbon, and particularly net carbon increases, it is difficult to disentangle all the factors to discern the specific functions.

Olsson and Johnson (2005) used both isotopic labeling and phospholipid fatty acid analysis (PLFA) to study changes in *Plantago lanceolata* L. in sandy loam soil in greenhouse pots, inoculated with *G. intraradices*. They incorporated ^{13}C in photosynthetic chambers at 8 weeks plant growth and replicates were sampled 5 times over 32 days. There was significantly more PLFA 16:1w5 in mycorrhizal roots, while neutral lipid fatty acids (NLFA) were 12 x greater in the soil with mycorrhizal colonization (Olsson and Johnson, 2005). In *G. intraradices*, 50-70% of neutral lipids were the fatty acid 16:1w5. Carbon can remain in AM fungal residues much longer than the reduction within a few days of finely branched hyphae or 5-6 days observed for intraradical vesicles and spores. Close correlation of ^{13}C of NLFA in extraradical and intraradical hyphae indicated lipids were translocated rapidly between fungal structures (Olsson and Johnson, 2005). The occurrence of rapid transport and turnover time of carbon in AM fungal hyphae concurs with results reviewed in Bago et al. (2000) and Staddon et al., (2003). Rapid transit of carbon in hyphae in the form of lipids followed by the deposition in fungal hyphae in forms resistant to decomposition may contribute to increased soluble carbon and also reduced mineralization potential observed.

CHANGES IN SOIL ORGANIC CARBON

The multiple effects of greater storage of carbon in hyphae, difference in carbon form in hyphae compared to root exudates and contribution of fungal hyphae to soil aggregation, may together affect the soil mineralization processes. The rapid mineralization of carbon from cover crops (Hassink, 1994), is not always associated with increases in total soil carbon. This is a controversial point. Studies that show consistent increases in soil carbon are associated with slow addition of carbon to soils such as surface residue (Clapp et al., 2000; Manns et al., 2007), clay soils (Angers et al., 1997) or cooler temperatures (Van den Bygaart et al., 2003). The resistance to decomposition was found to be more important than the amount of carbon decomposed in soil carbon increases in homogeneous plots growing oats with/without surface residue of chopped corn (*Zea mays* L) stalks compared to oat (*Avena sativa* L.) straw (Manns et al., 2009). The greatest benefit of fungal hyphae may result from an "afterlife" effect on soil that deters soil decomposition as found in forest soils (Staddon et al., 2003) and in sunflower (*Helianthus annuus*) rhizocosms (Langley et al., 2005).

In addition to fungal lipid storage and hyphae walls made of chitin, the secretion of the protein glomalin (Steinberg and Rilling, 2003) has also been studied in respect to increased residence time of carbon in soil. Glomalin is a glycoprotein secreted by AM hyphae that is insoluble in water (Wright and Upadhyaya, 1998). Glomalin has been found to correlate to soil aggregation (Rillig et al., 2002) and soil carbon (Rillig et al., 2003). Reports suggest that due to a high carbon content (Rillig and Allen, 1999), glomalin may contribute up to 5% to the soil carbon pool (Rillig et al., 2007). The effect of glomalin on soil aggregation and carbon may be greatest in sandy soils where fungal hyphae are prominent (Degens et al., 1996), although plant, soil and environmental factors must be taken into account too.

AM fungal hyphae may also influence nitrogen mineralization in soil by reducing soil moisture (Augé et al., 2001) or simply by the reduction of plant root exudates that increase mineralization. The resistance of aggregates to decomposition is also influenced by the water repellency of the soil. Carbon from plant residue and fungal decomposition has been found to increase water repellency of soil aggregates (Dekker and Ritsema, 1994) in association with aliphatic molecules which include lipids with long chain fatty acids (Marschner et al., 2008) that have a longer residence time in soil (Smith and Read, 2008).

The connection of AM hyphae to soil carbon through hydrophobicity is the subject of ongoing research (Rillig et al., 2005, 2010). An increase in hydrophobic forces that resist decomposition would allow soil aggregates to hold more moisture, increasing carbon input from photosynthesis without greater mineralization at the same time thus allowing net carbon gain. The degree of water repellency ranges from 10-30% (Simon et al., 2009) and is dependent on the increase in carbon from residue and fungal decomposition and the influence of aggregate size from roots, hyphae and fungal exudates. The contribution of AM fungal hyphae in water repellency is not yet distinguished.

In 3 years of monthly sampling of high organic matter soils growing oats (*A. sativa* cv Ida) in small plots, Manns and Emery (Under review) observed a strong correlation of soil carbon to AM colonization and water repellency when oat plants and surface residues were included in the analysis. The water repellency was increased from the residue alone, in support of many literature reports that detail organic molecules contributing to hydrophobic forces within aggregates (Dekker and Ritsema, 1994). With the combination of surface residue x AM fungal colonization from oat plants, and in optimum soil moisture conditions, the maximum resistance to decomposition occurred (Manns and Emery Under Review). AM could contribute to resistance to decomposition through the recalcitrance of glomalin, and slower decomposition of lipids and fungal hyphae and in addition, the residue and fungi could contribute to hydrophobicity to aid aggregates in resisting decomposition. With very high soil carbon (70-88 g kg^{-1}) the aggregate size reached a balance between moisture retention, hydrophobicity and decomposition (Manns and Emery, Under Review).

The interaction of decomposer and AM fungi may represent the optimum ecology for increasing soil carbon (Manns et al., 2009). A mutual association between microbial species has been observed to improve nutrient cycling (Lindahl et al., 1999; Ohtonen et al., 1999; Dighton, 2003) from increased carbon use efficiency (Anderson and Domsch, 1990). Although some studies cite competition for soil nitrogen between plant roots, bacteria and fungi (Schimel and Bennett, 2004; Raynaud, 2006), the complexity of the microbial food web can result in positive associations (Milleret et al., 2009) and is species dependent (Dighton et al., 1987). In the 21 year DOK (Bio-Dynamic, Bio-Organic and Konventionell) trial, aggregate stability increased 60% along with AM colonized root length (Mader et al., 2002). The report concluded that the enhanced diversity in microbial communities transformed carbon at lower energy costs resulting in more efficient resource utilization (Mader et al., 2002).

This review suggests that high carbon solubility in soils associated with AM colonization is not necessarily accompanied by increased nitrogen mineralization. To the contrary, nitrogen immobilization can take place, and may explain contrasting results in prairie and organic agriculture systems compared to field tillage. The initial carbon content, soil texture and the extent of AM colonization may contribute to high variability in study results on soil mineralization (Billings et al., 2010). AM colonization appears to be an important component of organic agriculture systems and studies that examine the link of AM colonization with Nmin and soluble carbon would define the mechanism of activity.

The interaction of carbon and nitrogen also made a significant difference in climate predictions in models testing the projected effects of nitrogen limitation (Sokolov et al., 2008). With nitrogen limitation of CO_2 uptake (that is assumed in most terrestrial carbon models), elevated CO_2 resulted in positive feedback on climate warming. When considering that increased carbon would stimulate nitrogen immobilization, as suggested also by this discussion, carbon uptake would not be limited, and a net carbon sequestration would result in negative feedback to global warming (Sokolov et al., 2008). Woodlands with high mycorrhizal fungi and nitrogen immobilization may verify the influence of carbon and nitrogen interaction on possible net carbon uptake. Future research is needed on the changes in nitrogen with level of carbon, method of carbon assimilation from plants, and interactions with saprophytic and AM fungi, in light of the importance of understanding organic matter in the soil for predicting carbon and nitrogen turnover in a high CO_2 environment (Billings et al., 2010).

Conclusion

This review has looked at changes in soil carbon solubility associated with AM colonization and the effect of AM on the soil carbon and nitrogen mineralization of soil that is a much larger question in light of increasing CO_2. Many field scale, glass house, and growth chamber experiments have found net increase in soil carbon with AM colonization when the total plant growth period is taken into consideration. Since field carbon changes slowly compared to variability and environmental conditions, net changes at field scale are usually found after long term management changes (Mader et al., 2002; Angers et al., 1997), with dry conditions (van den Bygaart et al., 2003) or with surface residue such as corn stalks with slow decomposition (Clapp et al., 2000). However, in organic agricultural experiments with high AM colonization, or studies on transition from prairie to cultivation, higher solubility of carbon does not necessitate a corresponding increase in carbon and nitrogen mineralization.

Studies to date suggest that where AM colonization changes the distribution, form, and turnover time of carbon, in addition, fungal hyphae modify the environment to reduce decomposition forces in general. Specifically, soil moisture is reduced, the solubility of nitrogen is greatly reduced and aggregates become increasingly hydrophobic. Increased plant growth thereby perpetuates the soil and plant environment through micro-scale changes in soil aggregates. In agriculture, understanding the importance of AM associations and fostering their presence, could lead to increasingly sustainable and productive agricultural systems without loss of nutrients to leaching or atmosphere.

Acknowledgments

Many thanks to Dr. Maren Oelbermann, Dr. Neil Emery and many anonymous reviewers for their time and efforts to mentor my writing and greatly clarify the message. I am grateful to the Symons Trust Fund for Canadian Studies for covering study expenses at Trent University, and to the Green Crop Network for funding during my studies at the University of Manitoba. This work was supported by the Natural Sciences and Engineering Research Council (NSERC) and industrial and government partners through the Green Crop Networks (GCN) Research Network.

References

Anderson, T.H.; Domsch, K.H. *Soil Biol Biochem.* 1990, 22, 251-255.

Angers, D.A.; Bolinder, M.A.; Carter, M.R.; Gregorich, E.G.; Drury, C.F.; Liang, B.C.; Voroney, R.P.; Simard, R.R.; Donald, R.G.; Beyaert, R.; Martel, J. *Soil Tillage Res.* 1997, 41, 191-201.

Augé, R.M.; Stodola, A. J. W.; Tims, J. E.; Saxton, A. M.; *Plant Soil* 2001, 230, 87-97.

Bago, B.; Pfeffer, P.E.; Shachar-Hill, Y. *Plant Physiol.* 2000, 124, 949-957.

Barrett, J. E.; Burke, I. C.; *Soil Biol. Biochem.* 2000, 32, 1701-1716.

Billings, S.A.; Lichter, J.; Ziegler, S.; Hungate, B.A.; de B. Richter, D. *Soil Biol. Biochem.* 2010, 42, 665-668.

Burford, J. R.; Bremner, J. M. *Soil Biol. Biochem* 1975, 7, 389-394.

Carlile, M. J.; Watkinson, S.; Gooday, G.W. *The Fungi*. Academic Press, London, 2001.

Clapp, C. E.; Allmaras, R. R.; Layese, M. R.; Linden, D. R.; Dowdy, R. H. *Soil Tillage Res.*55, 126-142.

Degens, B. P.; Sparling, G. P.; Abbott, L. K. *App. Soil Ecol.* 1996, 3, 149-159.

Dekker, L. W.; Ritsema, D. J. *Water Resour. Res.* 1994, 30, 2507-2517.

DeLuca, T. H.; Keeney, D. R. *Soil Sci.*1993, 155, 219-228.

Dighton, J. *Fungi in Ecosystem Processes*; Marcel Dekker Inc.: New York, NY, 2003.

Dighton, J.; Thomas, E.; Latter, P. *Biol. Fert. Soils* 1987, 4, 45-150.

Douds, D. D. Jr.; Johnson, C. R.; Koch, K. E. *Plant Physiol.* 1988, 86, 491-496.

Drury, C. F.; Zhang, T. Q.; Kay, B. D. *Soil Sci. Soc. Am. J.* 2003,67, 1388-1404.

Godbold, D.; Hoosbeek, M. R.; Lukac, M.; Cotrufo, F.; Janssens, I. A.; Ceulemans, R.;Polle, A.; Velthorst, E. J.; Scarascia-Mugnozza, G.; De Angelis, P.; Miglietta, F.; Peressotti, A. *Plant Soil* 2006, 281, 15-24.

Gosling, P.; Hodge. A.; Goodlass, G.; Bending, G. D. *Agric. Eco. Environ.* 2006, 113, 17-35.

Hanson, P. J.; Edwards, N. T.; Garten, C. T.; Andrews, J. A. *Biogeochem.* 2000, 48, 115-146.

Hassink, J. *Soil Biol. Biochem.* 1994, 26, 1221-1231.

Holland, E. A.; Coleman, D. C. *Ecology* 1987, 68, 425-433.

Hu, S.; Coleman, D.C.; Beare, M. H.; Hendrix, P. E. *Agric. Eco. Environ.* 1995, 54, 77-88.

Jackson, L. E.; Burger, M.; Cavagnaro, T. R. *Ann. Rev. Plant Biol.* 2008, 59, 341-363.

Jakobsen, I.; Rosendahl., L. *New Phytol.* 1990, 115, 77-83.

Johnson, D.; Leake, J. R.; Ostle, N.; Ineson, P.; Read, D. J. *New Phyt.* 2002, 153, 327-334.

Kabir, Z.; O'Halloran, I. P.; Fyles, J. W.; Hamel, C. *Plant Soil* 1997, 192, 285-293.

Kuzyakov, Y. *Soil Biol. Biochem.* 2006, 38, 425-448.
Langley, J. A.; Johnson, N. C.; Koch, G. W. *Plant Soil* 2005, 277, 335-344.
Li, C.; Frolking, S.; Butterbach-Bahl, K. *Clim. Change* 2005, 72, 321-338.
Liang, C.; Zhang, X.; Rubert IV, K.; Balser, T. C. *Biol. Fert. Soils* 2007, 43, 631-639.
Lindahl, B.; Stenlid, J.; Olsson, S.; Finlay, R. *New Phytol.* 1999, 144, 183-193.
Lundquist, E. J.; Jackson, L. E.; Scow, K. M. *Soil Biol. Biochem.* 1999a, 31, 1031-1038.
Lundquist, E. J.; Jackson, L. E.; Scow, K. M.; Hsu, C. *Soil Biol. Biochem.* 1999b, 31, 221-236.
Mader, P.; FlieBback, A.; Dubois, D.; Gunst, L.; Fried, P.; Niggli, U. *Science* 2002, 296, 1694-1697.
Manns, H. R. Under Review.
Manns, H. R.; Emery, N. R. J. Under Review.
Manns, H. R.; Maxwell, C. D.; Emery, N. R. J. *Soil Till. Res.* 2007, 96, 83-94.
Manns, H. R.; Maxwell, C. D.; Emery, N. R. J. *J. Plant Interact.* 2009, 4, 167-178.
Manzoni, S.; Porporato, A. *Soil Biol. Biochem.* 2009, 41, 1355-1379.
Marinari, S.; Liburdi, K.; Fliessbach, A.; Kalbitz, K. *Soil Tillage Res.* 2010, 106, 211-217.
Marschner, P.; Timonen, S. *App. Soil Ecol.* 2005, 28, 23-36.
Marulanda, A.; Azcón, R.; Ruiz-Lozano, J.M. *Physiol. Plantarum* 2003, 119, 526-533.
McCarty, G. W.; Bremner, J. M.; *Biol. Fert. Soils* 1992, 14, 219-222.
McGill, W. B.; Cannon, K. R.; Robertson, J. A.; Cook, R. D. *Can. J. Soil Sci.* 1986, 66, 1-19.
Millerett, R.; Le Bayon, R. C.; Gobat, J. M. *Plant Soil* 2009, 316, 1-12.
Moyano, F. E.; Kutsch, W. L.; Schulze, E. D. *Soil Biol. Biochem.* 2007, 39, 843-853.
Ohtonen, R.; Fritze, H.; Pennanen, T.; Jumpponen, A.; Trappe, J. *Oecologia* 1999, 119, 239 - 246.
Olsson, P. A.; Johnson, N. C. *Ecol. Lett.* 2005, 8, 1264-1270.
Parniske, M. *Nature Rev.*; *Microbiol.* 2008, 6, 763-775.
Paul, E. A.; Kucey, R. M. N. *Science* 1981, 213, 473-474.
Paul, E. A.; Collins, H. P.; Paustian, K.; Elliott, E. T.; Frey, S.; Juma, N.; Janzen, H. H.; Campbell, C. A.; Zentners, R. P.; Lafond, G. P; Moulin, A. P. *Can. J. Soil Sci.* 2004, 84, 49-61.
Pfeffer, K. P. E.; Douds, D. D. Jr.; Becard, G.; Shachar-Hill, Y. *Plant Physiol.* 1999, 120, 587-598.
Piotrowski, J. S.; Rillig, M. C. *Adv. Agron.* 2008, 97, 111-130.
Poudel, D. D.; Horwath, W. R.; Lanini, W. T.; Temple, S. R.; van Bruggen, A. H. C. *Agric. Eco. Environ.* 2002, 90, 125-137.
Raiesi, F.; Ghollarata, M. *Pedobiologia* 2006, 50, 413-425.
Raynaud, X.; Lata , J-C.; Leadley, P.W. *Plant Soil* 2006, 287, 95-116.
Rillig, M.C.; Allen, M. F. *Mycorrhiza* 1999, 9, 1-8.
Rillig, M. C.; Wright, S.; Eviner, V. *Plant Soil* 2002, 238, 325-333.
Rillig, M. C.; Ramsey, P. W.; Morris, S.; Paul, E. A. *Plant Soil* 2003, 253, 293-299.
Rillig, M.C. *Pedobiologia* 2005, 49; 395-399.
Rillig, M. C.; Caldwell, B. A.; Woosten, H. A. B.; Sollins, P. *Biogeochem.* 2007, 85, 25-44.
Rillig, M. C.; Mardatin, N. F.; Leifheit, E. F.; Antunes, P. M. *Soil Biol. Biochem. (*In Press).
Sakamoto, K.; Oba, Y. *Biol. Fertil. Soils* 1994, 17, 39-44.
Schimel, J. P.; Bennett, J. *Ecology* 2004, 85, 591-602.
Schwab, S. M.; Leonard, R. T.; Menge, J. A. *Can. J. Bot.* 1984, 62, 1227-1231.

Simon, T. S.; Javurek, M.; Mikanova, O.; Vach, M. *Soil Tillage Res*. 2009, 105, 44-48.
Smit, A.; Velthof, G. L. *Plant Soil* 2010, 331, 139-150.
Smith, S.; Read, D. J. *Mycorrhizal Symbiosis*, Academic Press, New York, NY, 1997.
Smith, S.; Read, D. J. *Mycorrhizal Symbiosis,* Academic Press, New York, NY, 2008.
Sokolov, A.P.; Kicklighter, D.W.; Melillo, J.M.; Felzer, B.S.; Schlosser, C.A.; Cronin, T.W. *J. Climate* 2008, 21, 3776-3796.
Staddon, P. L.; Ostle, N.; Dawson, L. A.; Fitter, A. H. *J. Exper. Bot.* 2003, 54, 1461-1469.
Stanford, G.; Vander Pol, R. A.; Dzienia, S. *Soil Sci. Soc. Am. Proc.* 1975, 39, 284-289.
Steinberg, P. D.; Rillig, M. C. *Soil Biol. Biochem.* 2003, 35, 191-194.
Swift, M.; Heal, O.; Anderson, J. *Decomposition in Terrestrial Ecosystems,* University of California Press, Berkeley, CA, 1979.
Talbot, J. M.; Allison, S. D.; Treseder, K. K. *Functional Ecology* 2008, 22, 955-963.
Thomsen, I. K.; Petersen, B. M.; Bruun, S.; Jensen, L. S.; Christensen, B. T. *Soil Biol. Biochem.* 2008, 40, 849-852.
Van den Bygaart, A. J.; Gregorich, E. G.; Angers, D. A. *Can J. Soil Sci.* 2003, 83, 363-390.
Wright, S. F.; Upadhyaya, A. *Plant Soil* 1998, 198, 97-107.
Zhang, X.; Amelung, W.; Yuan, Y.; Zech, W. *Soil Science* 1998, 163, 220-229.

Reviewed by Dr. Maren Oelbermann, Department of Environment and Resource Studies, University of Waterloo, Canada. moelbermann@uwaterloo.ca

In: Mycorrhizal Fungi
Editor: Susanne M. Fulton
ISBN 978-1-61122-659-1
© 2011 Nova Science Publishers, Inc.

Chapter 9

ARBUSCULAR MYCORRHIZAL INTERACTIONS WITH RHIZOBACTERIA OR SAPROTROPHIC FUNGI AND ITS IMPLICATIONS TO BIOLOGICAL CONTROL OF PLANT DISEASES

M.G.B. Saldajeno[1] and M. Hyakumachi[2]
[1]United Graduate School of Agricultural Sciences, Gifu University, Gifu City, Japan
[2]Laboratory of Plant Pathology, Faculty of Applied Biological Sciences, Gifu University, Gifu City, Japan

ABSTRACT

Persistent chemical usage in conventional agriculture aggravated the environmental contamination and paved the way to the conception of several approaches to integrated pest management. In this context, microbe-microbe interactions such as the mycorrhizal relationships with other soil micro biota in the rhizosphere like the rhizosphere bacteria and saprotrophic fungi are particularly important. A good understanding of the interactions between beneficial microbial groups is indispensable to the improvement of their management as biocontrol agents or biofertilizers. In recent years, our knowledge of the interaction effects between arbuscular mycorrhizal fungi (AMF) and rhizosphere bacteria or saprotrophic fungi in relation to plant growth and induction of plant resistance against diseases and nematodes have greatly increased. This chapter seeks to review the current research results on the effects of combined inoculation of the AMF and rhizosphere bacteria on plant growth, root colonization and induction of systemic resistance and compare it with those of AMF and saprotrophic fungi and to discuss their mechanisms of action and implications to biological control of plant diseases.

Keywords: arbuscular mycorrhiza, rhizobacteria, saprotrophic fungi, plant growth, induced systemic resistance, biocontrol, multitrophic interactions

INTRODUCTION

In conventional agriculture, biocontrol agents against plant pathogens or nematodes are usually applied separately, however, recent research outputs support the idea that mixtures of two or more biocontrol agents boost efficacy and consistency of the biocontrol than when applied alone (Raupach and Kloepper, 1998; Guetsky *et al.*, 2001; Roberts *et al.*, 2005). Failure of such mixtures in improving disease suppression (Hubbard *et al.*, 1983; Dandurand and Knudsen, 1993) often arises when co-inoculants inhibit each other, thus, compatibility of the co-inoculants is basic for successful development of mixtures of biocontrol agents (Baker, 1990). The increasing need for effective biocontrol agents has paved the way in understanding the co-operative activities among the rhizosphere microbial populations such as between arbuscular mycorrhizal fungi (AMF) and rhizobacteria or saprotrophic fungi.

AMF, which belongs to the genera *Glomales* (Morton and Benny, 1990) are ubiquitous components of both natural and agricultural ecosystems (Smith and Read, 2008) and known to form associations with roots of higher plants. AMF contribute greatly to sustainability of inherent soil structure and fertility, and as such, are considered a key component of sustainable plant-soil ecosystems (Jeffries and Barea, 2001; Jeffries *et al.*, 2003). Most agricultural plants benefit immensely from arbuscular mycorrhizal associations, principally in overcoming biotic and abiotic stresses (Nelsen, 1987; Linderman, 1994; Tisdall, 1994), so it is logical that AMF are a very significant group among the beneficial soil microbiota. Moreover, AMF are known to promote plant growth and control or alleviate impact of plant diseases in various plant hosts particularly diseases caused by various soil-borne pathogens (Dehne, 1982; Caron, 1989; Hooker *et al.*, 1994; Linderman, 1994; Azcón-Aguilar and Barea, 1996). AMF also interact with almost all soil microbiota in the rhizosphere (for review on the interactions, see Bagyaraj, 1984; Linderman and Paulitz, 1990; Linderman, 1992; Fitter and Garbaye, 1994; Gryndler, 2000; Johansson *et al.*, 2004; Barea *et al.*, 2005). We will discuss the AMF interaction with two special groups of soil microbiota in this Chapter.

A wide variety of beneficial free-living soil bacteria is often referred to as the plant growth-promoting rhizobacteria (PGPR) (Kloepper and Scroth, 1978; Kloepper *et al.*, 1980) because when grown in association with a host plant, they positively influence the growth of their hosts through a range of direct or indirect mechanisms. One classic example of PGPR is the nitrogen fixing *Rhizobia* which has been exploited for its ability to fix N_2 in leguminous plants. This review, however, will not deal with *Rhizobia* symbioses as this is a very vast and well studied topic with countless excellent up-to-date reviews (refer to Hansen, 1994; Schultze and Kondorosi, 1998; Gualtiere and Bisseling, 2000; Sessitsch *et al.*, 2002).

Saprotrophic fungi, examples of which are plant growth-promoting fungi (PGPF), live on the rhizosphere and generally procure their nutritional requirements from nonliving or decaying organic matter in the soil. PGPF are non-pathogenic soil inhabiting saprobes (Hyakumachi, 1994) reported to promote plant growth in various crops (Meera *et al.*, 1994; 1995; Shivanna *et al.*, 1994; 1996a; 2005) and effective in controlling several plant diseases such as anthracnose (Meera *et al.*, 1995; Chandanie *et al.*, 2006), Fusarium crown and root rot in tomatoes (Horinouchi, 2007), take-all and common root rot diseases in wheat (Shivanna *et al.*, 1996b) Pythium and Rhizoctonia damping off, Pythium foliar blight, Sclerotium blight, Fusarium wilt, and brown patch diseases of turf grasses (Hyakumachi, 1994; Hyakumachi and Kubota, 2004a).

Most of the earlier studies on AMF interaction with soil microbiota in the rhizosphere dealt with selected rhizobacteria (McAllister *et al.*, 1994) and soil-borne pathogens (Jalali and Jalali, 1991). But recently, studies on AMF interactions with beneficial saprotrophic fungi have been increasing. Since AMF, PGPR or saprotrophic fungi like the PGPF are all beneficial microorganisms, their synergistic or additive effect could be more valuable than their individual effects.

In this chapter, the outcomes of combined inoculation of AMF and beneficial rhizosphere microbiota such as PGPR or PGPF on plant growth, root or rhizosphere colonization, and disease suppression are reviewed and its implications to biological control of plant diseases are presented.

INTERACTION BETWEEN AMF AND BENEFICIAL RHIZOSPHERE MICROBIOTA

The beneficial soil microbiota is so diverse that a lot of researchers tentatively placed them into groups like diazotrophs, *Bacilli*, *Pseudomonads*, *Trichoderma*, *Rhizobia*, phosphate solubilizing, lignin degrading, chitin degrading, or cellulose degrading bacteria and fungi. Rather than grouping these PGPR and PGPF into groups mentioned above, they can be generally organized according to their goal of application (Lugtenberg *et al.*, 2002), viz., biofertilizers (such as *Rhizobia*), phytostimulators (such as growth hormone-producing strains), rhizoremediators (pollutant degrader strains) and biopesticides.

Table 1. Examples of AMF-PGPR interactions

AM fungus	PGPR	Interaction Effects	Reference
AMF mix	*A. chroococcum* *A. brasilense*	No effect on plant growth and PGPR population; No effect or decreased AMF root colonization; Decreased plant yield	Constantino *et al.*, 2008
Gigaspora rosea	*P. putida*	Decreased SDW; Reduced density of PGPR on the root system and AMF root colonization	Gamalero *et al.*, 2010
Gi. rosea	*P. putida*	Enhanced plant growth; Reduced density of PGPR on the root system; Enhanced AMF root colonization	Gamalero *et al.*, 2008
Glomus clarum	*Pseudomonas* spp.	No effect on plant dry matter production; No effect or reduced AMF root colonization	Walley and Gemida, 1997
G. constrictum *G. intraradices*	*B. megaterium*	Decreased or increased plant growth depending on AMF combination; No effect or increased AMF root colonization depending on AMF combination	Marulanda *et al.*, 2008
G. deserticola	*Azospirillum*	Increased SDW and RDW; No effect on PGPR population or AMF root colonization	MarVásquez *et al.*, 2000
G. etunicatum	*P. fluorescens*, *P. putida*	No effect on CFU of PGPR; No effect on AMF root colonization	Paulitz and Linderman, 1989
G. fasciculatum	*A. brasilense*	Improved plant growth; Increased population of PGPR in the rhizosphere	Biro *et al.*, 2000
G. fasciculatum	*Pseudomonas* sp.	Enhanced plant biomass; Enhanced AMF root colonization; Increased vigor index of plants inoculated with *F. oxysporum* or *R. solani*	Basu and Santhaguru, 2009
G. fasciculatus	*A. chroococcum*	Enhanced plant dry weight; Increased rhizosphere soil population of PGPF; Enhanced infection and spore production of AMF	Bagyaraj and Menge, 1978
G. intraradices *Gi. margarita*	*B. cepacia*, *B. subtilis*	Increased shoot dry mass; No effect on PGPR population; No effect or increased AMF root colonization; Decreased root galls and nematode multiplication	Siddiqui and Akhtar, 2009
G. intraradices	*B. subtilis*	Enhanced plant growth; Reduced root biomass; No effect on AMF root colonization	Kohler *et al.*, 2007

Table 1. (Continued)

AM fungus	PGPR	Interaction Effects	Reference
G. intraradices	*B. subtilis* *Enterobacter* sp.	Increased SDW and AMF root colonization; Decreased rhizosphere soil population of PGPR	Toro *et al.*, 1997
G. intraradices	*B. thuringiensis*	Enhanced root growth; Enhanced AMF root colonization and arbuscule richness	Marulanda *et al.*, 2006
G. intraradices *G. mosseae*	*Bacillus* spp.	No effect on shoot growth; No effect on AMF colonization frequency but enhanced colonization intensity and arbuscule abundance	Vivas *et al.*, 2003c
G. intraradices	*P. alcaligenes* *B. pumilus*	Increased shoot dry mass; Increased AMF root colonization; Reduced galling, nematode population and root rot in tomato plants	Akhtar and Siddiqui, 2008a
G. intraradices *G. mosseae*	*P. fluorescens* *P. brasilensis*	Increased or decreased plant growth and AMF root colonization depending on specific AMF-PGPR combination	Jäderlund *et al.*, 2008
G. intraradices	*P. fluorescens* *P. putida*	Reduced population of PGPR in the rhizosphere; No effect on AMF root colonization	Paulitz and Linderman, 1989
G. intraradices	*P. putida*	Enhanced plant growth, PGPR population in roots, and AMF root colonization; Decreased galling in roots and nematode population	Siddiqui and Akhtar, 2008b
G. intraradices	*P. straita* *Rhizobium*	Increased SDW; Enhanced AMF root colonization; Reduced galling and nematode population	Akhtar and Siddiqui, 2008b
G. intraradices	*S. coelicolor*	No significant effect in shoot and root dry matter; Increased AMF root colonization	Abdel-Fattah and Mohamedin, 2000
G. mosseae	*Azospirillum*	No effect on plant growth, PGPR population or AMF colonization	MarVásquez *et al.*, 2000
G. mosseae	*Azotobacter* spp. *Corynebacterium* spp. *Gram-negative non-motile rod* *Gram-positive non-motile rod* *Pseudomonas* spp.	No effect on SDW; Enhanced or reduced PGPR population on the rhizosphere soil depending on inoculated bacterial isolate	Ames *et al.*, 1984
G. mosseae	*B. subtilis*	Enhanced plant growth, PGPF population in rhizosphere soil and AMF root colonization; Reduced incidence of root rot	Haggag and Abd-El Latif, 2001
G. mosseae	*Brevibacillus* sp.	Increased shoot and root biomass and AMF root colonization	Vivas *et al.*, 2003b; 2003c
G. mosseae	*P. fluorescens*	Increased SDW; Increased PGPR population in the rhizosphere soil; No effect on AMF root colonization	Edwards *et al.*, 1998
G. mosseae	*P. fluorescens*	Promoted plant growth; Increased AMF root colonization	Gamalero *et al.*, 2004
G. mosseae	*Paenibacillus* sp.	Enhanced fresh leaf and root weights; Enhanced AMF root colonization; Decreased necrosis on tomato roots caused by *P. parasitica*	Budi *et al.*, 1999
G. mosseae *Gi. rosea*	*Various soil bacteria* *P. fluorescens* *C. fungivorans*	Varied effects on plant growth and AMF root colonization depending on the specific PGPR, plant or AMF combination	Pivato *et al.*, 2009
G. mosseae *G. deserticola* *G. intraradices*	*B. pumilus* *B. licheniformis*	Enhanced or no effect on shoot and root biomass and AMF root colonization	Medina *et al.*, 2003
Glomus spp.	*Bacillus* sp.	No effect on plant and root fresh weight; No effect on marketable fruit yield; No effect on protection against Verticillium wilt	Tahmatsidou *et al.*, 2006
Indigenous AMF *G. etunicatum*	*B. subtilis*	Decreased SDW; Decreased the frequency of indigenous AMF root colonization of maize under pot culture conditions	Xiao *et al.*, 2008
Indigenous AMF	*P. putida*	Increased SDW and RDW; No effect on PGPR population in the rhizosphere soil; Increased mycorrhizal colonization at 6 wk	Meyer and Linderman, 1986
Indigenous AMF	*Pseudomonas* spp.	Increased or no effect on plant height; No effect on AMF root colonization	Roesti *et al.*, 2006

SDW= shoot dry weight; RDW= root dry weight; CFU= colony forming unit

Only three outcomes of interaction between AMF and PGPR or PGPF could be possible: viz., synergistic, antagonistic and neutral. And most often than not, since microbial entities are highly diversified and form very complex and dynamic relationships, the observed phenomenon turns contradictory and its reproducibility is very low (Gryndler, 2000).

More recent studies on combinations of AMF and PGPR or PGPF not only deal with plant growth promotion or disease control under normal environmental conditions but under conditions of stress like salinity, drought or metal toxicity. Furthermore, recently developing areas of PGPR or PGPF usage include forest regeneration, phytoremediation of contaminated soils, and agrowaste detoxification.

The summary of the interaction effects between AMF and PGPR or PGPF reviewed in this chapter is presented in Tables 1 and 2, respectively.

Table 2. Examples of AMF-PGPF interactions

AM fungus	PGPF	Interaction Effects	Reference
Gigaspora rosea	*R. mucilaginosa*	Increased plant dry matter and AMF root colonization depending on time of inoculation of PGPF, type of PGPF inoculum carrier or amount of PGPF inoculum	Fracchia *et al.*, 2003
Gi. rosea	*T. pseudokoningii* (5 strains)	Reduced SDW, RDW and AMF root colonization; No effect on CFU of PGPF	Martinez *et al.*, 2004
Glomus coronatum	*F. oxysporum*	No effect on plant growth; Enhanced AMF root colonization; No effect on nematode control	Diedhiou *et al.*, 2003
G. deserticola	*F. oxysporum*	Increased SDW but no effect on RDW; No effect on CFU of PGPF; Enhanced AMF root colonization	Fracchia *et al.*, 2000
G. deserticola	*T. harzianum*	No effect on plant growth; No effect on PGPF or AMF colonization	MarVasquez *et al.*, 2000
G. etunicatum	*G. virens*	Enhanced or no effect on AMF root colonization depending on medium used	Paulitz and Linderman, 1991
G. intraradices	*A. niger*	No effect on plant growth, root biomass or AMF root colonization	Kohler *et al.*, 2007
G. intraradices *Gi. margarita*	*A. niger* *P. chrysogenum*	Increased shoot dry mass; No effect or decreased PGPR population; No effect on AMF root colonization; Decreased root galls and nematode multiplication	Siddiqui and Akhtar, 2009
G. intraradices	*C. rosea*	Enhanced plant growth; Reduced CFU of PGPF and AMF root colonization	Ravnskov *et al.*, 2006
G. intraradices	*P. lilacinus* *P. chlamydosporia* *T. harzianum* *G. virens*	Enhanced or no effect on plant growth; No effect on PGPF population isolated from eggs or female nematodes; No effect on AMF root colonization; No effect or decreased galling in roots	Siddiqui and Akhtar, 2008a
G. intraradices	*T. harzianum*	Enhanced SDW; Reduced RDW and AMF root colonization	Green *et al.*, 1999
G. intraradices	*T. harzianum*	No effect on CFU of PGPF; Increased AMF root colonization; Significantly lowered Fusarium wilt in melon plants	Martínez-Medina *et al.*, 2010
G. intraradices	*T. harzianum*	No reduction in disease severity or incidence relative to AMF or PGPF alone	Datnoff *et al.*, 1995
G. mosseae	*F. concolor, F. equiseti, F. graminearum, F. lateritium, F. moniliforme, F. oxysporum, F. solani, F. stilboide*	Varied effects on SDW, RDW and AMF root colonization depending on the type of soil where the plant was grown; No effect on CFU of PGPF	Garcia-Romero *et al.*, 1998
G. mosseae	*F. equiseti*	Enhanced SDW; Reduced PGPF population or AMF root colonization; Enhanced resistance against anthracnose; No effect or enhanced suppression of damping-off depending on time of pathogen inoculation	Saldajeno and Hyakumachi, 2010 (In press)
G. mosseae	*F. equiseti* *A. alternata*	No effect or enhanced plant growth depending on host plant or time of inoculation of PGPF; No effect on CFU of PGPF; No effect or reduced AMF root colonization depending on host plant or time of inoculation of PGPF	McAllister *et al.*, 1997
G. mosseae	*F. equiseti* *A. alternata*	No effect on plant growth and CFU of PGPF; No effect or reduced AMF root colonization depending on time of inoculation of PGPF; Reduced AMF spore germination	McAllister *et al.*, 1996

Table 2. (continued)

AM fungus	*PGPF*	Interaction Effects	Reference
G. mosseae	*F. oxysporum*	Increased SDW; No effect on RDW; No effect on CFU of PGPF; Enhanced AMF root colonization	Fracchia *et al.*, 2000
G. mosseae	*G. virens*	No effect on AMF root colonization	Paulitz and Linderman, 1991
G. mosseae	*P. simplicissimum*	No effect on SDW, PGPF population, or AMF root colonization; No effect on induced resistance against anthracnose; No effect or enhanced suppression of damping-off depending on time of pathogen inoculation	Chandanie *et al.*, 2006; 2009
G. mosseae	*Phoma* sp.	No effect on SDW and AMF root colonization; Reduced isolation frequency of PGPF in roots; Reduced level of disease protection	Chandanie *et al.*, 2005; 2006
G. mosseae	*R. mucilaginosa*	Increased plant dry matter and AMF root colonization depending on time of inoculation of PGPF, type of PGPF inoculum carrier, or amount of PGPF inoculum	Fracchia *et al.*, 2003
G. mosseae	*R. mucilaginosa, M. pulcherrima, T. cutaneum* var. *cutaneum, S. cerevisiae, C. laurentii, D. occidentalis* var. *occidentalis*	Increased SDW, RDW, plant biomass and yield; Increased yeast population in the rhizosphere soil; Enhanced AMF root colonization	Boby *et al.*, 2008
G. mosseae	*T. koningii* *F. solani*	Varied effects on plant growth, PGPF population in rhizosphere soil, and AMF root colonization depending on time of inoculation and host plant	McAllister *et al.*, 1994
G. mosseae	*T. aureoviride*	Enhanced plant biomass and foliar area; Increased AMF root colonization	Calvet *et al.*, 1993
G. mosseae	*T. harzianum*	Enhanced SDW and AMF root colonization; Reduced root and rhizosphere soil population of PGPF; No effect or enhanced suppression of damping-off depending on time of pathogen inoculation	Chandanie *et al.*, 2009
G. mosseae	*T. harzianum*	No effect on plant growth, PGPF population or AMF root colonization	MarVásquez *et al.*, 2000
G. mosseae	*T. harzianum* *P. oxalicum*	Enhanced plant growth, PGPF population in rhizosphere soil and AMF root colonization; Reduced incidence of root rot	Haggag and Abd-El Latif, 2001
G. mosseae	*T. pseudokoningii (5 strains)*	Increased SDW except a strain which has no effect on SDW; No effect on RDW and CFU of PGPF; No effect on AMF root colonization except a strain which reduced it	Martinez *et al.*, 2004

SDW= shoot dry weight; RDW= root dry weight; CFU= colony forming unit

In Vitro Interactions

An *in vitro* assay prior to further tests in the greenhouse or in a controlled growth chamber can give significant insight on the interaction between the two microbial partners. Most *in vitro* tests seek to determine effects of PGPR or PGPF on AMF spore or conidial germination, and hyphal growth or elongation.

Biocontrol PGPR or PGPF could produce antibiotics (Duffy *et al.*, 2004; Dwivedi *et al.*, 2009) or anti-fungal substances that can affect AMF spores and its growing hyphae where effects could be more pronounced during its pre-symbiotic phase (McAllister *et al.*, 1994; Garcia-Romera *et al.*, 1998). Calvet *et al.* (1992) in their study involving *Glomus mosseae* and some saprotrophic fungi isolated from organic substrates, found that germination rate of *G. mosseae* was hastened and development of mycelia from germinated spores were enhanced by the presence of *Trichoderma* spp. while the presence of *Penicillium decumbens* and *Aspergillus fumigatus* inhibited spore germination of *G. mosseae*. The two fungi might have produced antibiotic-like substances affecting *G. mosseae* spore germination under axenic conditions. Several pseudomonads inhibited germination of *Glomus clarum* spores when incubated either on membranes placed directly on bacterial lawns or on agarose block separated from the bacteria by a membrane but no inhibition was observed when the agarose was physically separated from the bacteria indicating that there is an involvement of a non-volatile diffusible substances inhibitory to *G. clarum*. It is possible that substances inhibitory to pathogenic fungi produced by pseudomonads are also inhibitory to *G. clarum* (Walley and Germida, 1997). Interaction of AMF with other beneficial soil microbiota showed similar enhancement effects. For example, Fracchia *et al.* (2003) demonstrated that the yeast *Rhodotorula mucilaginosa* significantly increased the hyphal length of *G. mosseae* and *Gigantea rosea* spores.

Some other PGPR or PGPF could produce volatiles, exudates or hormones that may significantly affect AMF spore germination or hyphal growth, and fungal extract from AMF could likewise affect growth of PGPR or PGPF. Xiao *et al.* (2008) found that *Bacillus subtilis* produces some unknown volatiles that inhibited the spore germination and hyphal growth of *Glomus etunicatum.* Soluble and volatile substances produced by *Alternaria alternata* and *Fusarium equiseti* were involved in the inhibition of spore germination of *G. mosseae* (McAllister *et al.*, 1996) On the contrary, Calvet *et al.* (1989) showed that germination of the resting spores of *G. mosseae* on water agar was not affected by the presence of *Trichoderma* spp. although the inoculation of any of the isolates strongly enhanced the production of vegetative spores. Likewise, the *in vitro* experiments by Martinez *et al.* (2004) in which several strains of the saprotrophic fungus *Trichoderma pseudokoningii* were paired with spores of *G. mosseae* and *Gi. rosea,* resulted in the inhibition of spore germination of both AMF depending upon the strains used. They speculated that soluble exudates and volatile substances produced by the saprobes caused the inhibition of AMF spore germination.

In vitro tests of Pivato *et al.* (2009) using several bacterial strains resulted to increased or decreased germination rate of spores and hyphal length of *G. mosseae*. Their results showed that mycorrhizal development could be influenced positively or negatively depending on the bacterial or AMF species. This observation equally holds true in saprotrophic fungi. For example, the effect of genus *Trichoderma* on AMF spore germination may also differ with the species used (Rousseau *et al.*, 1996; Green *et al.*, 1999).

The conidial germination of *Trichoderma harzianum* was enhanced in the presence of the fungal extract from AMF while that of *Fusarium oxysporum* f. sp. *chrysanthemi* was reduced and the measured effects were directly correlated with extract concentration. Filion *et al.* (1999) suggested that direct interactions exist between AMF and soil microorganisms.

INTERACTION EFFECT ON PLANT GROWTH

The growth promotion by AMF is generally thought to be due to the increased nutrient uptake, specifically P uptake, made possible through the arbuscular mycorrhizal symbiosis (Linderman, 1988; Barea and Jeffries, 1995; Azcón-Aguilar and Barea, 1997). The AMF mycelia have a higher absorptive capacity for water and mineral nutrients compared to the roots, thus improving the plants' mineral absorption capabilities. But still some confusion exists as to whether growth increase is due to AMF or other P solubilizing microbes in the rhizosphere (Linderman, 1992).

PGPR are well-known for their role in improving plant health. Various commercial products containing PGPR for agriculture and forestry are available in the market (Lucy *et al.*, 2004; Berg, 2009). *Rhizobia* is of course the most successful inocula marketed for over a century but with the pressing environmental and health contamination brought by continuous usage of synthetic chemical fertilizers, the market for the so called "bio-inoculants or biofertilizers" is still expanding. For in-depth discussion on the role of PGPR on plant growth promotion, several recent excellent review articles are available (Lugtenberg *et al.*, 1991; Rodríguez and Fraga, 1999; Bloemberg and Lugtenberg, 2001; Vessey, 2003; Kennedy *et al.*, 2004; Lucy *et al.*, 2004; Artursson *et al.*, 2006; Adesemoye and Kloepper, 2009; Berg, 2009) detailing the processes and mechanisms of PGPR-induced plant growth promotion. Several mechanisms which help explain how PGPR enhance plant growth can be divided into direct or indirect mechanisms. Direct mechanisms include biological nitrogen fixation, provision of bioavailable phosphorus for plant uptake, sequestration of iron for plants by siderophores, production of plant hormones like auxins, cytokinins, and gibberellins and lowering of plant ethylene levels (Lucy *et al.*, 2004; Vessey, 2003). Indirect mechanisms include production of antibiotics, reduction of iron available to the plant pathogens in the rhizosphere, synthesis of fungal cell wall-lyzing enzymes and competition for space within the roots with the deleterious microorganisms in the soil.

Like the PGPR, the PGPF can help enhance plant growth through various mechanisms which include mineralization, suppression of deleterious microorganisms, and hormone production (Hyakumachi and Kubota, 2004b). Plant growth promotion of PGPF can be attributed to the suppression of indigenous pathogenic *Pythium* spp. in the soil (Hyakumachi, 1994) which resulted in a notable growth promotion of field-grown cucumbers. In addition, there are certain fungal species that are capable of producing growth hormone in axenic culture (Ram, 1959; Contreras-Cornejo *et al.*, 2009). Some strains of *Phoma* sp. for instance have been found to produce abscissic acid which is reported to promote plant growth (Hyakumachi and Kubota, 2004b).

Fracchia *et al.* (2000) showed that dual inoculation of *G. mosseae* or *Glomus deserticola* and *F. oxysporum* led to enhanced growth of plants. Similarly, Calvet *et al.* (1993) concluded that the combined inoculation of *Trichoderma aureoviride* and *G. mosseae* has a synergistic effect on the growth of marigold plants. Haggag and Abd-El Latif (2001) found that combined inoculation of *G. mosseae* and *T. harzianum* or *Penicillium oxalicum* enhanced growth of geranium plants.

A field study by Diedhiou *et al.* (2003) showed that combined application of *Glomus coronatum* and the non-pathogenic strain of *F. oxysporum* did not increase plant growth. On the contrary, Garcia-Romera *et al.* (1998) noted that dual inoculation with *G. mosseae* and some strains of *Fusarium* sp. led to enhanced growth of soybean plants. The shoot dry weight of soybean plants cultivated in nonsterilized soils or sterile soils inoculated with *G. mosseae* were not negatively affected by the presence of *Fusarium*. Ravnskov *et al.* (2006) found that *Clonostachys rosea* and *Glomus intraradices* were mutually inhibitory, but promoted plant growth and altered the soil microbial communities.

Chandanie *et al.* (2005; 2006; 2009) noted that plant growth was stimulated when *Trichoderma* was combined with AMF; no stimulation when *Penicillium* was combined with AMF; while *Phoma* when combined with AMF was inhibitory to growth but when applied alone stimulated growth. They also noted that plants treated with *G. mosseae* alone did not increase shoot dry weight or root dry weight compared to the controls.

Kohler *et al.* (2007) observed that the highest growth in lettuce plants was achieved when co-inoculated by *G. intraradices* and *B. subtilis* although *G. intraradices* alone hardly enhanced lettuce seedling growth. The increase in lettuce plant growth may be due to enhanced supply of P and K to the crop by *B. subtilis.* Toro *et al.* (1997) also observed that *B. subtilis* co-inoculated with *G. intraradic*es could increase the shoot dry weight of onions. On the contrary, Medina *et al.* (2003) observed that dual inoculation of *Bacillus* spp. and AMF did not always increase shoot biomass of *Medicago sativa* plants compared with AMF alone inoculated plants.

Azospirillum sp. increased shoot and root growth in *G. deserticola* mycorrhizal plants while *Pseudomonas* sp. did not (MarVázquez *et al.*, 2000). In the gamma-sterilised substrate, co-inoculation of *Glomus fasciculatum* with *Azospirillum* was effective in improving plant growth and nutrient-uptake in the rhizosphere of alfalfa (Biró *et al.*, 2000). Basu and Santhaguru (2009) found a more pronounced impact on root and shoot length, plant vigor index, dry weight and total N and P content in *Vigna radiata* when the plants were dual inoculated with fluorescent *Pseudomonas* and *G. fasciculatum*. Marulanda *et al.* (2006) found that the maximum root development of *Retama sphaerocarpa* was obtained by co-inoculation of *Bacillus thuringiensis* and the indigenous *G. intraradices* while Tahmatsidou *et al.* (2006) did not observe any enhancement in plant and root fresh weight of strawberry co-inoculated with *Glomus* spp. and *Bacillus* sp.

Jäderlund *et al.* (2008) showed that very specific interactions do occur; for example, *G. intraradices* together with *Pseudomonas fluorescens* increased plant dry weight on *Microdochium nivale*-infested plants, but *Paenibacillus brasilensis* inhibited this positive effect. This is an example of two completely different reactions between the same AMF and two species of bacteria, which are known to enhance plant growth and inhibit pathogens. Similarly, Marulanda-Aguirre *et al.* (2008) found that *Bacillus megaterium* had contrasting effects on plant performance depending upon the AMF involved.

When searching for plant growth-promoting microorganisms, it is therefore important to test for the most suitable combination of plant, bacteria and/or fungi in order to achieve satisfactory plant growth benefits.

Interaction Effect on the Population or Colonization of Co-Inoculant

A. Impact on AMF Root Colonization

The effect of combined AMF-PGPR or AMF-PGPF inoculations on the population of each co-inoculant (AMF root colonization and rhizosphere or root population of PGPR or PGPF) is highly variable; in the case of PGPR, results differ depending on the host plant, bacterial isolate and the AMF-PGPR combination.

The ACC deaminase producing *Pseudomonas putida* promotes mycorrhization of *Gi. rosea* when inoculated into cucumber while its mutant impaired in ethylene production decreased mycorrhization (Gamalero *et al.*, 2008). When the cucumber plants were grown under salt stress conditions, *P. putida* decreased AMF root colonization and arbuscule abundance in dual inoculated plants (Gamalero *et al.*, 2010). Pivato *et al.* (2009) observed that the promotion by *P. fluorescens* of mycorrhization appears to be fungal rather than plant specific. In their experiment on maize, MarVázquez *et al.* (2000) observed that the biocontrol agent *Pseudomonas* sp. did not affect negatively *G. mosseae* and *G. deserticola* establishment in the roots. Basu and Santhaguru (2009) demonstrated that the percent colonization by *G. fasciculatum* increased in the presence of fluorescent *Pseudomonas*.

In their study, Kohler *et al.* (2007) confirmed that *B. subtilis* had no effect on the root colonization of *G. intraradices* and that the level of AMF infection was not related to the ability of the co-inoculated treatments to improve plant growth. Toro *et al.* (1997) discovered that *B. subtilis* could increase the AMF root colonization. Vivas *et al.* (2003a) also reported a *Bacillus* sp. isolated from a desert area that could enhance the root colonization of two AMF: *G. mosseae* and *G. intraradices.* On the contrary, Xiao *et al.* (2008) showed that *B. subtilis* identified as an antifungal strain decreased the frequency of root colonization in maize by indigenous AMF under pot culture conditions. PGPR that produce antifungal metabolites are potential threats to AMF but Dwivedi *et al.* (2009) demonstrated that antifungal metabolites-producing PGPR do not necessarily interfere and may even promote arbuscular mycorrhizal symbiosis. Similarly, a *Paenibacillus* sp. isolate from the mycorrhizosphere of *Sorghum bicolor* which has antagonistic activity, stimulated mycorrhization of *G. mosseae* (Budi *et al.*, 1999). Marulanda *et al.* (2006) reported that a drought-adapted bacterium, *B. thuringiensis*, was only effective in increasing arbuscule richness and activity when associated with the autochthonous AMF.

Any bacteria involved in mycorrhizal establishment and/or its functioning is referred to as mycorrhiza helper bacteria (MHB) (Garbaye, 1994) and is currently the most studied group among bacteria especially in symbiotic associations. Some PGPR investigated in association with AMF fall into the category of MHB especially those that promote mycorrhization. Some MHB species are responsible for multiple helper effects, because they influence both plants and the associated mycorrhizal fungi (Bonfante and Anca, 2009).

In the case of PGPF, the effect of combined inoculation on the population of each co-inoculant generally differs according to the host plant, and species of AMF, or PGPF. Various research results showed that it can be inconsistent between species of the same genus, or even within the strains of the same species of the PGPF involved. Results may also differ according to other factors such as time of inoculation of the AMF or PGPF, the amount of inoculum used, the type of medium or inoculum carrier used, or the type of soil where the plant was grown (sterilized, partially sterilized or unsterilized soil).

Dual inoculation with *G. mosseae* or *G. deserticola* and *F. oxysporum* resulted to additive AMF colonization of the roots of pea plants (Fracchia *et al.*, 2000). Similarly, Haggag and Abd-El Latif (2001) noted an increase in root colonization of *G. mosseae* in geranium when inoculated together with *T. harzianum* or *P. oxalicum*. Diedhiou *et al.* (2003) reported that the combined application of *G. coronatum* and the non-pathogenic strain of *F. oxysporum* enhanced mycorrhization of tomato roots. Garcia-Romera *et al.* (1998) showed that dual inoculation with *G. mosseae* and different strains of *Fusarium* sp. either led to additive or no effect on AMF colonization of soybean roots depending on the kind of soil where the plants were grown.

Fracchia *et al.* (2003) noted that percentage root length colonization of *G. mosseae* in soybean and of *Gi. rosea* in red clover was increased significantly when *R. mucilaginosa* was inoculated 2 weeks before the AMF, but was unaffected when the yeast was inoculated at the same time or 2 weeks after the AMF. They also found differences in plant growth and AMF colonization but not in PGPF population when the PGPF was inoculated as thin agar slice or as suspension of an aqueous solution. Likewise, McAllister *et al.* (1997) observed that *A. alternata* or *F. equiseti* has either no effect or decreased *G. mosseae* root colonization depending on the time of inoculation of the PGPF and the host plant; while the population of the PGPF in the rhizosphere soil was not affected by *G. mosseae*. They also noted that interaction effect differed between maize and lettuce plants. These results validated their earlier study (McAllister *et al.*, 1994) in lettuce and maize using a different set of PGPF. On the other hand, Paulitz and Linderman (1991) found differences in AMF colonization when the PGPF was added as either wheat bran or peatmoss-Czapek inoculum.

Chandanie *et al.* (2005; 2006; 2009) demonstrated that the percent root length colonized by *G. mosseae* was independent of the presence of *Phoma* or *Penicillium* isolates in the soil; while *Trichoderma* enhanced root colonization of *G. mosseae*.

Some studies however noted inhibitory effects of the saprotrophic co-inoculant on AMF. For example, Saldajeno and Hyakumachi (2010) reported that *F. equiseti* has an adverse effect on root colonization by *G. mosseae*. A similar observation was noted by Ravnskov *et al.* (2006) between *C. rosea* and the AMF *G. intraradices*.

B. Impact on Rhizosphere Soil or Root Colonization of PGPR or PGPF

The impact of AMF on the rhizosphere soil or root colonization of the PGPR or PGPF co-inoculant is generally measured by counting the population of the co-inoculant, and often expressed as colony forming units (CFU). And as with AMF colonization being highly inconsistent among genus or species of AMF, PGPR or PGPF involved or plant host under study, PGPR or PGPF population in the rhizosphere soil or in the roots behave in a similar fashion.

Numerous research reports hinted that the interaction between AMF and saprotrophic fungi depends on the species of the saprotrophic fungi or strains of AMF involved (Martinez *et al.*, 2004). For instance, McAllister *et al.* (1994) reported a synergistic interaction between *G. mosseae* and *T. aureoviridae*; Calvet *et al.* (1993) observed the opposite between *G. mosseae* and *Trichoderma koningii*. Green *et al.* (1999) and Rousseau *et al.* (1996) noted antagonistic interaction between *G. intraradices* and *T. harzianum* while synergistic reaction was observed by Datnoff *et al.* (1995).

Mycorrhizal plants are capable of producing compounds which can interfere with rhizosphere microorganisms and modify microbial community around the mycorrhizal roots (the mycorrhizosphere effect) (Linderman, 1988; Linderman and Paulitz, 1990). Moreover, although not well documented, the extraradical mycelium of AMF might also impact strongly on the microbial population around the mycorrhizal roots (Filion *et al.*, 1999). For instance, it was observed that substances released by the extraradical mycelia of *G. intraradices* either stimulated or inhibited conidial germination of non-pathogenic *T. harzianum* and root pathogen *F. oxysporum* f. sp. *chrysanthemi*. The reduction in root colonization of *Phoma* sp. and *T. harzianum* might possibly be related to the exudates secreted by the mycorrhizal roots and/or the extraradical mycelia of *G. mosseae* (Chandanie *et al.*, 2006). On the other hand, the strong sporulation ability of *Penicillium simplicissimum* under natural conditions and the inherent characteristics related to the species may explain the tolerance of the fungus to such exudates (Chandanie *et al.*, 2006; 2009).

Ravnskov *et al.* (2006) noted that inoculation of *G. intraradices* decreased CFU of *C. rosea* suggesting that *G. intraradices* is inhibitory to *C. rosea*. Chandanie *et al.* (2005; 2009) demonstrated that *G. mosseae* significantly reduced the root colonization ability of *Phoma* sp. and *T. harzianum* in cucumber plants. Conversely, Haggag and Abd-El Latif (2001) showed that population and survival of *T. harzianum* and *P. oxalicum* increased even beyond 90 days after inoculation in the presence of *G. mosseae* under natural conditions. The AMF had no significant effect on colonization of *Penicillium* isolate both in the rhizosphere soil and roots of cucumber (Chandanie *et al.*, 2009). Garcia-Romera *et al.* (1998) demonstrated that the dual inoculation with *G. mosseae* and some strains of *Fusarium* sp. did not influence the CFU of *Fusarium* in the rhizosphere soil.

These results demonstrated that the interaction effects between individual saprotrophic fungi and AMF depends on the species involved. While AMF adversely affected the population development of some saprotrophic fungi in the roots of some plants, it had slight or no effect on populations of other saprotrophic fungi.

The same trend could be generalized in the case of PGPR population in the root or in the rhizosphere soil. For example, MarVázquez *et al.* (2000) demonstrated that *G. mosseae* and *G. deserticola* induced changes in the *Azospirillum* and *Pseudomonas* populations in maize roots depending on the inoculant combination involved. The rhizosphere soil population of *B. subtilis* was decreased by *G. intraradices* (Toro *et al.*, 1997) while it was increased by *G. mosseae* (Haggag and Abd-El Latif, 2001).

Effects on Plant Disease Suppression

PGPR are known to enhance plant growth through an indirect way, by helping control pathogenic microorganisms (for review on PGPR as biocontrol agent, please refer to Lugtenberg *et al.*, 1991; Whipps, 2001; Zehnder *et al.*, 2001; Compant *et al.*, 2005). PGPR like *Bacillus* (Kloepper *et al.*, 2004), *Pseudomonas*, *Streptomyces* (Schrey and Tarkka, 2008), and non-streptomycete actinomycetes (El-Tarabily and Sivasithamparam, 2006) are just a few of the PGPR that have gained a lot of attention because of their potential as biocontrol agents.

On the other hand, disease suppression capability of fungi like the AMF and the PGPF have been discussed and reviewed in detail (Dehne, 1982; Bagyaraj, 1984; Paulitz and Linderman, 1991; Hooker *et al.*, 1994; Hyakumachi, 1994; Linderman, 1994; Azcón-Aguilar and Barea, 1996; Xavier and Boyetchko, 2002; Hyakumachi and Kubota, 2004b; Harman *et al.*, 2004; Vinale *et al.*, 2008) and has been thought to be due to several mechanisms stimulated by the presence of the inoculum on the plant, although much of these mechanisms are still poorly understood.

It is generally accepted that antagonistic capability of PGPR and PGPF include the actions of hyperparasitism, antibiosis and competition or a combination of these actions. Another important mechanism is the induction of systemic resistance and this has been observed both in PGPR (Wei *et al.*, 1991; Pieterse *et al.*, 2003; van Loon, 2007) and in PGPF (Shivanna *et al.*, 1996c).

Studies on the interaction effects of AMF and beneficial soil microorganisms on disease suppression is scanty, as most studies done were on the direct interaction of AMF with the pathogen itself (Kaye *et al.*, 1983; Krishna and Bagyaraj, 1983; Caron *et al.*, 1986; Garcia-

Garrido and Ocampo, 1988; Rosendahl and Rosendahl, 1990; Trotta *et al.*, 1996; Bødker *et al.*, 2002; Karagiannidis *et al.*, 2002; de la Peña *et al.*, 2006; Garmendia *et al.*, 2006; Alejo-Iturvide *et al.*, 2008).

In general, there is a tendency of intensified development of leaf disease symptoms in arbuscular mycorrhizal plants (Gernns *et al.*, 2001) and this is correlated to improved nutrition and higher physiological activities compared to non-arbuscular mycorrhizal plants (Dehne, 1982).

Using the commercial formulations of *G. intraradices* and *T. harzianum* on their study on tomatoes, Datnoff *et al.* (1995) demonstrated that combined *G. intraradices* and *T. harzianum* inoculation was effective in reducing both the incidence and severity of Fusarium crown and root rot under field conditions. Similarly, Haggag and Abd-El Latif (2001) noted that the dual application of *G. mosseae* and *T. harzianum*, *P. oxalicum* or *B. subtilis* reduced Geranium root rot caused by *Fusarium solani* and *Macrophomina phaseolina* either in artificially- or naturally-infested soils.

Chandanie *et al.* (2009) tested *P. simplicissimum* and *T. harzianum* against Rhizoctonia damping-off. Pre-treatment of the cucumber seedlings for seven or twelve days with AMF and PGPF before transplanting into pathogen infested soil protected plants from the damping-off disease caused by *Rhizoctonia solani* more than the protection conferred by individual AMF or PGPF. The degree of protection provided by pre-treating seedlings with *G. mosseae* for seven days increased when the treatment duration was increased to twelve days, but such an increase was not found in case of *P. simplicissimum* or its combined treatment with *G. mosseae*. Saldajeno and Hyakumachi (2010) similarly demonstrated that inoculating the cucumber seedlings with *F. equiseti* and *G. mosseae* 6 or 12 days prior to damping-off pathogen inoculation suppressed damping-off symptom development. Taken all together, the protective capability of pre-inoculated *G. mosseae* could be a result of combined mechanisms including competition, altered root exudation, anatomical and morphological changes in the root system and local activation of plant defense properties (Azcón-Aguilar and Barea, 1996; Pozo *et al.*, 1998; 1999; 2002; Pozo and Aguilar, 2007).

In another study of Chandanie *et al.* (2006), inoculation of PGPF isolates (*Phoma* sp. or *P. simplicissimum*) into the root system of cucumber showed considerable disease protection against the anthracnose pathogen *Colletotrichum orbiculare*. Although the treatment of *G. mosseae* had no significant effect on the disease development, combined inoculation of *G. mosseae* with *Phoma* sp. reduced the level of disease protection. However, the level of protection induced by *P. simplicissimum* or *T. harzianum* was not altered by combining it with *G. mosseae*. Saldajeno and Hyakumachi (2010) likewise found that combined inoculation of *F. equiseti* and *G. mosseae* significantly decreased the total number of anthracnose lesions per leaf and total lesion area in cucumber plants. As the resistance against the *C. orbiculare* was achieved when there was no direct contact between the pathogen and the inducer within the plant, the induction of systemic resistance is implicated as the mechanism of disease suppression (Chandanie *et al.*, 2006; Saldajeno and Hyakumachi, 2010).

PGPF-mediated ISR has been demonstrated by increased lignin deposition at the point of attempted penetration by the pathogen *C. orbiculare* in the epidermal tissues of cucumber hypocotyls (Shivanna *et al.*, 1996c) and also conspicuous superoxide generation by culture filtrates of respective PGPF isolates (Koike *et al.*, 2001). Biochemical analyses have revealed systemic accumulation of salicylic acid and increased activities of chitinase, ß-1,3-glucanases and peroxidase in cucumber plants induced by PGPF (Shivanna *et al.*, 1996c; Hyakumachi and Kubota, 2004a; Yedidia *et al.*, 1999). Hossain *et al.* (2007) hypothesized that multiple defense mechanisms are involved in *P. simplicissimum*-mediated ISR in *Arabidopsis* plants. Additional studies on the expression of pathogenesis-related genes and the signaling pathways involved in PGPF-mediated ISR are currently being pursued.

Dual inoculation of *V. radiata* with an efficient biocontrol strain of fluorescent *Pseudomonas* and *G. fasciculatum* in pathogen-infested soil alleviated the adverse effects of *F. oxysporum* and *R. solani* pathogens in *V. radiata* by increasing its vigor index. It is hypothesized that the production of plant growth promoting substances by the AMF and antibiotics by the PGPR were responsible for the alleviation of adverse effects of the pathogens on the plants (Basu and Santhaguru, 2009).

EFFECTS AGAINST PLANT NEMATODES

There are a considerable number of studies on the AMF-PGPR or AMF-PGPF combination against plant nematodes especially by the group of Siddiqui and Akhtar in India. Their research results show that several PGPR or saprotrophic fungi when applied together with AMF with or without organic amendments or fertilizer in the presence or absence of a pathogen (*M. phaseolina*) could decrease galling and nematode multiplication in tomato or chickpea plants. AMF have been demonstrated as effective biocontrol agents against plant parasitic nematodes, and its combination with PGPR or PGPF help enhanced its efficiency in plant parasitic nematode biocontrol. On the other hand, Diedhiou *et al.* (2003) studied interactions between the AMF *G. coronatum* and the non-pathogenic *F. oxysporum* in the control of *Meloidogyne incognita* on tomato. They found that pre-inoculation of tomato plants with either *G. coronatum* or *F. oxysporum* stimulated plant growth and reduced *M. incognita* infestation but combined application of the AMF and *F. oxysporum* only enhanced mycorrhization of tomato roots and did not increase overall nematode control or plant growth.

FIELD APPLICATION

A number of commercial AMF, PGPR, or PGPF bioinoculants have been successful in the field. For instance, in a field study of Nemec *et al.* (1996), they demonstrated that these bioinoculants were successful in controlling diseases of tomato, celery, bell pepper and citrus in Florida. Bioinoculants are traditionally applied singly and mixing two or more bioinoculants in a single field application is rare. Datnoff *et al.* (1995) proved that a mixture of two bioinoculants could work well in the field. Their combination of commercial formulations of AMF *G. intraradices* and PGPF *T. harzianum* proved to be efficient in controlling Fusarium crown and root rot of tomato in Florida. Recently, more field studies are being conducted using mixtures of two or more beneficial microorganisms like the AMF-PGPR or AMF-PGPF or AMF-PGPR-PGPF combinations. In the rain-fed wheat fields in India for instance, Roesti *et al.* (2006) using different *Pseudomonad* strains and AMF consortia, found that the AMF and PGPR synergistically improve grain quality of wheat. They also found that the rhizobacterial community structure in a wheat field is dependent on the fertilizer management and plant's growth stage. Tahmatsidou *et al.* (2006) in their study with Verticillium wilt of strawberry in raised beds in the field found that dual inoculation of AMF and PGPR inoculants did not give greater protection than single inoculation. The AMF and PGPR may have additive beneficial effects but was not confirmed in their study. In a corporate banana plantation setting, Phirke *et al.* (2008) demonstrated that replacing 50% of chemical fertilizers with a consortium of biofertilizers (*Azotobacter* and *Aspergillus*) together with the native AMF population in the soil in the matrix of soil conditioner enabled proper nourishment and protection of bananas in an IPNM (Integrated Pest and Nutrient Management) treatment as compared to the traditional way. Their study revealed that exploitation of PGPR through judiciously designed IPNM system could relatively increase banana productivity with the least occurrence of fusarial wilt and negligible evidence of Sigatoka, save 50% of chemical fertilizers and permitted control over soil fertility in producer's favor over traditional cultivation practices.

CURRENT APPLICATION OF AMF AND PGPR OR PGPF

At present, researchers are trying to utilize beneficial effects of AMF-PGPR, AMF-PGPF, or AMF-PGPR-PGPF combinations on plant growth promotion or disease control not only under normal environmental conditions but under stress conditions of salinity, drought or metal toxicity (for PGPR role in stress agriculture, please refer to Saleem *et al.*, 2007). Marulanda *et al.* (2006) showed that a drought-adapted *B. thuringiensis* and an indigenous AMF isolate could reduce by 42% the water required to produce 1mg of shoot biomass in *R. sphaerocarpa*, a positive AMF-PGPR effect on plants grown under drought environments. Alguacil *et al.* (2009) found differential effects of *Pseudomonas mendocina* and *G. intraradices* on lettuce plants under elevated atmospheric CO_2 and drought conditions. Vivas *et al.* (2003a) demonstrated that a PGPR *Bacillus* sp. and two species of AMF could improve plant and fungal parameters under polyethylene glycol-induced drought stress. In saline soil, root colonization by ACC deaminase-producing bacteria and AMF can improve plant tolerance to such stressful condition (Gamalero *et al.*, 2010). An indigenous cadmium (Cd)-tolerant AMF *G. mosseae* and Cd-adapted strain of rhizobacteria *Brevibacillus* sp. were found to improve plant tolerance to Cd contamination (Vivas *et al.*, 2003b; 2003c). An improvement of growth of *Eucalyptus globulus* grown in soil amended with sewage sludge contaminated with heavy metals and inoculated with some AMF and saprotrophic fungi was observed (Arriagada *et al.*, 2009).

Moreover, newly emerging areas of PGPR or PGPF usage such as forest regeneration, endangered plant species conservation, phytoremediation of contaminated soils, and agrowaste detoxification are gaining momentum. An AMF and rhizobacteria, singly or in combination, were inoculated to three endangered plant species and they found that the plants were dependent on AMF for their growth. Therefore, AMF could be used to improve *ex situ* conservation projects (Zubek *et al.*, 2009). Aranda *et al.* (2009) found that the phytotoxicity of an agrowaste (olive mill residue) could be reduced by incubation with saprotrophic fungi and the application of the AMF *G. deserticola* in the agrowaste-saprotrophic fungi mix could further enhance the beneficial effect of the organic olive mill residue on growth of tomato plants.

Conclusion and Implications to Biological Control of Plant Diseases

In recent years, biological control of plant pathogens is accepted as a major practice in agriculture which seeks to veer away from the environmental and health contamination brought about by persistent pesticide and chemical fertilizer usage. Biological control is regarded as a directed and accurate management of common ecosystems components, like certain antagonistic rhizosphere organisms, to protect plants against pathogens (Azcón-Aguilar and Barea, 1996). Since interactions of biological control agents with non-target beneficial organisms in the rhizosphere might stimulate or inhibit the beneficial effects of both introduced and non-target organisms in an agro ecosystem (Calvet *et al.*, 1993; Paulitz and Linderman, 1991; Brimmer and Boland, 2003), it is very crucial to study interactions of biological control agents with non-target beneficial organisms in the soil to ensure the successful development, commercialization and usage of biological control strategies. Also, mixtures of two or more beneficial soil microorganisms may contribute an additive effect to the biocontrol more than the effectiveness of its individual components. Still, even with the advent of more advance techniques nowadays, our knowledge about interactions between beneficial soil microorganisms like the AMF and biological control agents is still very limited (Paulitz and Linderman, 1991).

Beneficial soil microorganisms like the AMF, PGPR and PGPF have attracted much attention in their role in reducing plant diseases. Although the full potential has not been reached yet, the work to date is very promising and offers organic growers effective control of some serious plant diseases. Aside from the biocontrol potential, these microorganisms can also help enhance plant growth and hence, increase productivity. The large scale or commercial application of AMF, PGPR or PGPF individually or in combination to crops as bioinoculants is very attractive as it will substantially reduce the use of hazardous chemical fertilizers and pesticides.

References

Abdel-Fattah, G.M., and Mohamedin, A.H. (2000). Interactions between a vesicular-arbuscular mycorrhizal fungus (*Glomus intraradices*) and *Streptomyces coelicolor* and their effects on sorghum plants grown in soil amended with chitin of brawn scales. *Biol Fertil Soils, 32,* 401-409.

Adesemoye, A.O., and Kloepper, J.W. (2009). Plant–microbes interactions in enhanced fertilizer-use efficiency. *Appl Microbiol Biotechnol, 85,* 1-12.

Akhtar, M.S., and Siddiqui, Z.A. (2008a). *Glomus intraradices*, *Pseudomonas alcaligenes*, and *Bacillus pumilus*: effective agents for the control of root-rot disease complex of chickpea (*Cicer arietinum* L.). *J Gen Plant Pathol, 74*, 53-60.

Akhtar, M.S., and Siddiqui, Z.A. (2008b). Biocontrol of a root-rot disease complex of chickpea by *Glomus intraradices*, *Rhizobium* sp. and *Pseudomonas straita*. *Crop Prot, 27,* 410-417.

Alejo-Iturvide, F., Márquez-Lucio, M.A., Morales-Ramírez, I., Vázquez-Garcidueñas, M.S., and Olalde-Portugal, V. (2008). Mycorrhizal protection of chili plants challenged with *Phytophthora capsici*. *Eur J Plant Pathol, 120*, 13-20.

Alguacil, M.M., Kohler, J., Caravaca, F., and Roldán, A. (2009). Differential effects of *Pseudomonas mendocina* and *Glomus intraradices* on lettuce plants physiological response and aquaporin PIP2 gene expression under elevated atmospheric CO_2 and drought. *Microb Ecol, 58*, 942-951.

Ames, R.N., Reid, C.P.P., and Ingham, E.R. (1984). Rhizosphere bacterial population responses to root colonization by a vesicular-arbuscular mycorrhizal fungus. *New Phytol, 96,* 555-563.

Aranda, E., Sampedro, I., Díaz, R., García-Sánchez, M., Arriagada, C.A., Ocampo, J.A., and García-Romera, I. (2009). The effects of the arbuscular mycorrhizal fungus *Glomus deserticola* on growth of tomato plants grown in the presence of olive mill residues modified by treatment with saprophytic fungi. *Symbiosis, 47*, 133-140.

Arriagada, C., Sampedro, I., Garcia-Romera, I., and Ocampo, J. (2009). Improvement of growth of *Eucalyptus globulus* and soil biological parameters by amendment with sewage sludge and inoculation with arbuscular mycorrhizal and saprobe fungi. *Science Total Environ, 407*, 4799-4806.

Artursson, V., Finlay, R.D., and Jansson, J.K. (2006). Interactions between arbuscular mycorrhizal fungi and bacteria and their potential for stimulating plant growth. *Environ Microbiol, 8*, 1-10.

Azcón-Aguilar, C., and Barea, J.M. (1997). Applying mycorrhiza biotechnology to horticulture: significance and potentials. *Scientia Horticulturae, 68*, 1-24.

Azcón-Aguilar, C., and Barea, J.M. (1996). Arbuscular mycorrhizas and biological control of soil-borne plant pathogens – an overview of the mechanisms involved. *Mycorrhiza, 6*, 457-464.

Bagyaraj, D.J. (1984). *Biological interactions with VA mycorrhizal fungi*. In C.L. Powell, and D. J. Bagyaraj (Eds.), *VA mycorrhiza* (pp. 132-153). Boca Raton, Florida: CRC Press.

Bagyaraj, D.J., and Menge, J.A. (1978). Interaction between a VA mycorrhiza and *Azotobacter* and their effects on rhizosphere microflora and plant growth. *New Phytol, 80*, 567-573.

Baker, R. (1990). An overview of current and future strategies and models for biological control. In D. Hornby, R.J. Cook, Y. Henis, W.H. Ko, A.D. Rovira, B. Schippers, and P.R. Scott (Eds.), *Biological control of soil-borne plant pathogens* (pp. 375-388). Wallingford: CAB International.

Barea, J.M., and Jeffries, P. (1995). *Arbuscular mycorrhizas in sustainable soil-plant systems*. In H. Varma, and B. Hock (Eds.), *Mycorrhiza* (pp. 341-408). Berlin: Springer-Verlag.

Barea, J.M., Pozo, M.J., Azcón, R., and Azcón-Aguilar, C. (2005). Microbial co-operation in the rhizosphere. *J Exp Bot, 56,* 1761-1778.

Basu, M.J., and Santhaguru, K. (2009). Impact of *Glomus fasciculatum* and fluorescent pseudomonads on growth performance of *Vigna radiata* (L.) Wilczek challenged with phytopathogens. *J Plant Prot Res, 49,* 190-194.

Berg, G. (2009). Plant-microbe interactions promoting plant growth and health: perspectives for controlled use of microorganisms in agriculture. *Appl Microbiol Biotechnol, 84,* 11-18.

Biró, B., Köves-Péchy, K., Vörös, I., Takács, T., Eggenberger, P., and Strasser, R.J. (2000). Interrelations between *Azospirillum* and *Rhizobium* nitrogen-fixers and arbuscular mycorrhizal fungi in the rhizosphere of alfalfa in sterile, AMF-free or normal soil conditions. *Appl Soil Ecol, 15*, 159-168.

Bloemberg, G.V., and Lugtenberg, B.J.J. (2001). Molecular basis of plant growth promotion and biocontrol by rhizobacteria. *Curr Opinion Plant Biol, 4,* 343-350.

Boby, V.U., Balakrishna, A.N., and Bagyaraj, D.J. (2008). Interaction between *Glomus mosseae* and soil yeasts on growth and nutrition of cowpea. *Microbiol Res, 163*, 693-700.

Bødker, L., Kjøller, R., Kristensen, K., and Rosendahl, S. (2002). Interactions between indigenous arbuscular mycorrhizal fungi and *Aphanomyces euteiches* in field-grown pea. *Mycorrhiza, 12*, 7-12.

Bonfante, P., and Anca, I.A. (2009). Plants, mycorrhizal fungi, and bacteria: a network of interactions. *Annu Rev Microbiol, 63*, 363-383.

Brimmer, T.A., and Boland, G.J. (2003). A review of the non-target effects of fungi used to biologically control plant diseases. *Agric Ecosyst Environ, 100*, 3-16.

Budi, S.W., van Tuinen, D., Martinotti, G., and Gianinazzi, S. (1999). Isolation from the *Sorghum bicolor* mycorrhizosphere of a bacterium compatible with arbuscular mycorrhiza development and antagonistic towards soil borne fungal pathogens. *Appl Environ Microbiol, 65*, 5148-5150.

Calvet, C., Barea, J.M., and Pera, J. (1992). *In vitro* interactions between the vesicular-arbuscular mycorrhizal fungus *Glomus mosseae* and some saprophytic fungi isolated from organic substrates. *Soil Biol Biochem, 24*, 775-780.

Calvet, C., Pera, J., and Barea, J.M. (1989). Interactions of *Trichoderma* spp. with *Glomus mosseae* and two wilt pathogenic fungi. *Agric Ecosys Environ, 29*, 59-65.

Calvet, C., Pera, J., and Barea, J.M. (1993). Growth response of marigold (*Tagetes erecta* L.) to inoculation with *Glomus mosseae*, *Trichoderma aureoviride* and *Pythium ultimum* in a peat-perlite mixture. *Plant Soil, 148*, 1-6.

Caron, M. (1989). Potential use of mycorrhizae in control of soil-borne diseases. *Can J Plant Pathol, 11*, 177-179.

Caron, M., Fortin, J.A., and Richard, C. (1986). Effect of inoculation sequence on the interaction between *Glomus intraradices* and *Fusarium oxysporum* f. sp. *radicis-lycopersici* in tomatoes. *Can J Plant Pathol, 8,* 12-16.

Chandanie, W.A., Kubota, M., and Hyakumachi, M. (2005). Interaction between arbuscular mycorrhizal fungus *Glomus mosseae* and plant growth promoting fungus *Phoma* sp. on their root colonization and growth promotion of cucumber. *Mycoscience, 46*, 201-204.

Chandanie, W.A., Kubota, M., and Hyakumachi, M. (2006). Interactions between plant growth promoting fungi and arbuscular mycorrhizal fungus *Glomus mosseae* and induction of systemic resistance to anthracnose disease in cucumber. *Plant Soil, 286,* 209-217.

Chandanie, W.A., Kubota, M., and Hyakumachi, M. (2009). Interactions between the arbuscular mycorrhizal fungus *Glomus mosseae* and plant growth-promoting fungi and their significance for enhancing plant growth and suppressing damping-off of cucumber (*Cucumis sativus* L.). *Appl Soil Ecol, 41*, 336-341.

Compant, S., Duffy, B., Nowak, J., Clément, C., and Barka, E.A. (2005). Use of plant growth-promoting bacteria for biocontrol of plant diseases: principles, mechanisms of action, and future prospects. *Appl Environ Microbiol, 71,* 4951-4959.

Constantino, M., Gómez-Álvarez, R., Álvarez-Solís, J.D., Geissen, V., Huerta, E., and Barba, E. (2008). Effect of inoculation with rhizobacteria and arbuscular mycorrhizal fungi on growth and yield of *Capsicum chinense* Jacquin. *J Agric Rural Dev Tropics Subtropics, 109*, 169-180.

Contreras-Cornejo, H.A., Macías-Rodríguez, L., Cortés-Penagos, C., and López-Bucio, J. (2009). *Trichoderma virens*, a plant beneficial fungus, enhances biomass production and promotes lateral root growth through an auxin-dependent mechanism in *Arabidopsis*. *Plant Physiol, 149*, 1579-1592.

Dandurand, L.M., and Knudsen, G.R. (1993). Influence of *Pseudomonas fluorescens* on hyphal growth and biocontrol activity of *Trichoderma harzianum* in the spermosphere and rhizosphere of pea. *Phytopathology, 83*, 265-270.

Datnoff, L.E., Nemec, S., and Pernezny, K. (1995). Biological control of Fusarium crown and root rot of tomato in Florida using *Trichoderma harzianum* and *Glomus intraradices*. *Biological Control, 5,* 427-431.

Dehne, H.W. (1982). Interaction between vesicular-arbuscular mycorrhizal fungi and plant pathogens. *Pytopathology, 72*, 1115-1119.

de la Peña, E., Rodríguez-Echeverría, S., van der Putten, W.H., Freitas, H., and Moens, M. (2006). Mechanism of control of root-feeding nematodes by mycorrhizal fungi in the dune grass *Ammophila arenaria*. *New Phytol, 169*, 829-840.

Diedhiou, P.M., Hallmann, J., Oerke, E.C., and Dehne, H.W. (2003). Effects of arbuscular mycorrhizal fungi and a non-pathogenic *Fusarium oxysporum* on *Meloidogyne incognita* infestation of tomato. *Mycorrhiza, 13*, 199-204.

Duffy, B., Keel, C., and Défago, G. (2004). Potential role of pathogen signaling in multitrophic plant-microbe interactions involved in disease protection. *Appl Environ Microbiol, 70*, 1836-1842.

Dwivedi, D., Johri, B.N., Ineichen, K., Wray, V., and Wiemken, A. (2009). Impact of antifungals producing rhizobacteria on the performance of *Vigna radiata* in the presence of arbuscular mycorrhizal fungi. *Mycorrhiza, 19*, 559- 570.

Edwards, S.G., Young, J.P.W., and Fitter, A.H. (1998). Interactions between *Pseudomonas flourescens* biocontrol agents and *Glomus mosseae*, an arbuscular mycorrhizal fungus, within the rhizosphere. *FEMS Microbiol Lett, 166*, 297-303.

El-Tarabily, K.A., and Sivasithamparam, K. (2006). Non-streptomycete actinomycetes as biocontrol agents of soil-borne fungal plant pathogens and as plant growth promoters. *Soil Biol Biochem, 38*, 1505-1520.

Filion, M., St-Arnaud, M., and Fortin, J.A. (1999). Direct interaction between the arbuscular mycorrhizal fungus *Glomus intraradices* and different rhizosphere microorganisms. *New Phytol, 141*, 525-533.

Fitter, A.H., and Garbaye, J. (1994). Interactions between mycorrhizal fungi and other soil microorganisms. *Plant Soil, 159*, 123-132.

Fracchia, S., Garcia-Romera, I., Godeas, A., and Ocampo, J.A. (2000). Effect of the saprophytic fungus *Fusarium oxysporum* on arbuscular mycorrhizal colonization and growth of plants in greenhouse and field trials. *Plant Soil, 223*, 175-184.

Fracchia, S., Godeas, A., Scervino, J.M., Sampedro, I., Ocampo, J.A., and García-Romera, I. (2003). Interaction between the soil yeast *Rhodotorula mucilaginosa* and the arbuscular mycorrhizal fungi *Glomus mosseae* and *Gigaspora rosea*. *Soil Biol Biochem, 35*, 701-707.

Gamalero, E., Trotta, A., Massa, N., Copetta, A., Martinotti, M.G., and Berta, G. (2004). Impact of two fluorescent pseudomonads and an arbuscular mycorrhizal fungus on tomato plant growth, root architecture and P acquisition. *Mycorrhiza, 14*, 185-192.

Gamalero, E., Berta, G., Massa, N., Glick, B.R., and Lingua, G. (2008). Synergistic interactions between the ACC deaminase-producing bacterium *Pseudomonas putida* UW4 and the AM fungus *Gigaspora rosea* positively affect cucumber plant growth. *FEMS Microbiol Ecol, 64*, 459-467.

Gamalero, E., Berta, G., Massa, N., Glick, B.R., and Lingua, G. (2010). Interactions between *Pseudomonas putida* UW4 and *Gigaspora rosea* BEG9 and their consequences for the growth of cucumber under salt-stress conditions. *J Appl Microbiol, 108*, 236-245.

Garbaye, J. (1994). Helper bacteria: a new dimension to the mycorrhizal symbiosis. *New Phytol, 128*, 197-210.

Garcia-Garrido, J.M., and Ocampo, J.A. (1988). Interaction between *Glomus mosseae* and *Erwinia carotovora* and its effect on the growth of tomato plants. *New Phytol, 110*, 551-555.

Garcia-Romera, I., Garcia-Garrido, J.M., Martin, J., Fracchia, S., Mujica, M.T., Godeas, A., and Ocampo, J.A. (1998). Interaction between saprophytic *Fusarium* strains and arbuscular mycorrhizas of soybean plants. *Symbiosis, 24*, 235-246.

Garmendia, I., Aguirreolea, J., and Goicoechea, N. (2006). Defence-related enzymes in pepper roots during interactions with arbuscular mycorrhizal fungi and/or *Verticillium dahliae*. *Biocontrol, 51*, 293-310.

Gernns, H., von Alten, H., and Poehling, H.M. (2001). Arbuscular mycorrhiza increased the activity of a biotrophic leaf pathogen – is a compensation possible? *Mycorrhiza, 11*, 237-243.

Green, H., Larsen, J., Olsson, P.A., Jensen, D.F., and Jakobsen, I. (1999). Suppression of the biocontrol agent *Trichoderma harzianum* by mycelium of the arbuscular mycorrhizal fungus *Glomus intraradices* in root-free soil. *Appl Environ Microbiol, 65*, 1428-1434.

Gryndler, M. (2000). Interactions of arbuscular mycorrhizal fungi with other soil organisms. In Y. Kapulnik, and D.D. Douds Jr (Eds.), *Arbuscular mycorrhizas: Physiology and function* (pp. 239-262). The Netherlands: Kluwer Academic Publishers.

Gualtiere, G., and Bisseling, T. (2000). The evolution of nodulation. *Plant Mol Biol, 42*, 181-194.

Guetsky, R., Shtienberg, D., Elad, Y., and Dinoor, A. (2001). Combining biocontrol agents to reduce the variability of biological control. *Phytopathology, 91*, 621-627.

Haggag, W.M., and Abd-El Latif, F.M. 2001. Interaction between vesicular arbuscular mycorrhizae and antagonistic biocontrol microorganisms on controlling root rot disease incidence of geranium plants. *OnLine J Biol Sci, 1*, 1147-1153.

Hansen, A.P. (1994). *Symbiotic N_2 fixation of crop legumes*. Weikershein, Germany: Margraf Verlag.

Harman, G.E., Howell, C.R., Viterbo, A., Chet, I., and Lorito, M. (2004). *Trichoderma* species – opportunistic avirulent plant symbionts. *Nature Reviews, 2*, 43-56.

Hooker, J.E., Jaizme-Vega, M., and Atkinson, D. (1994). Biocontrol of plant pathogens using arbuscular mycorrhizal fungi. In S. Gianinazzi, and H. Schuepp (Eds.), *Impact of arbuscular mycorrhizas on sustainable agriculture and natural ecosystems* (pp. 191-200). Basel: Birkhauser.

Horinouchi, H., Muslim, A., Suzuki, T., and Hyakumachi, M. (2007). *Fusarium equiseti* GF191 as an effective biocontrol agent against Fusarium crown and root rot of tomato in rock wool systems. *Crop Prot, 26*, 1514-1523.

Hossain, M.M., Sultana, F., Kubota, M., Koyama, H., and Hyakumachi, M. (2007). The plant growth-promoting fungus *Penicillium simplicissimum* GP17-2 induces resistance in *Arabidopsis thaliana* by activation of multiple defense signals. *Plant Cell Physiol, 48*, 1724-1736.

Hubbard, J.P., Harman, G.E., and Hadar, Y. (1983). Effect of soilborne *Pseudomonas* spp. on the biological control agent, *Trichoderma hamatum*, on pea seeds. *Phytopathology, 78*, 655-659.

Hyakumachi, M. (1994). Plant growth promoting fungi from turfgrass rhizosphere with potential for disease suppression. *Soil Microorganisms, 44*, 53-68.

Hyakumachi, M., and Kubota, M. (2004a). Biological control of plant diseases by plant growth promoting fungi. *Proc Int Sem Biological Cont Soilborne Plant Dis* (pp. 87-123). Japan-Argentina Joint Study.

Hyakumachi, M., and Kubota, M. (2004b). Fungi as plant growth promoter and disease suppressor. In D.K. Arora (Ed.), *Fungal biotechnology in agricultural, food, and environmental applications* (vol. 21, pp. 101-110). New York: Marcel Dekker, Inc.

Jalali, B.L., and Jalali, I. (1991). Mycorrhiza in plant disease control. In D.K. Arora, B. Rai, K.G. Mukerji, and G.R. Knudsen (Eds.), *Handbook of applied mycology: Soil and plants* (vol. 1, pp. 131-154). New York: Marcel Dekker, Inc.

Jäderlund, L., Arthurson, V., Granhall, U., and Jansson, J.K. (2008). Specific interactions between arbuscular mycorrhizal fungi and plant growth-promoting bacteria: as revealed by different combinations. *FEMS Microbiol Lett, 287*, 174-180.

Jeffries, P., and Barea, J.M. (2001). Arbuscular mycorrhiza: a key component of sustainable plant–soil ecosystems. In B. Hock (Ed.), *The mycota: Fungal associations* (vol. 9, pp. 95-113). Berlin: Springer.

Jeffries, P., Gianinazzi, S., Perotto, S., Turnau, K., and Barea, J.M. (2003). The contribution of arbuscular mycorrhizal fungi in sustainable maintenance of plant health and soil fertility. *Biol Fertil Soils, 37*, 1-16.

Johansson, J.F., Paul, L.R., and Finlay, R.D. (2004). Microbial interactions in the mycorrhizosphere and their significance for sustainable agriculture. *FEMS Microbiol Ecol, 48*, 1-13.

Karagiannidis, N., Bletsos, F., and Stavropoulos, N. (2002). Effect of Verticillium wilt (*Verticillium dahliae* Kleb.) and mycorrhiza *(Glomus mosseae*) on root colonization, growth and nutrient uptake in tomato and eggplant seedlings. *Scientia Horticulturae, 94*, 145-156.

Kaye, J.W., Pfleger, F.L., and Stewart, E.L. (1984). Interaction of *Glomus fasciculatum* and *Pythium ultimum* on greenhouse-grown poinsettia. *Can J Bot, 62*, 1575-1579.

Kennedy, I.R., Choudhury, A.T.M.A., and Kecskés, M.L. (2004). Non-symbiotic bacterial diazotrophs in crop-farming systems: can their potential for plant growth promotion be better exploited? *Soil Biol Biochem, 36*, 1229-1244.

Kloepper, J.W., Leong, J., Tientze, M., and Schroth, M.N. (1980). Enhanced plant growth by siderophores produced by plant growth-promoting rhizobacteria. *Nature, 286*, 885-886.

Kloepper, J.W., Ryu, C.M., and Zhang, S. (2004). Induced systemic resistance and promotion of plant growth by *Bacillus* spp. *Phytopathology, 94*, 1259-1266.

Kloepper, J.W., and Schroth, M.N. (1978). Plant growth promoting bacteria on radishes. *Proc IV Int Conf Plant Pathogenic Bacteria* (vol. 2, pp. 879-882). Angers, France.

Kohler, J., Caravaca, F., Carrasco, L., and Roldán, A. (2007). Interactions between a plant growth-promoting rhizobacterium, an AM fungus and a phosphate-solubilising fungus in the rhizosphere of *Lactuca sativa*. *Appl Soil Ecol, 35*, 480-487.

Koike, N., Hyakumachi, M., Kageyama, K., Tsuyumu, S., and Doke, N. (2001). Induction of systemic resistance in cucumber against several diseases by plant growth promoting fungi: lignification and superoxide generation. *Eur J Plant Pathol, 107*, 523-533.

Krishna, K.R., and Bagyaraj, D.J. (1983). Interaction between *Glomus fasciculatum* and *Sclerotium rolfsii* in peanut. *Can J Bot, 61*, 2349-2351.

Linderman, R.G. (1988). Mycorrhizal interactions with the rhizosphere microflora: the mycorrhizosphere effect. *Phytopathology, 78*, 366-371.

Linderman, R.G. (1992). Vesicular-arbuscular mycorrhizae and soil microbial interactions. In G.J. Bethlenfalvay, and R.G. Linderman (Eds.), *Mycorrhizae in sustainable agriculture* (ASA Special Publication No. 54, pp. 45-70). Madison, Wi.

Linderman, R.G. (1994). Role of VAM fungi in biocontrol. In F.L. Pfleger, and R.G. Linderman (Eds.), *Mycorrhizae and plant health* (pp. 1-26). St Paul: APS.

Linderman, R.G., and Paulitz, T.C. (1990). Mycorrhizal-rhizobacterial interactions. In D. Hornby, R.J. Cook, Y. Henis, W.H. Ko, A.D. Rovira, B. Schippers, and P.R. Scott (Eds.), *Biological control of soil-borne plant pathogens* (pp. 261-283). Wallingford: CAB International.

Lucy, M., Reed, E., and Glick, B.R. (2004). Applications of free living plant growth-promoting rhizobacteria. *Antonie van Leeuwenhoek, 86*, 1-25.

Lugtenberg, B.J.J., Chin-A-Woeng, T.F.C., and Bloemberg, G.V. (2002). Microbe–plant interactions: principles and mechanisms. *Antonie van Leeuwenhoek, 81*, 373-383.

Lugtenberg, B.J.J., de Weger, L.A., and Bennett, J.W. (1991). Microbial stimulation of plant growth and protection from disease. *Curr Opinion Biotechnol, 2*, 457-464.

MarVázquez, M., César, S., Azcón, R., and Barea, J.M. (2000). Interactions between arbuscular mycorrhizal fungi and other microbial inoculants (*Azospirillum*, *Pseudomonas*, *Trichoderma*) and their effects on microbial population and enzyme activities in the rhizosphere of maize plants. *Appl Soil Ecol, 15*, 261-272.

Martinez, A., Obertello, M., Pardo, A., Ocampo, J.A., and Godeas, A. (2004). Interactions between *Trichoderma pseudokoningii* strains and the arbuscular mycorrhizal fungi *Glomus mosseae* and *Gigaspora rosea*. *Mycorrhiza, 14*, 79-84.

Martínez-Medina, A., Pascual, J.A., Pérez-Alfocea, F., Albacete, A., and Roldán, A. (2010). *Trichoderma harzianum* and *Glomus intraradices* modify the hormone disruption induced by *Fusarium oxysporum* infection in melon plants. *Phytopathology, 100*, 682-688.

Marulanda, A., Barea, J.M., and Azcón, R. (2006). An indigenous drought-tolerant strain of *Glomus intraradices* associated with a native bacterium improves water transport and root development in *Retama sphaerocarpa*. *Microb Ecol, 52*, 670-678.

Marulanda-Aguirre, A., Azcón, R., Ruiz-Lozano, J.M., and Aroca, R. (2008). Differential effects of a *Bacillus megaterium* strain on *Lactuca sativa* plant growth depending on the origin of the arbuscular mycorrhizal fungus coinoculated: physiologic and biochemical traits. *J Plant Growth Regul, 27*, 10-18.

McAllister, C.B., Garcia-Garrido, J.M., García-Romera, I., Godeas, A., and Ocampo, J.A. (1997). Interaction between *Alternaria alternata* or *Fusarium equiseti* and *Glomus mosseae* and its effects on plant growth. *Biol Fertil Soils*, *24*, 301-305.

McAllister, C.B., Garcia-Garrido, J.M., García-Romera, I., Godeas, A., and Ocampo, J.A. (1996). *In vitro* interactions between *Alternaria alternata*, *Fusarium equiseti* and *Glomus mosseae*. *Symbiosis*, *20*, 163-174.

McAllister, C.B., García-Romera, I., Godeas, A., and Ocampo, J.A. (1994). Interactions between *Trichoderma koningii*, *Fusarium solani* and *Glomus mosseae*: effects on plant growth, arbuscular mycorrhizas and the saprophyte inoculants. *Soil Biol Biochem*, *26*, 1363-1367.

Medina, A., Probanza, A., Gutierrez-Mañero, F.J., and Azcón, R. (2003). Interactions of arbuscular-mycorrhizal fungi and *Bacillus* strains and their effects on plant growth, microbial rhizosphere activity (thymidine and leucine incorporation) and fungal biomass (ergosterol and chitin). *Appl Soil Ecol*, *22*, 15-28.

Meera, M.S., Shivanna, M.B., Kageyama, K., and Hyakumachi, M. (1994). Plant growth promoting fungi from zoysiagrass rhizosphere as potential inducers of systemic resistance in cucumbers. *Phytopathology*, *84*, 1399-1406.

Meera, M.S., Shivanna, M.B., Kageyama, K., and Hyakumachi, M. (1995). Responses of cucumber cultivars to induction of systemic resistance against anthracnose by plant growth promoting fungi. *Eur J Plant Pathol*, *101*, 421-430.

Meyer, J.R., and Linderman, R.G. (1986). Response of subterranean clover to dual inoculation with vesicular-arbuscular mycorrhizal fungi and a plant growth-promoting bacterium, *Pseudomonas putida*. *Soil Biol Biochem*, *18*, 185-190.

Morton, J.B., and Benny, G.L. (1990). Revised classification of arbuscular mycorrhizal fungi (Zygomycetes): a new order, Glomales, two new suborders, Glomineae and Gigasporineae, and two new families, Acaulosporaceae and Gigasporaceae, with an emendation of Glomaceae. *Mycotaxon*, *37*, 471- 491.

Nelsen, C.E. (1987). The water relations of vesicular-arbuscular mycorrhizal systems. In G. Safir (Ed.), *Ecophysiology of VA mycorrhizal plants* (pp. 71-91). Boca Raton, Florida: CRC Press.

Nemec, S., Datnoff, L.E., and Strandberg, J. (1996). Efficacy of biocontrol agents in planting mixes to colonize plant roots and control root diseases of vegetables and citrus. *Crop Prot*, *15*, 735-742.

Paulitz, T.C., and Linderman, R.G. (1989). Interactions between fluorescent pseudomonads and VA mycorrhizal fungi. *New Phytol*, *113*, 37-45.

Paulitz, T.C., and Linderman, R.G. (1991). Lack of antagonism between the biocontrol agent *Gliocladium virens* and vesicular arbuscular mycorrhizal fungi. *New Phytol*, *117*, 303-308.

Phirke, N.V., Kothari, R.M., and Chincholkar, S.B. (2008). Rhizobacteria in mycorrhizosphere improved plant health and yield of banana by offering proper nourishment and protection against diseases. *Appl Biochem Biotechnol*, *151*, 441-451.

Pieterse, C.M.J., van Pelt, J.A., Verhagen, B.W.M., Ton, J., van Wees, S.C.M., Léon-Kloosterziel, K.M., and van Loon, L.C. (2003). Induced systemic resistance by plant growth promoting rhizobacteria. *Symbiosis*, *35*, 39-54.

Pivato, B., Offre, P., Marchelli, S., Barbonaglia, B., Mougel, C., Lemanceau, P., and Berta, G. (2009). Bacterial effects on arbuscular mycorrhizal fungi and mycorrhiza development as influenced by the bacteria, fungi, and host plant. *Mycorrhiza*, *19*, 81-80.

Pozo, M.J., Azcón-Aguilar, C., Dumas-Gaudot, E., and Barea, J.M. (1998). Chitosanase and chitinase activities in tomato roots during interactions with arbuscular mycorrhizal fungi or *Phytophthora parasitica*. *J Expt Bot*, *49*, 1729-1739.

Pozo, M.J., Azcón-Aguilar, C., Dumas-Gaudot, E., and Barea, J.M. (1999). β-1,3-glucanase activities in tomato roots inoculated with arbuscular mycorrhizal fungi and/or *Phytophthora parasitica* and their possible involvement in bioprotection. *Plant Sci*, *141*, 149-157.

Pozo, M.J., and Azcón-Aguilar, C. (2007). Unravelling mycorrhiza-induced resistance. *Curr Opinion Plant Biol*, *10*, 393-398.

Pozo, M.J., Cordier, C., Dumas-Gaudot, E., Gianinazzi, S., Barea, J.M., and Azcón-Aguilar, C. (2002). Localized versus systemic effect of arbuscular mycorrhizal fungi on defence responses to *Phytophthora* infection in tomato plants. *J Exp Bot*, *53*, 525-534.

Ram, C.S.V. (1959). Production of growth-promoting substances by *Fusarium* and their action on root elongation in *Oryzae sativa* L. *Proc Indian Acad Sci*, *49*, 167-182.

Raupach, G.S., and Kloepper, J.W. (1998). Mixtures of plant growth-promoting rhizobacteria enhance biological control of multiple cucumber pathogens. *Phytopathology*, *88*, 1158-1164.

Ravnskov, S., Jensen, B., Knudsen, I.M.B., Bødker, L., Jensen, D.F., Karliński, L., and Larsen, J. (2006). Soil inoculation with the biocontrol agent *Clonostachys rosea* and the mycorrhizal fungus *Glomus intraradices* results in mutual inhibition, plant growth promotion and alteration of soil microbial communities. *Soil Biol Biochem*, *38*, 3453-3462.

Roberts, D.P., Lohrke, S.M., Meyer, S.L.F., Buyer, J.S., Bowers, J.H., Baker, C.J., Li, W., de Souza, J.T., Lewis, J.A., and Chung, S. (2005). Biocontrol agents applied individually and in combination for suppression of soilborne diseases of cucumber. *Crop Prot*, *24*, 141-155.

Rodríguez, H., and Fraga, R. (1999). Phosphate solubilizing bacteria and their role in plant growth promotion. *Biotechnol Advances*, *17*, 319-339.

Roesti, D., Gaur, R., Johri, B.N., Imfeld, G., Sharma, S., Kawaljeet, K., and Aragno, M. (2006). Plant growth stage, fertilizer management and bio-inoculation of arbuscular mycorrhizal fungi and plant growth promoting rhizobacteria affect the rhizobacterial community structure in rain-fed wheat fields. *Soil Biol Biochem*, *38*, 1111-1120.

Rosendahl, C.N., and Rosendahl, S. (1990). The role of vesicular-arbuscular mycorrhiza in controlling damping-off and growth reduction in cucumber caused by *Pythium ultimum*. *Symbiosis*, *9*, 363-366.

Rousseau, A., Benhamou, N., Chet, I., and Piché, Y. (1996). Mycoparasitism of the extramatrical phase of *Glomus intraradices* by *Trichoderma harzianum*. *Phytopathology*, *86*, 434-443.

Saldajeno, M.G.B., and Hyakumachi, M. (2010). The plant growth-promoting fungus *Fusarium equiseti* and the arbuscular mycorrhizal fungus *Glomus mosseae* stimulate plant growth and reduce severity of anthracnose and damping-off diseases in cucumber (*Cucumis sativus* L.) seedlings. *Ann Appl Biol* (in press).

Saleem, M., Arshad, M., Hussain, S., and Bhatti, A.S. (2007). Perspective of plant growth promoting rhizobacteria (PGPR) containing ACC deaminase in stress agriculture. *J Ind Microbiol Biotechnol*, *34*, 635-648.

Schrey, S.D., and Tarkka, M.T. (2008). Friends and foes: streptomycetes as modulators of plant disease and symbiosis. *Antonie van Leeuwenhoek*, *94,* 11-19.

Schultze, M., and Kondorosi, A. (1998). Regulation of symbiotic root nodule development. *Annu Rev Gen*, *32*, 33-57.

Sessitsch, A., Howieson, J.G., Perret, X., Antoun, H., and Martinez-Romero, E. (2002). Advances in *Rhizobium* research. *Crit Rev Plant Sci*, *21*, 323-378.

Shivanna, M.B., Meera, M.S., and Hyakumachi, M. (1994). Sterile fungi from zoysiagrass rhizosphere as plant growth promoters in spring wheat. *Can J Microbiol*, *40*, 637-644.

Shivanna, M.B., Meera, M.S., Kageyama, K., and Hyakumachi, M. (1996a). Growth promotion ability of zoysiagrass rhizosphere fungi in consecutive plantings of wheat and soybean. *Mycoscience*, *37*, 163-168.

Shivanna, M.B., Meera, M.S., and Hyakumachi, M. (1996b). Role of root colonization ability of plant growth promoting fungi in the suppression of take-all and common root rot of wheat. *Crop Prot*, *15*, 497-504.

Shivanna, M.B., Meera, M.S., Kageyama, K., and Hyakumachi, M. (1996c). Plant growth promoting fungi induce systemic disease resistance in cucumber. In T. Wenhua, R.J. Cook, and A. Rovira (Eds.), *Advances in biological control of plant diseases* (pp. 175-184). Beijing, China: China Agricultural University Press.

Shivanna, M.B., Meera, M.S., Kubota, M., and Hyakumachi, M. (2005). Promotion of growth and yield in cucumber by zoysiagrass rhizosphere fungi. *Microbes Environ*, *20*, 34-40.

Siddiqui, Z.A., and Akhtar, M.S. (2008a). Synergistic effects of antagonistic fungi and a plant growth promoting rhizobacterium, an arbuscular mycorrhizal fungus, or composted cow manure on populations of *Meloidogyne incognita* and growth of tomato. *Biocontrol Sci Tech*, *18*, 279-290.

Siddiqui, Z.A., and Akhtar, M.S. (2008b). Effects of fertilizers, AM fungus and plant growth promoting rhizobacterium on the growth of tomato and on the reproduction of root-knot nematode *Meloidogyne incognita*. *J Plant Interact*, *3*, 263-271.

Siddiqui, Z.A., and Akhtar, M.S. (2009). Effects of antagonistic fungi, plant growth-promoting rhizobacteria, and arbuscular mycorrhizal fungi alone and in combination on the reproduction of *Meloidogyne incognita* and growth of tomato. *J Gen Plant Pathol*, *75*, 144-153.

Smith, S.E., and Read, D.J. (2008). *Mycorrhizal symbiosis* (3rd ed.). Cambridge, UK: Academic Press.

Tahmatsidou, V., O'Sullivan, J., Cassells, A.C., Voyiatzis, D., and Paroussi, G. (2006). Comparison of AMF and PGPR inoculants for the suppression of *Verticillium* wilt of strawberry (*Fragaria* x *ananassa* cv. Selva). *Appl Soil Ecol*, *32*, 316-324.

Tisdall, J.M. (1994). Possible role of soil microorganisms in aggregation in soils. *Plant Soil*, *159*, 115-121.

Toro, M., Azcón, R., and Barea, J.M. (1997). Improvement of arbuscular mycorrhiza development by inoculation of soil with phosphate-solubilizing rhizobacteria to improve rock phosphate bioavailability (^{32}P) and nutrient cycling. *App Environ Microbiol*, *63*, 4408-4412.

Trotta, A., Varese, G.C., Gnavi, E., Fusconi, A., Sampò, S., and Berta, G. (1996). Interactions between the soilborne root pathogen *Phytophthora nicotianae* var. *parasitica* and the arbuscular mycorrhizal fungus *Glomus mosseae* in tomato plants. *Plant Soil*, *185*, 199-209.

van Loon, L.C. (2007). Plant responses to plant growth promoting rhizobacteria. *Eur J Plant Pathol*, *119*, 243-254.

Vessey, J.K. (2003). Plant growth promoting rhizobacteria as biofertilizers. *Plant Soil*, *255*, 571-586.

Vinale, F., Sivasithamparam, K., Ghisalberti, E.L., Marra, R., Woo, S.L., and Lorito, M. (2008). *Trichoderma*–plant–pathogen interactions. *Soil Biol Biochem*, *40*, 1-10.

Vivas, A., Marulanda, A., Ruiz-Lozano, J.M., Barea, J.M., and Azcón, R. (2003a). Influence of a *Bacillus* sp. on physiological activities of two arbuscular mycorrhizal fungi and on plant responses to PEG-induced drought stress. *Mycorrhiza*, *13*, 249-256.

Vivas, A., Vörös, I., Biró, B., Barea, J.M., Ruiz-Lozano, J.M., and Azcón, R. (2003b). Beneficial effects of indigenous Cd-tolerant and Cd-sensitive *Glomus mosseae* associated with a Cd-adapted strain of *Brevibacillus* sp. in improving plant tolerance to Cd contamination. *App Soil Ecol*, *24*, 177-186.

Vivas, A., Vörös, I., Biró, B., Campos, E., Barea, J.M., and Azcón, R. (2003c). Symbiotic efficiency of autochthonous arbuscular mycorrhizal fungus (*G. mosseae*) and *Brevibacillus* sp. isolated from cadmium polluted soil under increasing cadmium levels. *Environ Pol*, *126*, 179-189.

Walley, F.L., and Germida, J.J. (1997). Response of spring wheat (*Triticum aestivum*) to interactions between *Pseudomonas* species and *Glomus clarum* NT4. *Biol Fertil Soils*, *24*, 365-371.

Wei, G., Kloepper, J.W., and Tuzun, S. (1991). Induction of systemic resistance of cucumber to *Colletotrichum orbiculare* by select strains of plant growth-promoting rhizobacteria. *Phytopathology*, *81*, 1508-1512.

Whipps, J.M. (2001). Microbial interactions and biocontrol in the rhizosphere. *J Exp Bot*, *52*, 487-511.

Xavier, L.J.C., and Boyetchko, S.M. (2002). Mycorrhizae as biocontrol agents. In K.G. Mukerji, C. Manoharachary, and B.P. Chamola (Eds.), *Techniques in mycorrhizal studies* (pp. 493 - 536). Dordrecht, The Netherlands: Kluwer Academic Publishers.

Xiao, X., Chen, H., Chen, H., Wang, J., Ren, C., and Wu, L. (2008). Impact of *Bacillus subtilis* JA, a biocontrol strain of fungal plant pathogens, on arbuscular mycorrhiza formation in *Zea mays*. *World J Microbiol Biotechnol*, *24*, 1133-1137.

Yedidia, I., Benhamou, N., and Chet, I. (1999). Induction of defense responses in cucumber plants (*Cucumis sativus* L.) by the biocontrol agent *Trichoderma harzianum*. *Appl Environ Microbiol*, *65*, 1061-1070.

Zehnder, G.W., Murphy, J.F., Sikora, E.J., and Kloepper, J.W. (2001). Application to rhizobacteria for induced resistance. *Eur J Plant Pathol*, *107*, 39-50.

Zubek, S.,Turnau, K., Tsimilli-Michael, M., and Strasser, R.J. (2009). Response of endangered plant species to inoculation with arbuscular mycorrhizal fungi and soil bacteria. *Mycorrhiza*, *19*, 113-123.

In: Mycorrhizal Fungi
Editor: Susanne M. Fulton
ISBN 978-1-61122-659-1
© 2011 Nova Science Publishers, Inc.

Chapter 10

ARBUSCULAR MYCORRHIZAL FUNGI IN HEAVY METAL POLLUTED SOILS AND TOLERANCE MECHANISMS

Mehdi Zarei and Jamal Sheikhi
Department of Soil Science, College of Agriculture, University of Shiraz, Shiraz, Iran

ABSTRACT

There are spacious studies concerning nutritional and physiological effects of arbuscular mycorrhizal fungi (AMF) on the plant and microbial communities of natural and agricultural ecosystems, but there are less information about importance and tolerance mechanisms of AMF in the heavy metals (HMs) polluted soils that all limited to two last decades. Totally, potentially toxic metals influence on the abundance, species richness and soil microbial and plant diversity. HM pollution does not affect on all organisms in an equal manner. Many researchers were reported the existence of AMF and colonization of root plants in metals polluted sites. It was described the identified and unidentified species of one or more AMF genus in the studies. *Glomus* was a common genus in the polluted soils. This chapter reviewed and discussed the results of some scientists regarding the presence and diversity of AMF and their role in imparting metal tolerance to plants in the HMs polluted soils.

Keywords: Arbuscular mycorrhiza, Abundance, Diversity, Tolerance and Heavy metals pollution

INTRODUCTION

Heavy metals (HMs) are elements with specific weight higher than 5 g/cm^3 (Holleman and Wiberg, 1985) and have the ability to formation of sulphides (Adriano, 2001). Some metals such as zinc, copper and nickel are essential and useful for plants and animals physiology. They play role in enzyme catalyzed, redox reactions, in electron transfer and metabolism of nucleic acids (Göhre and Paszkowski, 2006), conversely there are not any benefit biological role for some elements such as lead, mercury, arsenic and cadmium (Mertz,1981). HMs belong to non-organic pollutants and they occur in the nature, terrestrial and aquatic ecosystems and even in the atmosphere. Origin of these metals is diverse in the soils and often there are a mixture of metals with different toxic potential (Shalaby, 2003). HMs may be effective as a strong selective factor in the development of compatible microorganisms under stress condition. Among soil microorganisms, arbuscular mycorrhizal fungi (AMF) are known as a major microsymbiont of symbiosis with most plants in the polluted soils. These fungi can tolerate the stress of HMs with different mechanisms. In this chapter, the presence, abundance, diversity and tolerance mechanisms of AMF in the polluted soils were studied and discussed.

Arbuscular Mycorrhizal Fungi in Heavy Metal Polluted Soils

Many researchers have been reported the presence of AMF and roots colonization of plants with these fungi in the HMs polluted soils (Bohn and Liberta, 1982; Diaz and Honrubia, 1994; Pawlowska et al., 1996; Gaur and Adholeya, 2006). There are different identified and non-identified species, belonging to one or more AMF genera. Most common identified species of AMF in the HMs polluted areas are given in Table 1.

Table 1. Identified species of AMF in the polluted areas with heavy metals

Acaulospora bireticulata	*Glomus intraradices*
Acaulospora delicata	*Glomus macrocarpum*
Acaulospora nicolsonii	*Glomus mosseae*
Entrophospora sp.	*Glomus occultum*
Gigaspora gigantea	*Glomus pansihalos*
Glomus aggregatum	*Glomus pubescens*
Glomus albidum	*Glomus rubiforme*
Glomus claroideum	*Glomus tenue*
Glomus constrictum	*Glomus tortusum*
Glomus deserticola	*Scutellospora dipurpurescens*
Glomus fasciculatum	*Scutellopspora hetrogama*
Glomus geosporum	*Scutellospora weresubiae*

Griffioen, 1994; Raman and Sambandan, 1998; Shetty et al., 1995; Gaur and Adholeya, 2004.

Some research about presence, abundance and diversity of AMF in the HMs polluted soils was briefed at below:

Gildon and Tinker (1981) reported that clover was colonized by AMF up to 35 percent had normal growth in the strongly polluted soils around the mine contain zinc and cadmium . Dueck et al. (1986) observed AMF propagules in *Festuca rubra* and *Calamagrostis epigejos* plants that grown in the polluted soils with atmospheric deposits resulted from an explosions Furnace.

Sambandan et al. (1992) identified fifteen AMF species, belonging to *Glomus* (10 species), *Gigaspora* (2 species), *Acaulospora* (2 species) and *Sclerocystis* (1 species) genera in the polluted soils by HMs in India. The spore numbers were 622 per 100 g soil, the percentage of root colonization varied from 22 to 71 percent and among fungal species, *Glomus geosporum* is observed in all areas. *Glomus* and *Gigaspora* genera in symbiosis with fourteen plant species grown in the soils around a magnesium carbonate mine, were identified by Raman et al. (1993). Diaz and Honrubia (1993) showed that plant roots of alfalfa grown in the soils with 236 and 456 mg kg^{-1} DTPA extractable zinc and lead respectively, colonized with arbuscular mycorrhiza and number of AMF spores was less than non-polluted soils. Griffioen (1994) reported the presence of *Scutellospora dipurpurascens* species in the rhizosphere of *Agrostis capillaris* grown around the soils a refinery in West Virginia.

Weissenhorn et al. (1995) measured root colonization in the polluted soils with 1220 and 895 mg cadmium and lead per kg, respectively up to 40 percent. Also, the spore numbers were reported 100 spores per 50 g soil in the agricultural soils near to melting factory of zinc and lead. Leyval et al. (1995) reported density of AMF in the polluted soils with HMs in Norway, between 3 to 46 spores per g soil. *Glomus constrictum* identified in the soils of around zinc mine in the U.S. Kansas (Shetty et al., 1995). Pawlowska et al. (1996) studied AMF symbiosis status in relation to 100 m^2 vegetations of disturbed and undisturbed areas, in the southern Poland. The soils were contained with zinc, lead and cadmium with total concentration of 49000, 4560 and 180 mg kg^{-1}, respectively and concentrations of DTPA extractable of 1270, 323 and 37.4 mg kg^{-1}, respectively. *Glomus aggregatum*, *Glomus constrictum*, *Glomus fasciculatum*, *Glomus pansihalos* spores, an unidentified *Glomus* species and *Entrophospora* with average spore numbers of 25 per 100 grams substrate were observed in the undisturbed region. While in the undisturbed region, *Glomus fasciculatum* spores and an unidentified species of *Entrophospora* with average of 20 spores per 100 grams substrate were found. Ramn and Sambandan (1998) studied arbuscular mycorrhizal symbiosis of 22 plant species in the HMs polluted soils in three regions of India, Tamil Nadu. Arbuscular mycorrhizal organs including of arbuscule and hyphae, were observed within the roots of 19 plant species. Younger roots were contained more arbuscules and mature roots were contained more vesicles and complex hyphae networks. Spores of 15 AMF species belong to four genera *Glomus*, *Scutellospora*, *Scelerocystis* and *Gigaspora* were observed in the rhizospheric soils of this region. Many of AMF species were observed commonly in the rhizospheric soil of plant species. *Glomus fasciculatum*, *Glomus geosporum* and *Gigaspora gigantea* were associated with most plant species, and were introduced as common endophyte in the polluted areas. Del Val et al. (a 1999) evaluated diversity of AMF in the long-term experiments on soils treated with sewage sludge containing of zinc, cadmium, copper, nickel and lead. Six AMF ecotypes belonging to *Glomus* include *Glomus claroideum*, *Glomus mosseae* and four unidentified species with different tolerance characteristics were observed in the rhizospheric soil of all plants that used for the trap culture and in different soil treatments. With addition of sewage sludge, the spore numbers were decreased but did not disappear completely, this subject showed adaptation of these fungi to the environmental stresses. Average spore numbers in control, with addition of 100 cm^3 $ha^{-1}y^{-1}$ and in the treatment of 300 cm^3 $ha^{-1}y^{-1}$, were 300 and 100 spores per 100 g dry soil, respectively. Among plant species, *Sorghum bicolor* has the greatest effects on the spore numbers. *Glomus claroideum* and an unidentified of *Glomus* were common species in the rhizosphere of all plants grown with the 300 cm^3 ha^{-1} y^{-1}. *Glomus claroideum* was identified and purified as a species adapted ecotype with conditions of plot treatment with 300 cm^3 ha^{-1} y^{-1}. The total spore numbers had negative correlation with amount of HMs and had the highest correlation coefficients with zinc and cadmium cations in the soil solution. Increased species diversity and richness of native AMF in the medium levels of pollution may be due to the presence of species with competitive capacity and/or compatible to these conditions that are able to improve plant root colonization and to complete their life cycle. But at high pollution levels, intensify effects of heavy metals toxicity on root colonization and proliferation of fungi, decreased the diversity of AMF species. Chaudhry et al. (1999) reported the fungal species belonging to *Glomus* and *Gigaspora* genera in symbiosis with most plants grown in the polluted soils with the waste materials consisting of zinc and iron. Pawlowska et al. (2000) reported the *Glomus constrictum*, *Glomus mosseae* and an unidentified species of *Glomus* in soils with moderate contamination by heavy metals in the around of Minnesota. Tonin et al. (2001) studied AMF in relation to a metallophyte plant, *Viola calaminaria* that grown in highly polluted soils in Belgium, containing total and extractable zinc and cadmium, 20961, 934 and 41, 1.7 mg kg^{-1}. It was identified at least four different *Glomus* species in the rhizosphere of this plant based on spores morphology and using of Terminal-RFLP technique. This absolutely metallophyte plant, usually grows in the rich soils of lead and zinc in Belgium. It was not measured HMs in the plant tissues even in the pollution conditions or only very small amounts of these elements accumulated in the plant tissues (Baker, 1981). Hildebrandt et al. (1999) showed high mycorrhizal colonization of these plants in the highly polluted conditions. Turnau et al. (2001) studied community of AMF in the rhizosphere and roots of strawberry (*Fragaria vesca*) grown in a sandy soil of Chrzanow in southern Poland which was polluted by zinc, lead, cadmium, arsenic and copper. Molecular analysis of the stained roots with tryptophan blue has shown that *Glomus gerdemannii* colonized the roots up to 54 percent. Abundance of *Glomus claroideum* and *Glomus occultum* were 32 and 34 percent respectively, and abundance of *Glomus mosseae* and *Glomus intraradices* were reported up to 5-7 percent within the roots. At least *Glomus mosseae* always associated with other fungi, but there was no report about tolerance of this fungus. Orlowska et al. (2002) reported effective AMF symbiosis in roots of plants cultivated in the rich soils of zinc, lead and cadmium in the southern Poland. Mycorrizal colonization was measured in *Plantago lanceolata* up to 90 percent and in *Biscutella laevigata* up to 40 percent. In addition to vesicles and hyphae coils, arbuscules were observed. Gonzalez-Chavez et al. (2002) reported the fungal species belonging to genera of *Glomus*, *Acaulospora* and *Entrophospora* in the rhizospheric soil of *Holcus lanatus* around an arsenic mine.

Whitfield et al. (2004) investigated diversity of AMF in the rhizosphere of *Thymus polytrichus* plants growing in the soils with high, medium and low pollution in northern England. Total concentration of cadmium, lead and zinc, were 9.43, 1358 and 3641 mg kg^{-1} and EDTA extractable concentrations of these metals were 3.84, 517.7 and 937.3 mg kg^{-1} respectively; medium polluted soils contained total concentration of cadmium, lead and zinc 5.08, 966.7 and 2024 mg kg^{-1} and EDTA extractable concentrations of these elements, 2.50, 525.4 and 616.5 mg kg^{-1}, respectively and a low polluted soils contained total concentration of cadmium, lead and zinc 2.68, 573.8 and 912.8 mg kg^{-1} and EDTA extractable concentrations of these elements, 1.55, 299.9 and 286.4 mg kg^{-1}, respectively. The results of molecular analysis using primers AM1-NS31 showed that AMF strains found within the roots of this plant at low and high polluted soils were different, and *Glomus* was dominant in the polluted soils. *Glomus mosseae* and *Glomu fasciculatum* and some unidentified species of

Glomus genus and fewer species of *Acaulospora* were exclusively found in the highly polluted soils. Vallino et al. (2006) studied AMF diversity using primers AM1-NS31 and ARCH1311-NS8 within the roots of *Solidago gigantea* in a HM polluted area in northern Italy. Phylogenic analysis of the fungal sequences obtained from the roots, separated 14 groups. 12 sequences were belonged to *Glomus* genus, one *Gigaspora* and the other *Paraglomus*.

Zarei et al. (2008a), Zarei (2008) and Zarei et al. (2010) in order to study the variations in spore abundance, root colonization parameters, morphological and molecular diversity of arbuscular mycorrhizal fungi (AMF) in a naturally HMs polluted site and their relationships with soil properties, 35 plots in the Anguran Zn and Pb mining region (Zanjan province, Iran) were selected along a transect from the mine to 4500 m away. Within each plot, a composite sample of root and rhizospheric soil from a dominant indigenous plant was collected. The soil samples were analyzed for their physico-chemical characteristics. The roots were examined for colonization, arbuscular and vesicular abundance, mycorrhizal frequency and intensity. Spores were extracted, counted and after trap culturing, identified at species level. The results showed that along the transect, the total and available (DTPA - extractable) Zn decreased from 6472 to 45 mg kg^{-1} and 75 to 5 mg kg^{-1}, respectively. For Pb the values varied from 5203 to 0 mg kg^{-1} for total and 32 to 0 mg kg^{-1} for available form. In contrast, root colonization rate in the dominant native plants (except *Alyssum* sp.) increased from 35 to 85 % and the spore numbers from 80 to 1306 per 200 g dry soil along the transect. Ten AMF species taxon were morphologically identified, which nine of them belonged to *Glomus* (*Glomus mosseae*, *G. intraradices*, *G. versiforme*, *G. ambisporum*,*G. constrictum*, *G. fasciculatum*, *G. geosporum*, *G. sinosum* and *Glomus* sp.) and one of them belonged to *Acaulospora* (*Acaulospora* sp.) (See figure 1). *Glomus* was the dominant genus in all plots. Isolation frequency, spore density and relative abundance of AMF species were different in HMs polluted and non-polluted sites. *G. mosseae* was the taxon most commonly observed in different plots, with higher spore density and relative abundance at high level of HMs pollution. *G. mosseae*, *G. intraradices* and *G. versiforme* were more abundant than other AMF species in the soils with moderate and low levels of Zn and Pb contents. Isolation frequency, spore density and relative abundance of *G. intraradices* were higher than those of other AMF in the non-polluted soils. The Shannon-Weiner index of AMF decreased from 1.55 to 0.52 as the HMs contents of soils increased. The Sorenson similarity index of AM fungi and plant community between non-polluted soils with those of the other pollution levels decreased with increasing soil HMs contents and was higher for AM fungal populations than plant community. By molecular methods, different sequence types which belonged to *Glomus* group Aa, Ab and B, *Glomus versiforme* and *Acaulospora* sp. were identified (Figure 2). Rarefaction analysis curves confirmed exhaustive molecular characterization of the AMF diversity present within root samples. Correspondence analyses showed that with increasing the HMs contcentrations, the number of AMF sequence types colonizing the roots decreased. Some AMF sequence types were only found at sites with the highest or the lowest soil HMs contents.

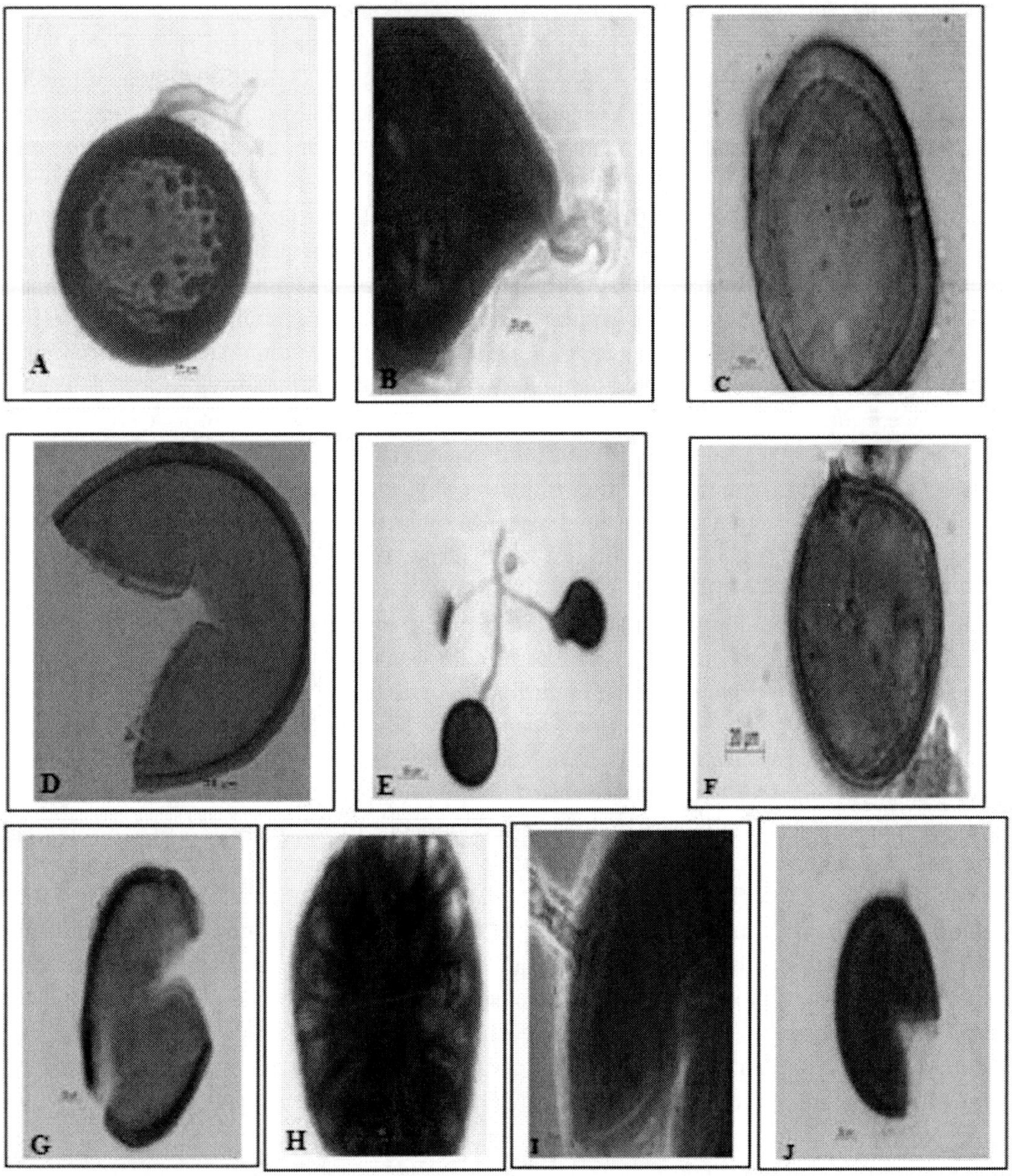

Figure 1. Spore with branching hyphae of *Glomus ambisporum* (A), spore with constriction of hypha at attachment point of *Glomus constrictum* (B), spores of *Glomus fasciculatum* (C), *Glomus geosporum* (D), *Glomus intraradices* (E), *Glomus mosseae* (F), *Glomus sinuosum* (H), *Glomus* sp. (I) and *Acaulospora* sp. (J) (Zarei, 2008).

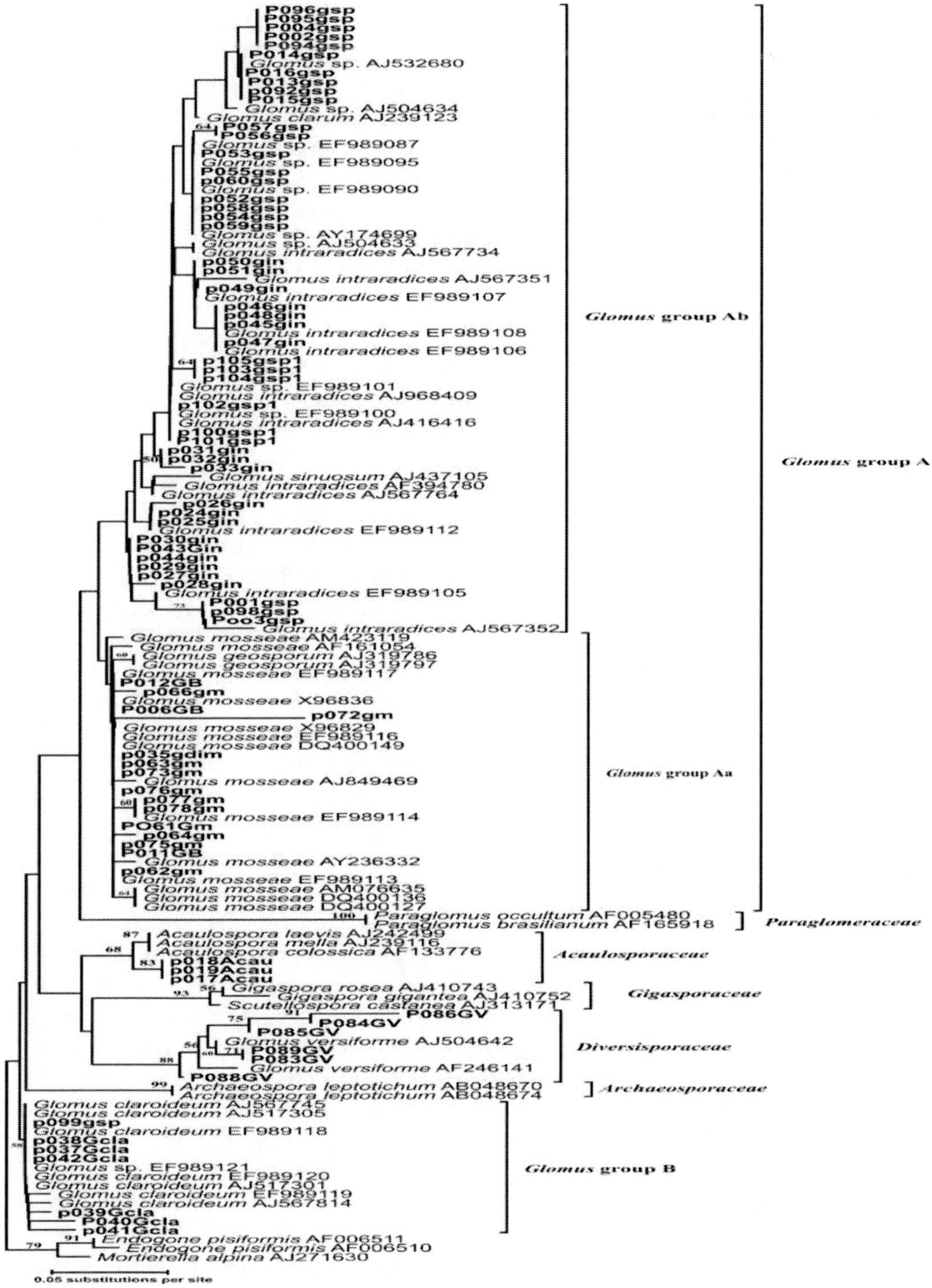

Figure 2. Rooted Neighbor Joining phylogram of sequences belonging to AMF group from 5.8S- rDNA sequences based on distance analysis (Kimura-2-parameter). Bootstrap supports above 50% are given on the branches (1000 replicates). Sequences obtained in study of Zarei (2008) are shown in bold.

Overall, the studies have been indicated that AMF are exist in the HMs polluted soils. As HM pollution increases abundance and diversity of AMF decrease. However, such plants especially metallophytes may be colonized more. Some species and sequences types may be exclusively found in the high HMs pollution levels. In the most studies about AMF in soils polluted with potentially toxic elements, due to lack of information about how the effects of host plants and soil physical and chemical characteristics, the determination of relationship between the presence and tolerance of these fungi to HMs is difficult because of the interactions of between these factors are complex and cannot completely separate the effects of HMs from other factors (Leyval et al., 1997; Gonzalez-Chavez et al., 2006).

Arbuscular Mycorrhizal Fungi and Tolerance to Heavy Metals

Many researchers have been reported that AMF ecotypes originated from the polluted soils are more resistance and tolerance to HMs contamination than native fungal strains of the non-polluted soils (Diaz et al., 1996). Gildon and Tinker (1981) isolated a strain of Glomus mosseae from HMs contaminated soils that could tolerate 100 mg Zn kg-1. Dueck et al. (1986) reported the presence of some strains of Glomus fasciculatum as tolerant strains to HMs in the several HMs contaminated areas in the Netherlands. Weissenhorn et al. (1993) isolated the spores belonging to Glomus mosseae group from contaminated soils with HMs in France. They showed that the two Glomus mosseae strains isolated from the polluted soils with cadmium had the ability to tolerate the cadmium concentrations from 50 to 70 and 200 to 500 mg l-1, respectively. Weissenhorn et al. (1994) reported the tolerance of AMF species to zinc and cadmium in the soils treated with these elements for one year. They assumed that rapid observation the AMF tolerance to elements attempting to relatively long time of renewable generation and high numbers of nuclear in the fungal spores, was based on phenotypic flexibility of these fungi than selectivity of tolerant genotypes. Leyval et al. (1995) used the spores germination of AMF as a biological indicator for the determination of fungi tolerance to HMs. They reported that the originated spores from soils that naturally polluted by HMs, had higher germination than spores from non-polluted soils, and its reasons were known to the natural presence of tolerant native AMF., They also showed that the spores of some Glomus mosseae isolates could germinate in the total concentrations of 6060 mg Pb kg-1, 1630 mg Cu kg-1, 24410 mg Zn kg-1and 3766 mg Zn (extractable with ammonium nitrate) kg-1, that probably was due to phenotypic flexibility or compatibility specific isolates of this species. Del Val et al. (1999b) isolated the different ecotypes of fungi consisting of two isolates of Glomus claroideum, an unidentified Glomus species and Glomus mosseae from the plots containing different amounts of HMs, and were studied their tolerance and compatibility in an experiment. They used two host plants species Allium porrum (Leek) and Sorghum bicolor for the study of structural aspects and colonization ability of different fungal isolates obtained polluted and non-polluted soils. Total root colonization and the number of arbuscules were different in the polluted and non-polluted soils and were strongly effected from the soil pollution by HMs. Glomus mosseae and an unidentified species of Glomus obtained from non-polluted plots and/or with moderate pollution, were formed the less arbuscules and their activities were largely limited to the production of extracellular mycelia in the presence of HMs. The results of these researchers showed that Glomus claroideum isolated from plots with high pollution, showed more compatibility potential to HMs and this isolate in comparison with other isolates originated from non-polluted soils, significantly increased the root colonization levels of leek and sorghum plants up to 20 and 15 percent, respectively in the HMs polluted soils and was introduced as an effective fungus in improving two plants growth under HMs pollution. Del val et al. (1999a) reported that development of hyphae and pre-symbiosis stages of Glomus mosseae were more sensitive to HMs and it was caused reduction of plant root colonization by this fungus. Hildebrandt et al. (1999) isolated a Glomus species from the rhizosphere of Viola calaminaria that grown in the HMs polluted soils. This isolate was improved corn growth in a polluted soil and reduced the concentration of HMs in plant roots and shoots in comparison with a common species of *Glomus* or control treatment (Kaldorf et al., 1999). Gonzalez-Chavez et al. (2002) evaluated the spore germination of *Glomus mosseae* and *Glomus caledonium* species isolated from rhizosphere of *Holcus lanatus* that grown in the polluted and non-polluted soils of around a arsenic mine. The results showed that spore germination of isolates originated from non-polluted soils was strongly stopped at concentration 50 mg arsenate kg^{-1}, while, spore germination of isolates from polluted soils was not affected or even sometimes was stimulated. Shalaby (2003) studied capability of spore germination and development of hyphae growth of isolates originated from non-polluted and polluted soils with zinc, lead and cadmium in in-vitro. The results showed that spores and pre-symbiosis phase were more sensitive to HMs. Among fungal strains effective concentration at which 50 percent of the spore germination and/or hyphae growth decreased were different and negative effects were observed at higher concentrations of HMs. The isolated spores from the polluted soils in comparison with isolated spores from the non-polluted soils were more tolerant to HMs. The natural resistance has been attributed more to phenotypic flexibility than genetic changes within spores because of the tolerance feature was lost after regeneration in the absence of HMs. Malcova et al. (2003) reported that AMF species and even different isolates of a species differed in terms of sensitivity to HMs. Pawlowska and Charvat (2004) studied the effects of zinc, lead and cadmium on the pre-symbiosis and symbiosis phases of the *Glomus intraradices* and *Glomus etunicatum* and reported that *Glomus intraradices* in comparison with *Glomus etunicatum* was more tolerance to each metals and they expressed that tolerance was depended on fungal genotypes. Extra to this, AMF propagules may have different responses to different elements. Zarei (2008) demonstrated spore numbers were more affected by Zn and Pb concentrations than root colonization. The variations of AM fungi propagules were more related to available than total concentration of both metals.

It was noticed, AMF fungal species isolated from the HMs polluted soil could be more adapted and tolerated to HMs pollution. Sensitivity of different AM species or isolates and even propagules to different HMs may be varied. This can be depended on phenotypes or genotypes characteristics of fungal species and HM type. However, HMs occurs in mixture, not singly in the environment, therefore, interactive effects of HMs should be considered.

Tolerance Mechanisms of Arbuscular Mycorrhizal Fungi to Heavy Metals

Because of be compulsive of AMF symbiosis with plants and lack of its growth in conventional culture media, less information is available about the tolerance mechanisms of these fungi to HMs. The more information was based on plant response to HMs and observation of fungal structures within colonized roots, which were difficult to separate of fungal and plant responses. Different mechanisms were proposed for explanation of plant responses to the high concentrations of HMs. Primary effects can be diagnosed in the molecular, biochemical and cellular levels and next effects are visible in the physiological and organelle levels. Levitt (1980) concerning with the behavior of organisms at high concentrations of HMs was suggested the three mechanisms of avoidance, detoxification and physiological tolerance. Avoidance mechanisms include limitation of uptake and transport of HMs to plant organs and include the reduction of absorption, extracellular increasing of metal diffusion, extracellular complex formation with metals by exudation of organic acids and other chelating agents. Physiological tolerance is because of the presence of specific metabolic pathways and enzyme adaptation. Tolerance involves mechanisms such as intracellular chelation of these metals due to formation of ligands such as metallothioneins, polyphosphates and/or the distribution of elements in the vacuole (Leyval et al., 1997). Tolerance, is creation a severe changes in sensitivity of some individual to a particular toxicant that can be structural and/or adaptive and due to phenotypic flexibility against inducers (Khade and Adholeya, 2007). Generally, suggested tolerance mechanisms of AMF to HMs are reviewed by Göhre and Paszkowski (2006) and Gonzalez-Chavez et al. (2006). In these fungi, the tolerance does not have a general pattern and may be different among species in the response to a particular metal. There are also large changes in the rate of tolerance to HMs among different populations of a species or ecotype.

Mechanisms number 1 is extracellular chelation. Mainly, efficient chelating agents in this stage include organic metabolites like organic acids (oxalic and citric acid), siderophores, riboflavins and glomalin that play a role in the detoxification via complexation and precipitation. Glomalin is a glycoprotein that is produced and released by colonized roots and extracellular mycelia of AMF. Chemical structure of glomalin was not fully determined and it known as a series of repeated monomer structures that inter connected hydrophobic bonds. Presence of iron proteins (0.04 to 8.8 percent), and oligosaccharides linked with nitrogen and being insoluble are its other properties (Gonzalez-Chavez et al. 2004). Glomalin play a significantly role in sequestering soil carbon and is as an important source soil carbon accompanied by humic substances (up to 25 percent). Gonzalez-Chavez et al. (2004) studied the role of glomalin in sequestering of HMs in three experiments. Glomalin was extracted from polluted soils or from *Gigaspora rosea* and of two AMF species that grown in three copper levels. Their results showed that glomalin sequestered high concentrations of lead, zinc and cadmium in the polluted soils by these elements. Glomalin extracted from *Gigaspora rosea* have ability for removing of copper, zinc, cobalt and nickel from a solution containing of these metals, but there was no effects on the calcium, magnesium and potassium. The highest sequestered copper was 28 mg g^{-1} glomalin. Mechanisms number 2 includes binding of HM to the cell wall components of fungi. Cell wall is the first place that metals contact with cell and is a protective barrier for absorption control of potentially toxic metals.

Different structural components and compositions of cell wall consisting of carboxyl, amine, hydroxyl, phosphate and sulfate groups can create potential locations to sequestering of metals in the cell wall surface. Also, binding of metals with polysaccharide such as chitosan and chitin has been reported. From other cell wall components such as melanin, glucan and manan can also play the role in binding with metals. Chitins and β (1-3)-glucans have been reported as a major cell wall component of *Glomus* species (Smith and Read, 1997). HMs binding with cell wall surface of hyphae and extracellular mycelia of AMF were reported. Turnau (1998) using transmission electron microscopy and rhodizoniate histochemical staining localized HMs in the colonized roots of *Euphorbia cyparissias* L. that sampled from Zn wastes. This author reported that about 80 percent of total intraradical mycelium contained high amounts of HMs. Amorphous or crystalloid sediments were recognizable at surface of this mycelium. Sulfur, phosphorus and arsenic in the amorphous sediments and calcium, iron, zinc and lower levels of arsenic in the crystalloid sediments and silicium, copper and zinc in the both sediments were detected. Additionally, metals residuals were observed in the internal layer of spore wall of some AMF morphotypes. Spores of AMF are involved in binding with HMs. The researchers were reported that *Glomus mosseae* spores sequestered copper between 470 to 680 $\mu g\ g^{-1}$ dry weight (Gonzalez-Chavez et al., 2006). Extraradical mycelium is considered a main structural component in arbuscular mycorrhizal symbiosis that contributed in the nutrient cycle and to formation an extensive network that connects the roots to soil matrix. The studies have shown that this mycorrhizal structure due to having glomalin plays an important role in tolerance of plant to HMs. Extraradical mycelium of AMF, increase soil available volume for plant roots (Marschner and Romheld, 1998). Thus, mycorrhizal plants in comparison with non-mycorrhizal plants access to more volume of soil and therefore can affect more on the absorption and transfer of elements and plant tolerance to HMs. Mechanisms 3 is in relation to the plasma membrane as a living and selective barrier against metals influx to fungal cells. Mechanisms number 4 used through metals specific and non specific carriers and also by pores in the plasma membrane of these fungi for control of HMs transferring to the cell. Mechanisms number 5 includes chelation in the cytoplasm as an intracellular buffer system that controls HMs toxicity by decreasing free ions concentration. In this system metallothioneins, the polypeptides attached to the metals, organic and amino acids and/or proteins of similar metallothioneins have been contributed. The presence of linked thioles with cadmium and higher concentrations of nitrogen and sulfur has been reported in arbuscular mycorrhiza propagules (Gonzalez-Chavez et al., 2006). Galli et al. (1995) in copper polluted substrates have found high amounts of cysteine thioles, glutamyl cysteine and glutathione in maize colonized roots by *Glomus intraradices*. Tyrosinase activity of extraradical mycelium of AMF in copper polluted substrates was more than non-polluted substrates. Activity of this enzyme in the polluted substrate, was higher in the presence of *Glomus caledonium* than *Glomus mosseae* and *Glomus claroideum*. Polyphosphates are phosphate storage organs within the fungal cells that may contribute in regulation of concentrations of HMs ions in the cells. Polyphosphate granules have an important role in ionic compartmentalization of HMs and maintaining of this state. Sequestering of HMs by intracellular polyphosphates of the fungal may be important in the reducing of metal transport to plants (Gonzalez-Chavez et al., 2006). Mechanisms numbers 6 and 7 include HM export via specific or non-specific active or passive transport from cells and metals sequestration in the vacuoles. Mechanisms number 8 is transport of HMs in the hyphae of the fungus. Mechanisms number 9 is active and passive transport of metals from the arbuscule to plant cells (Göhre and Paszkowski, 2006). Physiological interactions between the AMF and host cells in the presence of the HMs may be other mechanisms in relation to the reduction of HMs toxicity. In the molecular level, root colonization with the mycorrhizal fungi can regulate the genes expression of coding of HMs carriers in the plasma membrane that through influence on the absorption and transport of HMs. Göhre and Paszkowski (2006) expressed that *Glomus intraradices* in the symbiosis and pre-symbiosis phases is tolerant to HMs. They have also discovered *GintZnT1* gene in *Glomus intraradices* that is effective in the export of zinc to extracellular and thus in the decreasing of zinc toxicity. Similarly, Hildebrandt et al. (2007) expressed the preferential expression of this gene in the intraradical structures of *Glomus intraradices* and it gene products may play an important role in the zinc toxicity.

Effects of AMF on the plant nutrition, root exudations, rhizosphere microbial communities, soil structure and protection against environmental stresses can be considered indirect mechanisms in the increasing tolerance to HMs. Chen et al. (2003) showed that mycorrhizal infection improved phosphorus nutrition and increased plants growth especially shoots and these cases helped to reduce of zinc toxicity. One of the decreasing mechanisms of zinc toxicity may be improving of the phosphorus absorption in the result of mycorrhizal colonization. Because of metallic elements absorb and transfer via aqueous solutions, the transport of these elements to shoots increases following the increasing of plants transpiration. There is no doubt that mycorrhizal colonization is effective in the water relations of plants (Smith and Read, 1997). The reports were shown that the rate of transpiration significantly was higher in the mycorrhizal plants than non-mycorrhizal plants. The roots of mycorrhizal plants usually have more branches, even with the equal biomass with non-mycorrhizal roots, so they contain high absorption surfaces. The increased leaf area is an important factor in the increasing of transpiration rate. However, even in the plants with same leaf area and root system length, transpiration rates of mycorrhizal plants is higher due to the reduced stomatal resistance. Stomatal behavior is affected by hormonal changes which can dependent on phosphorus nutrition and mycorrhizal colonization. Based on the results of a study, root colonization of *Boueteloua gracilis* plants inoculated with *Glomus fasciculatum* reduced due to concentrations of abscisic acid (ABA) hormone in the plant leaves. However, there are conflicting reports in this field, that should be attend to combination of plant-fungus species, growth conditions and methods of measurement (Turnau et al., 2006). Glomalin is a glue-like protein which contains 30-40 percent of its weight of carbon that form a secondary root system and also it has ability for connection to soil particles that caused play an effective role in the improving physical and chemical characteristics of soils and aggregation and through way can improve plants growth in the polluted soils which confront poor fertility, high compaction, low permeability and weak vegetations (Del Val et al., (1999a)). Synergistic effects of other microorganisms such as plants growth promoting rhizobactaria (PGPR) and saprophytic fungi with AMF, can influence on the plants growth and HMs tolerance and play an important role in the phytoremediation. Some of PGPRs can increase the availability of HMs and their accumulation in plants. Also, Synergistic effects, has been reported between AMF and some useful microorganisms. A strains of *Brevibacillus* sp. isolated from polluted soils by nickel could enhance grow clover plant in the nickel polluted soil and increase mycorrhizal efficiency in the stress conditions of nickel toxicity (Vivas et al., 2003). Similar to this study in the zinc polluted soils was reported by Vivas et al. (2006). Different microorganisms with different mechanisms may have beneficial effects on the plant growth and its tolerance to HMs, so it is necessary to used multipurpose beneficial microbial inocula with AMF to enhance plant growth and improve phytoremediation efficiency.

ACKNOWLEDGMENT

We would like to thank Dr. Srefan Hempel for reviewing the chapter and for valuable comments.

REFERENCES

Adriano, D. C. (2001). *Trace elements in terrestrial environments: biogeochemistry, bioavailability and risks of metals* (pp. 867). New York, Springer-Verlag.

Baker, A. J. M. (1981). Accumulators and excluders-strategies in the response of plants to heavy metals. *Journal of Plant Nutrition, 3,* 643-654.

Bohn, K. S. & Liberta, A. E. (1982). In: D. H., Graves (Eds.), *Symposium on surface mining hydrology, sedimentology and reclamation.* Lexington, University of Kentucky.

Chaudhry, T. M., Hill, L., Khan, A. G. & Kuek, C. (1999). Colonization of iron and zinc contaminated dumped filter cake waste by microbes, plants and associated mycorrhizae. In: M. H. Wong, J. W. C Wong. & A. J. M. Baker (Eds.), *Remediation and management of degraded land*, (pp. 275-283). Boca Raton, FL: CRC Press.

Chen, B. D., Li, X. L., Tao, H. Q., Christie, P. & Wong, M. H. (2003). The role of arbuscular mycorrhiza in zinc uptake by red clover growing in a calcareous soil spiked with various quantities of zinc. *Chemosphere, 50,* 839-846.

del Val, C., Barea, J. M. & Azcon-Aguilar, C. (1999a). Diversity of arbuscular mycorrhizal fungus populations in heavy-metal-contaminated soils. *Applied and Environmental Microbiology*, *65,*718-723.

del Val, C., Barea, J. M. & Azcon-Aguilar, C. (1999b). Assessing the tolerance to heavy metals of arbuscular mycorrhizal fungi from sewage sludge-contaminated soils. *Applied Soil Ecology, 11,* 261-269.

Diaz, G. & Honrubia, M. (1993). Infectivity of mine soils from southeast Spain. 2. Mycorrhizal population levels in spoilt sites. *Mycorrhiza, 4,* 85-88.

Diaz, G. & Honrubia, M. (1994). A mycorrhizal survey of plants growing on mine wastes in southeast Spain. *Arid Soil Res. Rehab., 8,* 59-68.

Diaz, G., Azcon-Aguilar, C. & Hornubia, M. (1996). Influence of arbuscular mycorrhizae on heavy metal (Zn and Pb) uptake and growth of *Lygeum spartum* and *Anthyllis cystoides. Plant and Soil, 180,* 241-249.

Dueck, T. A., Visser, P., Ernst, W. H. O. & Schat, H. (1986). Vesicular-arbuscular mycorrhizae decrease zinc toxicity to grasses growing in zinc polluted soils. *Soil Biology and Biochemistry*, *18,* 331-333.

Galli, U., Schuepp, H. & Brunold, C. H. (1995). Thiols of Cu treated maiz plant inoculated with the AM fungus Glomus intraradices. *Physiologia Plantarm, 94,* 247.

Gaur, A. & Adholeya, A. (2004). Prospects of arbuscular mycorrhizal fungi in phytoremediation of heavy metal contaminated soils. *Current Science*, 86, 528-534.

Gildon, A & Tinker, P. B. (1981). A heavy metal-tolerant strain of mycorrhizal fungus. *Transactions of the British Mycological Society, 77,* 648 -649.

Gohre, V. & Paszkowski, U. (2006). Contribution of the arbuscular mycorrhizal symbiosis to heavy metal phytoremediation. *Planta, 223,* 1115-1122.

Gonlzalez- Chavez, M. D. C. A., Vangronsveld, J., Colpaert, J. & Leyval, C. (2006). Arbuscular mycorrhizal fungi and heavy metals: tolerance mechanisms and potential use in bioremediation. Chapter 12. In M. N.V., Prasad, K. S. Sajwan, & R. Naidue (Eds.), *Trace Elements in Environment Biogeochemistry, Biotechnology and Bioremediation.* Taylor and Francis group LLC, CRC press.

Gonzalez, M. C. (2000). *AMF As/Cu polluted soils, contribution to plant tolerance and importance of the external mycelium.* PhD thesis. Reading University, UK.

Gonzalez-Chavez, C., Harris, P. J., Dodd, J. & Meharg, A. A. (2002). Arbuscular mycorrhizal fungi confer enhanced arsenate resistance on Holcus lanatus. *New Phytologist, 155,* 163-171.

Griffioen, W. A. J. (1994). Characterization of a heavy metal-tolerant endomycorrhizal fungus from the surroundings of a zinc refinery. *Mycorrhiza*, 4, 197-200.

Hildebrandt, U., Regvar, M. & Bothe, H. (2007). Arbuscular mycorrhiza and heavy metal tolerance. *Phytochemisrty, 68,* 139-146.

Hildebrandt, U., Kaldorf, M. & Bothe, H. (1999). The zinc violet and its colonization by arbuscular mycorrhizal fungi. *Journal of Plant Physiology*, 154, 709-717.

Holleman, A. & Wiberg, E. (1985). *Lehrbuch der Anorganischen Chemie*. Academic Press, Berlin.

Kaldorf, M., Kuhn, A., Schroder, W. H., Hildebrandt, U. & Bothe, H. (1999). Selective element deposits in maize colonized by a heavy metal tolerance conferring arbuscular mycorrhizal fungus. *Journal of Plant Physiology, 154,* 718-728.

Khade, S. W. & Adholeya, A. (2007). Feasible bioremediation through arbuscular mycorrhizal fungi imparting heavy metal tolerance: a retrospective. *Bioremediation Journal, 11(1),* 33-43.

Levitt, J. (1980). *Responses of Plants to Environmental Stresses*. (Vol. 2), (2nd Ed.). New York, Academic Press.

Leyval, C., Singh, B. R. & Joner, E. J. (1995). Occurrence and infectivity of arbuscular mycorrhiza fungi in some Norwegian soils influenced by heavy metals and soil properties. *Water, Air and Soil Pollution, 84,* 203-216.

Leyval, C., Turnau, K. & Haselwandter, K. (1997). Effect of heavy metal pollution on mycorrhizal colonization and function: physiological, ecological and applied aspects. *Mycorrhiza*, 7, 39-153.

Malcova, R., Rydlova, J. & Vosatka, M. (2003). Metal free cultivation of *Glomus* sp. BEG isolated from Mn-contaminated soil reduces tolerance to Mn. *Mycorrhiza, 13,* 151-157.

Marschner, H. & Romheld, D. J. (1998). Role of root growth, arbuscular mycorrhiza and root exudates for the efficiency in nutrient acquisition. *Field Crops Reserach, 56,* 203-207.

Mertz, W. (1981). The essential trace elements. *Science,* 213, 1332-1338.

Orlowska, E., Zubek, S. Z., Jurkiewicz, A., Szarek-Lukaszewska, G. & Turnau, K. (2002). Influence of restoration of arbuscular mycorrhiza of *Biscutella laevigata* L. (Brassicaceae) and *Plantago lanceolata* L. (Plantaginaceae) from calamine spoil mounds. *Mycorrhiza, 12,* 153-60.

Pawlowska, T. E. & Charvat, I. (2004). Heavy-metal stress and developmental patterns of arbuscular mycorrhizal fungi. *Applied and Environmental Microbiology, 70(11),* 6643-6649.

Pawlowska, T. E., Blaszkowski, J. & Rühling, A. (1996). The mycorrhizal status of plants colonizing a calamine spoil mound in southern Poland. *Mycorrhiza, 6,* 499-505.

Pawlowska, T. E., Chaney, R. L., Chin, M. & Charvat, I. (2000). Effects of metal phytoextraction practices on the indigenous community of arbuscular mycorrhizal fungi at a metal-contaminated landfill. *Applied and Environmental Microbiology, 66,* 2526-2530.

Raman, N. & Sambandan, K. (1998). Distribution of VAM fungi in tannery effluent polluted soils of Tamil Nadu, India. *Bulletin of Environmental Contamination and Toxicology, 60,* 142-50.

Raman, N., Nagarajan, N., Gopinathan, S. & Sambandan, K. (1993). Mycorrhizal status of plant species colonizing a magnesite mine spoil in India. *Biology and Fertility of Soil, 16,* 76-78.

Sambandan, K., Kannan, K. & Raman, N. (1992). Distribution of vesicular-arbuscular mycorrhizal fungi in heavy metal polluted soils of Tamil-Nadu, Indian. *Journal of Environmental Biology, 13,* 159-67.

Shalaby, A. M. (2003). Responses of arbuscular mycorrhizal fungal spores isolated from heavy metal polluted and non-polluted soil to Zn, Cd, Pb and their interactions in vitro. *Pakistan Journal of Biological Science, 6,* 1416-1422.

Shetty, P. K., Hetrick, B. A. D., Figge, D. A. H. & Schwab, A. P. (1995). Effects of mycorrhizae and other soil microbes on revegetation of heavy metal contaminated mine spoil. *Environmental Pollution, 86,* 181-188.

Smith, S. E. & Read, D. J. (1997). *Mycorrhizal Symbiosis.* (2rd Ed.) London, Academic Press.

Tonin, C., Vandenkoornhuyse, P., Joner, E. J., Straczek, J. & Leyval, C. (2001). Assessment of arbuscular mycorrhizal fungi diversity in the rhizosphere of *Viola calaminaria* and effect of these fungi on heavy metal uptake by clover. *Mycorrhiza, 10,* 161-168.

Turnau, K. (1998). Heavy metal content and localization in mycorrhizal *Euphorbia cyparissias* from zinc wastes in southern Poland. *Acta Societatis Botanicorum Poloniae. 67,* 105.

Turnau, K., Jurkiewicz, A., Lingua, G., Barea, J. M. & Gianinazzi-Pearson, V. (2006). Role of arbuscular mycorrhiza and associated microorganisms in phytoremediation of heavy metal-polluted sites. Chapter 13. In M. N.V. Prasad, K. S. Sajwan & R. Naidue (Eds.). *Trace Elements in Environment Biogeochemistry, Biotechnology and Bioremediation.* Taylor and Francis group LLC, CRC press.

Turnau, K., Ryszka, P., Gianinazzi, P. V. & van Tuinen, D. (2001). Identificaltion of arbucular mycorrhizal fungi in soils and roots of plant colonizing zinc wastes in southern Poland. *Mycorrhiza, 10,* 169-174.

Vallino, M., Massa, N., Lumini, E., Bianciotto, V., Berta, G. & Bonfante, P. (2006). Assessments of arbuscular mycorrhizal fungal diversity in roots of *Solidago gigantea* growing in a polluted soil in Northern Italy. *Environmental Microbiology, 8,* 971-983.

Vivas, A., Barea, J. M., Biro, B. & Azcon, R. (2006). Effectiveness of autochtonous bacterium and mycorrhizal fungus on *Trifolium* growth, symbiotic development and soil enzymatic activities in Zn contaminated soil. *Journal of Applied Microbiology, 100,* 587-598.

Vivas A., Voros I., Biro B., Campos E., Barea J. M. & Azcon R. (2003). Symbiotic efficiency of autochthonous arbuscular mycorrhizal fungus (*Glomus mosseae*) and *Brevibacillus* sp. isolated from cadmium polluted soil under increasing cadmium levels. *Environmental Pollution, 126,* 179-189.

Weissenhorn, I., Glashoff, A., Leyval, C. & Berthelin, J. (1994). Differential tolerance to Cd and Zn of arbuscular mycorrhizal (AM) fungal spores isolated from heavy metal polluted and non-polluted soils. *Plant and Soil, 167,* 189-196.

Weissenhorn, I., Leyval, C. & Berthelin, J. (1993). Cd-tolerant arbuscular mycorrhizal (AM) fungi from heavy metal polluted soil. *Plant and Soil, 157,* 247-256.

Weissenhorn, I., Leyval, C., Belgy, G. & Berthelin, J. (1995). Arbuscular mycorrhizal contribution to heavy metal uptake by maize (*Zea mays* L.) in pot culture with contaminated soil. *Mycorrhiza, 5,* 245-252.

Whitfield, L., Richards, A. J. & Rimmer, D. L. (2004). Relationships between soil heavy metal concentration and mycorrhizal colonization in *Thymus polytrichus* in Northen England. *Mycorrhiza, 14,* 55-62.

Zak, J. C., Danielson, R. M. & Parkinson, D. (1982). Mycorrhizal fungal spore numbers and species occurrence in two amended mine spoils in Alberta, Canada. *Mycologia, 74,* 785-792.

Zarei, M. (2008). *Diversity of arbuscular mycorrhizal fungi in heavy metal pollution soils and their roles in phytoremediation.* Ph.D. dissertation in Soil Science (Soil Biology and Biotechnology), Agricultural Faculty, University of Tehran, Tehran, Iran, 219 pp (In Farsi).

Zarei M., Hempel S., Wubet T., Schäfer T., Savaghebi Gh., Salehi Jouzani Gh., Khayyam Nekouei M. & Buscot F. (2010) Molecular diversity of arbuscular mycorrhizal fungi in relation to soil chemical properties and heavy metal contamination. *Environmental Pollution 158,* 2757-2765.

Zarei M., Saleh-Rastin N., Salehi Jouzani Gh., Savaghei Gh. & Buscot F. (2008a). Arbuscular mycorrhizal abundance in contaminated soils around a zinc and lead deposit. *European Journal of Soil Biology, 44,* 381-391.

Zarei M., König S., Hempel S., Khayyam Nekouei M., Savaghebi Gh. & Buscot F. (2008b) Community structure of arbuscular mycorrhizal fungi associated to *Veronica rechingeri* at Anguran zinc and lead mining region. *Environmental Pollution, 156,* 1277-1283.

In: Mycorrhizal Fungi
Editor: Susanne M. Fulton

ISBN 978-1-61122-659-1
© 2011 Nova Science Publishers, Inc.

Chapter 11

PLANT DROUGHT TOLERANCE ENHANCEMENT BY ARBUSCULAR MYCORRHIZAL SYMBIOSIS

Ricardo Aroca, Rosa Porcel and Juan Manuel Ruiz-Lozano***
Estación Experimental del Zaidín (CSIC), Profesor Albareda 1,
18008 Granada, Spain

ABSTRACT

Since dry land becoming arid or semi-arid is rising in the last decades, more agricultural practices in accordance with the principles of sustainable agriculture are needed. One practice could be the use of arbuscular mycorhizal (AM) fungi as inoculants to render the plants more tolerant to drought stress. In the present chapter is described how AM symbiosis improves three aspects of plant drought tolerance. The three aspects described here are root water uptake capacity, water use efficiency (WUE), and antioxidant systems. Regarding water uptake capacity it is clear that AM plants are able to take up more water than non AM plants under drought conditions. This higher capacity of water extraction is linked in several cases to a more apoplastic water flow through roots and the water uptake by the mycelium itself. The precise role of aquaporins on this water uptake capacity enhancement has not been solved yet. WUE it is usually enhanced in AM plants under drought conditions when it is calculated as the ratio between biomass or yield and the water consumed. This fact means that AM plants can produce more biomass with the same amount of water consumed, that is, they use water more efficiently. These higher WUE could be linked to a better photosystem II efficiency. But how AM symbiosis affects dark respiration or CO_2 assimilatory enzymes activities under drought conditions is far to be known. Finally, although general speaking, AM symbiosis diminishes oxidative damage caused by drought in terms of lipid peroxidation, the question if this is a consequence of a better water status or if it is directly caused by a better antioxidant system is still a matter of debate.

* E-mail: raroca@eez.csic.es

INTRODUCTION

Dry land becoming arid or semi-arid has been increasing progressively in the last decades (Ben-Hur, 2006). So, even more natural and cultivated growing plants are being subjected to periods of water starvation. Plants have developed several mechanisms in order to cope with drought stress including morphological adaptations, optimization of water resources and antioxidant systems able to diminish the harmful effects of reactive oxygen species (ROS) linked to drought (Farooq et al., 2009). Thus, under drought conditions, plants trend to increase their root:shoot ratio in order to provide sufficient water to their aerial parts (Grzesiak et al., 1999). At the same time, when plants sense a diminution of the soil water availability, immediately they close their stomata in order to prevent leaf water lost. However, this stomatal closure also causes a restriction in the CO_2 uptake, diminishing the photosynthesis rate and plant growth at the same time. Therefore, one physiological trait that has been explored in order to obtain plants more adapted to drought periods is the water use efficiency (WUE). Thus, plants able to decrease transpiration rate but minimizing the consequent descent of photosynthesis rate are the ones desirable in order to save soil water resources without affecting growth (Wang et al., 2007).

As consequence of stomatal closure and the followed CO_2 uptake diminution, several ROS accumulate in leaf tissues. Hence, while CO_2 assimilation decrease under drought periods, the same amount of light is absorbed by light harvesting pigments and transformed to heat and/or reducing power. However if the reducing power is not used in assimilating CO_2, the excess of electrons generated in the water hydrolysis is absorbed by O_2 molecules, generating several ROS as superoxide anion (O_2^-), hydroxyl radical (HO•), or hydrogen peroxide (H_2O_2). So, plants with an efficient antioxidant system able to remove properly ROS generated under drought episodes usually are more tolerant to water deficit stress (Vanrensburg and Kruger, 1994).

Besides the mechanisms described above, plants also interact with several soil microorganisms that render the plant more tolerant to drought (Aroca and Ruiz-Lozano, 2009). Among such microorganisms one of the most spread are the arbuscular mycorrhizal (AM) fungi. Around 90% of natural growing plants are colonized by AM fungi. By this symbiosis plants receive from the fungus water and mineral nutrients not accessible to the plant roots (especially P). In return, the fungus receives from the plant C resources and a niche to complete their life cycle (Parniske, 2008). Hence, plants inoculated with AM fungi usually are able to take up more water under drought conditions (Marulanda et al., 2003), present higher WUE (Kaya et al., 2003) and have a more efficient antioxidant system (Wu et al., 2006a). In the present chapter a description of the changes caused by the AM symbiosis in such processes and the possible mechanisms behind these changes will be described.

AM SYMBIOSIS IMPROVES ROOT WATER UPTAKE UNDER DROUGHT CONDITIONS

The first place where soil drought is sensed by plants is in the roots. While soil water potential is decreasing progressively, plants have more difficult in absorbing water. In fact, when drought is becoming more intense root hydraulic conductance (L) is also decreasing in the same way (Aroca et al., 2006). L determines the capacity of the plant roots to take up water from the soil independently of the water potential gradient. In fact, plants with higher L also growth better under drought conditions (Matsuo et al., 2009), linked to a better ability to take up water. At the same time, as has been commented before, it is proven that AM plants are able to take up more water from the soil under drought conditions than non AM plants (Marulanda et al., 2003), most probably because AM fungi hyphae can transport water by itself from soil water reservoirs not accessible to plant roots (Ruiz-Lozano and Azcón, 1995).

It is worth to mention that water can follow three main paths across the roots prior to reach xylem vessels. Water can go following the cell interspaces or the cell walls; this path named apolastic. Apolastic water flow is mainly governed by the transpiration rate and most of the water flows by this path under optimal growing conditions. The other two paths include the water flowing by the symplastic route (living cells across plasmodesmata) and water flowing across vacuoles and plasma membranes. Since these two paths can not be separated empirically, the sum of both are named cell-to cell path (Steudle and Peterson, 1998). Cell-to-cell path may dominate under conditions where the stomata are closed like during the night or under several abiotic stresses.

Thus, several studies addressed how AM symbiosis regulates plant host L under drought stress. Apparently the results found in the literature are contradictory. Hence, Dell'Amico et al. (2002) in tomato plants using *Glomus clarum* as AM fungus and Sánchez-Blanco et al. (2004) in *Rosmarinus officinalis* plants using *G. deserticola* as inoculum found an increase in L of AM inoculated plants under drought conditions. On the contrary, Aroca et al. (2007) in bean, Aroca et al. (2008) in lettuce, and Ruiz-Lozano et al. (2009) in maize, did not find any difference in L values between AM and non AM plants under drought conditions. However, two different methods were used to measure L. Dell'Amico et al. (2002) and Sánchez-Blanco et al. (2004) used a pressure chamber by which both apoplastic and cell-to-cell paths are measured. By contrary, Aroca et al. (2007, 2008) and Ruiz-Lozano et al. (2009) measured L by the free exudation method, where only cell-to-cell path is measured. Therefore, and based on the transpiration measurements, it is possible that AM symbiosis favours water flowing by the apoplastic path (Aroca et al., 2007).

On the other hand, there are several evidences supporting the idea that aquaporins regulate ultimately plant L. Aquaporins are membrane intrinsic proteins present in all living organisms that facilitate the pass of water and other small and neutral molecules through them always following and osmotic gradient (Maurel et al., 2008). In plants, aquaporins are subdivided in five subgroups based on their amino acid sequence similarity. The five subgroups are plasma membrane intrinsic proteins (PIPs), tonoplast intrinsic proteins, nodulin-like intrinsic proteins, small and basic intrinsic proteins, and uncharacterized intrinsic proteins (Gupta and Sankararamakrishnan, 2009). Several studies showed how plants with one PIP gene silenced had also lower L values (Javot et al., 2003; Postaire et al., 2010).

In the last years, several reports pointed out the regulation of PIP aquaporins by AM symbiosis. Thus, Porcel et al. (2006) found that *G. mosseae* soybean inoculated plants decreased earlier the root expression of one PIP gene under drought conditions compared to non inoculated plants. This down-regulation was interpreted as a mechanism of avoiding water flow from roots to soil as soil water potential was decreasing progressively. Later on, Aroca et al. (2007) analyzed the expression of four different PIP genes in bean roots subjected to drought and inoculated or not with *G. intraradices* (Figure 1). Authors found that each PIP aquaporin gene responded different to drought and in almost all cases depending on the AM presence. Thus, *PvPIP1;1* expression increased by drought stress in AM plants but it remained unchanged in non AM ones. On the contrary, *PvPIP2;1* expression increased by drought in non AM plants, but it did not change in AM ones. On the other hand, *PvPIP1;3* expression was opposite regulated in AM and non AM plants. While *PvPIP1;3* expression increased in non AM plants, its expression was down regulated in AM ones. Finally, *PvPIP1;2* expression was down regulated in both kind of plants, but in a greater extent in AM ones. These results highlighted the specific function of each PIP gene in the plant response to drought stress.

Most recently, Ruiz-Lozano et al. (2009) analyzed the expression and abundance of several PIP genes under drought conditions in maize plants inoculated or not with the AM fungus *G. intrarradices*. In this research was found an up regulation of several PIP genes by drought stress in AM plants, which was confirmed at the protein level. So, all the above results together indicate that the regulation of aquaporin genes by AM symbiosis under drought stress depends on the plant specie and on the specific PIP gene analyzed. In the next future, a precise localization of these genes in AM and non AM roots is needed in order to understand their specific role in the enhancement of drought tolerance by AM symbiosis.

However, most recently Aroca et al. (2009) cloned the first aquaporin from any AM fungus (*GintAQP1*). Authors found some evidences supporting the idea that fungi aquaporins could compensate the down regulation of host plant aquaporins caused by drought. Also, they found that *GintAQP1* expression was up regulated in the mycelium part do not osmotically stressed when the other mycelium part was stressed by NaCl. So, a possible communication between un-stressed and stressed mycelium could be taken place. Obviously, more fungal aquaporins cloned are needed to completely understand this possible communication.

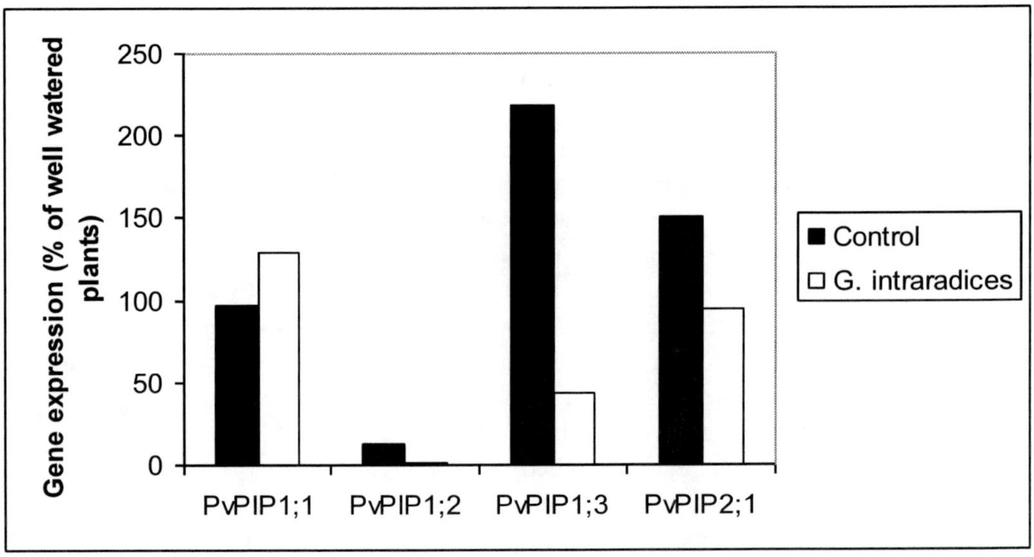

Figure 1. Percentage of changes in several aquaporin PIP gene expressions by drought stress in non inoculated (Control) and AM *Phaseolus vulgaris* plants (*G. intraradices*). Adapted from Aroca et al. (2007).

AM Symbiosis, Drought and Water Use Efficiency

As it has been stated above, plants able to achieve a better growth with less soil water used are needed in order to maximize soil water resources. These plants should have higher water use efficiency (WUE), that is, they should produce more biomass with the same amount of water consumed. However, conciliation between reductions in stomatal opening without affecting plant growth is not always achieved (Blum, 2009). In Table 1 is summarized some of the data available about the effect of AM symbiosis in WUE under drought conditions. In two out of three cases were AM symbiosis did not increase WUE, it also did not increase plant growth. At the same time, in two out of three non effect cases, WUE was estimated by isotopic discrimination. On the other hand, when AM symbiosis enhanced WUE of the host plant, it also enhanced plant growth. Finally, in almost all situations where AM symbiosis caused an increment of WUE, it was measured as the ratio between "biomass" gain and "water consumed". These result matched with the proposal of Blum (2009) who discouraged the use of the isotopic discrimination as a tool to estimate WUE under drought conditions in the field.

So, AM plants are not only able to take up more water under drought conditions (Marulanda et al., 2003), but also to use these water more efficiently. In fact, in the study of Kaya et al. (2003), colonized or non colonized watermelon plants consumed the same amount of water, but the AM plants produced more and bigger fruits, showing a better effective use of water (Blum, 2009). Hence, net photosynthesis should be enhanced under drought conditions in AM plants (Ruiz-Lozano et al., 1995; Khalvati et al., 2005). Net photosynthesis can be enhanced by several ways like diminishing dark respiration, increasing photosystem II (PSII) efficiency, or raising the activity of CO_2 assimilatory enzymes (Niu et al., 2003).

There are very few data about how AM symbiosis affects leaf dark respiration under drought conditions. Thus, Klingeman et al. (2005) found higher leaf dark respiration in AM rose plants than in non AM ones under drought conditions. In accordance, in this study no differences in WUE between inoculated and non inoculated plants were observed. Therefore, more work is needed in order to establish the possible role of dark respiration in changes of WUE caused by AM symbiosis.

The more spread method used to estimate PSII efficiency is the measurement of chlorophyll a fluorescence (Baker, 2008). By using this technique it has been proven that AM plants had more PSII efficiency than non AM ones under drought conditions (Pinior et al., 2005; Ruiz-Sánchez et al., 2010). More precisely, Pinior et al. (2005) found that AM symbiosis improved the activity of the oxygen evolving complex at the donor site of PSII and the electron flow between PSII and PSI in rose plants subjected to drought. This better PSII efficiency of AM plants under drought conditions could be involved in their higher WUE. Beside PSII efficiency, WUE efficiency can be also enhanced by a raise in the CO_2 assimilatory enzymes activities. Hence, Valentine et al. (2008) found that AM grapevines had higher Rubisco activity and higher WUE than non AM grapevine plants under drought conditions. However, much more research is needed to address the possible involvement of the CO_2 assimilatory enzymes in the enhancement of WUE by AM symbiosis under drought stress.

Table 1. Summary of the effects of AM inoculation on water use efficiency (WUE) and growth under drought conditions

Plant Species	Fungus Species	Method used	WUE effect	Growht effect	Reference
Olea europaea	mixed Glomus spp.	Isotopic discriminartion	Positive	Positive	Querejeta et al. (2006)
Rhamnus lyioides	mixed Glomus spp.	Isotopic discriminartion	Non effect	Positive	Querejeta et al. (2006)
Allium cepa	Several Glomus spp.	Yield vs Evapotranspiration	Positive	Positive	Bolandnazar et al. (2007)
Triticum durum	Glomus monosporum	Biomass vs Evapotranspiration	Positive	Positive	Al-Karaki (1998)
Lactuca sativa	Several Glomus spp.	Net Phtosynthesis vs Transpiration	Positive	Positive	Ruiz-Lozano et al. (1995)
Capsicum annum	Several Glomus spp.	Isotopic discrimination	Non effect	Non effect	Davies et al. (2002)
Rosa spp.	Glomus intraradices	Carbon gain vs Water consumed	Non effect	Non effect	Klingeman et al. (2005)
Citrullus lanatus	Glomus clarum	Yield vs Water consumed	Positive	Positive	Kaya et al. (2003)

As summary, it is evident that there are strong lacks in the knowledge of how and when AM symbiosis increases WUE of the host plant under drought conditions. More research is needed at the biochemical level to elucidate the mechanisms behind such possible raise in host plant WUE.

AM SYMBIOSIS, DROUGHT AND ANTIOXIDANT CAPACITY

As has been mentioned above, drought episodes could cause a raise in ROS production and as a consequence, an oxidative stress (Miller et al., 2010). Plants have evolved several antioxidants mechanisms to avoid oxidative damage linked to stressful conditions. These mechanisms can be divided as enzymatic or non-enzymatic. Non-enzymatic mechanisms include compounds able to scavenge directly several ROS as are ascorbic acid (AA), glutathione (GSH), or □-tocopherol. Enzymatic antioxidants include superoxide dismutase (SOD), glutathione reductase (GR), catalase (CAT, which converts H_2O_2 to water), and several peroxidases being the most studied the ascorbate peroxidase (APX). For a recent review of antioxidant systems in plants see Jaleel et al. (2009). Here will be revised in which degree AM symbiosis diminishes oxidative damage caused by drought and which antioxidant systems could be involved.

Table 2. Summary of the effects of AM inoculation in different components of plant antioxidant system

Plant species	Lipid peroxidation	SOD	APX	GR	CAT	POX	AA	GSH	Reference
Olea europaea	n.d.	↑	↑	=	↑	n.d.	n.d.	n.d.	Alguacil et al. (2003)
Phyllirea angustifolia	n.d.	↓	n.d.	n.d.	n.d.	↓	n.d.	n.d.	Caravaca et al. (2005)
Juniperus oxycedrus	n.d.	↓	n.d.	n.d.	n.d.	↓	n.d.	n.d.	Alguacil et al. (2006)
Poncirus trifoliata	↓	=	↑	↑	↑	↑	↑	↑	Wu et al. (2006b)
Citrus sinensis	↓	↑	n.d.	n.d.	↑	↑	↑	↑	Wu and Zou (2009)
Glycine max	↓	↑	↓	=	=	n.d.	n.d.	n.d.	Porcel and Ruiz-Lozano (2004)
Oryza sativa	↓	n.d.	n.d.	n.d.	n.d.	n.d.	↓	↑	Ruiz-Sánchez et al. (2010)

Superoxide dismutase (SOD), ascorbate peroxidase (APX), glutathione reductase (GR), other peroxidases (POX), ascorbic acid (AA), glutathione (GSH). N.d. means not detected. ↑or ↓ means increase or decrease, respectively. = means no changes.

In Table 2 is summarized some of the results found in the literature about how AM symbiosis affects antioxidant system of the host plant. All the data in Table 2 are from shoot measurements and in plants were growth was promoted by the symbiosis. Unfortunately, there are very few data about how AM colonization decreases shoot lipid peroxidation caused by drought (Porcel and Ruiz-Lozano, 2004; Wu et al., 2006b; Wu and Zou, 2009; Ruiz-Sánchez et al., 2010). In the studies of Wu and co-workers, the diminution of lipid peroxidation was correlated with a raise in the antioxidant enzymes activities and compounds levels; however such correlation was not found by Porcel and Ruiz-Lozano (2004). These last authors hypothesized that since AM symbiosis prevented leaf dehydration; also diminished the production of ROS, and therefore the host plant did not need to enhance the activity of the antioxidant enzymes. These results pointed out the multiple factors involved in the drought tolerance enhancement by AM symbiosis. The same hypothesis was also claimed by other authors (Caravaca et al., 2005; Alguacil et al., 2006).

At the same time, there are very few data about how AM symbiosis regulates host antioxidant compounds under drought stress (Wu et al., 2006b; Wu and Zou, 2009; Ruiz-Sánchez et al., 2010). In these three reports an increase in GSH content was found, and only Wu and co-workers also found an increase of AA contents. Tausz et al. (2004) claimed that GSH contents raise in plants acclimated to abiotic stresses, which could be the case of AM plants. GSH has not only the functions of scavenging peroxides or regenerating AA pool, but also it regulates the expression of photosynthetic genes and may keep the cell pools of reducing power (NADPH) under necessary conditions for plant living cells (May et al., 1998). No data about other antioxidant compounds regulated by AM symbiosis like a-tocopherol are available in the literature and more efforts are needed in this direction. More practical results were found by Subramanian et al. (2006), who found an increase in the AA pool content in tomato fruits of AM colonized plants under several water regimes (from well watered plants to severe droughted ones), without almost affecting plant yield. These results are really interesting in order to improve food quality since the high importance of AA in human nutrition (Ginter, 2007).

PERSPECTIVES

Although there are in the literature some negative results (Weber et al., 1993; Franzini et al., 2010), it can be concluded that AM symbiosis enhances plant host tolerance to drought. In fact, much progress has been made in understanding how AM symbiosis regulates L and PIP aquaporin expression under drought conditions in the last years. However, deeper studies are needed to completely understand the importance of this regulation. At the same time, it is clear that more research is needed to understand how AM symbiosis regulates plant host WUE, especially at biochemical and molecular levels (dark respiration and CO_2 assimilatory enzymes). On the other hand, studies to solve which signals regulate at the molecular level the antioxidant systems of AM plants under drought conditions are necessary. All these fields of study should be addressed in the near future.

REFERENCES

Alguacil, MM; Hernández, JA; Caravaca, F; Portillo, B; Roldán, A. Antioxidant enzyme activities in shoot from three mycorrhizal shrub species afforested in a degraded semi-arid soil. *Physiologia Plantarum*, 2003 118, 562-570.

Alguacil, MM; Caravaca, F; Díaz-Vivancos, P; Hernández, JA; Roldán, A. Effect of arbuscular mycorrhizae and induced drought stress on antioxidant enzyme and nitrate reductase activities in *Juniperus oxycedrus* L. grown in a composted sewage sludge-amended semi-arid soil. *Plant and Soil*, 2006 279, 209-218.

Al-Karaki, GN. Benefit, cost and water-use efficiency of arbuscular mycorrhizal durum wheat grown under drought stress. *Mycorrhiza*, 1998 8, 41-45.

Aroca, R; Ruiz-Lozano, JM. Induction of plant tolerance to semi-arid environments by beneficial soil microorganisms – A review. In: Lichtfouse E editor. *Sustainable agriculture reviews 2: climate change, intercropping, pest control and beneficial microorganisms*. Dordrecht: Springer; 2009; 121-136.

Aroca, R; Ferrante, A; Vernieri, P; Chrispeels, MJ. Drought, abscisic acid, and transpiration rate effects on the regulation of aquaporin gene expression and abundance in *Phaseolus vulgaris* plants. *Annals of Botany*, 2006 98, 1301-1310.

Aroca, R; Porcel, R; Ruiz-Lozano, JM. How does arbuscular mycorrhizal symbiosis regulate root hydraulic properties and plasma membrane aquaporins in *Phaseolus vulgaris* under drought, cold or salinity stresses? *New Phytologist*, 2007 173, 808-816.

Aroca, R; Vernieri, P; Ruiz-Lozano, JM. Mycorrhyzal and non-mycorrhizal *Lactuca sativa* plants exhibit contrasting responses to exogenous ABA during drought stress and recovery. *Journal of Experimental Botany*, 2008 59, 2029.2041.

Aroca, R; Bago, A; Sutka, M; Paz; JA; Cano, C; Amodeo, G; Ruiz-Lozano, JM. Expression analysis of the first arbuscular mycorrhizal fungi aquaporin described reveals concerted gene expression between salt-stressed and nonstressed mycelium. *Molecular Plant-Microbe Interactions*, 2009 22, 1169-1178.

Baker, NR. Chlorophyll fluorescence: a probe of photosynthesis *in vivo*. *Annual Review of Plant Biology*, 2008 59, 89-113.

Bem-Hur, M. Using synthetic polymers as soil conditioners to control runoff and soil loss in arid and semi-arid regions – a review. *Australian Journal of Soil Research*, 2006 44, 191-204.

Blum, A. Efective use of water (EUW) and not water-use efficiency (WUE) is the target of crop yield improvement under drought stress. *Field Crop Research* 2009 112, 119-123.

Bolandnazar, S; Aliasgarzad, N; Neishabury, MR; Chaparzadeh, N. Mycorrhizal colonization improves onion (*Allium cepa* L.) yield and water use efficiency under water deficit conditions. *Scientia Horticulturae*, 2007 114, 11-15.

Caravaca, F; Alguacil, MM; Hernández, JA; Roldán, A. Involvement of antioxidant enzyme and nitrate reductase activities during water stress and recovery of mycorrhizal *Myrtus communis* and *Phillyrea angustifolia* plants. *Plant Science*, 2005 169, 191-197.

Davies Jr, FT; Olalde-Portugal, V; Aguilera-Gomez, L; Alvarado, MJ; Ferrera-Cerrato, RC; Boutton, TW. Alleviation of drought stress of Chile ancho pepper (*Capsicum annum* L. cv. San Luis) with arbuscular mycorrhiza indigenous to Mexico. *Scientia Horticulturae*, 2002 92, 347-359.

Dell'Amico, J; Torrecillas, A; Rodríguez, P; Morte, A; Sánchez-Blanco, MJ. Responses of tomato plants associated with the arbuscular mycorrhizal fungus *Glomus clarum* during drought and recovery. *Journal of Agricultural Science*, 2002 138, 387-393.

Farooq, M; Wahid, A; Kobayashi, N; Fujita, D; Basra, SMA. Plant drought stress: effects, mechanisms and management. *Agronomy for Sustainable Development*, 2009 29, 185-212.

Franzini, VI; Azcón, R; Mendes, FL; Aroca, R. Interactions between *Glomus* species and *Rhizobium* strains affect the nutritional physiology of drought-stressed legume hosts. *Journal of Plant Physiology*, 2010 167, 614-619.

Ginter, E. Chronic vitamin C deficiency increases the risk of cardiovascular diseases. *Bratislava Medical Journal*, 2007 108, 417-421.

Grzesiak, S; Hura, T; Grzesiak, MT; Pienkowski, S. The impact of limited soil moisture and waterlogging stress conditions on morphological and anatomical root traits in maize (*Zea mays* L.) hybrids of different drought tolerance. *Acta Physiologiae Plantarum*, 1999 21, 305-315.

Gupta, AB; Sankararamakrishnan, R. Genome-wide analysis of major intrinsic proteins in the tree plant *Populus trichocarpa*: characterization of XIP subfamily of aquaporins from evolutionary perspective. *BMC Plant Biology*, 2009 9, 134.

Jaleel, CA; Riadh, K; Gopi, R; Manivannan, P; Inés, J; Al-Juburi, HJ; Chang-Xin, Z; Hong-Bo, S; Panneerselvam, R. Antioxidant defence responses: physiological plasticity in higher plants under abiotic constraints. *Acta Physiologiae Plantarum*, 2009 31, 427-436.

Javot, H; Lauvergeat, V; Santoni, V; Martin-Laurent, F; Guclu, J; Vinh, J; Frank, KI; Schäffner, AR; Bouchez, D; Maurel, C. Role of a single aquaporin isoform in root water uptake. *The Plant Cell*, 2003 15, 509-522.

Kaya, C; Higgs, D; Kirnak, H; Tas, I. Mycorrhizal colonisation improves fruit yield and water use efficiency in watermelos (*Citrllus lanatus* Thunb.) grown under well-watered and water-stressed conditions. *Plant and Soil*, 2003 253, 287-292.

Khalvati, MA; Hu, Y; Mozafar, A; Schmidhalter, U. Quantification of water uptake by arbuscular mycorrhizal hyphae and its significance for leaf growth, water relations, and gas exchange of barley subjected to drought stress. *Plant Biology*, 2005 7, 706-712.

Klingeman, WE; van Iersel, MW; Kang, JG; Augé, RM; Moore, JL; Flanagan, PC. Whole-plant gas exchange measurements of mycorrhizal 'Iceberg' roses exposed to cyclic drought. *Crop Protection*, 2005 24, 309-317.

Marulanda, A; Azcón, R; Ruiz-Lozano, JM. Contribution of six arbuscular mycorrhizal fungal isolates to water uptake by *Lactuca sativa* L. plants under drought stress. *Physiologia Plantarum*, 2003 119, 526-533.

Matsuo, N; Ozawa, K; Mochizuki, T. Genotypic differences in root hydraulic conductance of rice (*Oryza sativa* L.)in response to water regimes. *Plant and Soil*, 2009 316, 25-34.

Maurel, C; Verdoucq, L; Luu, DT; Santoni, V. Plant aquaporins: membrane channels with multiple integrated functions. *Annual Review of Plant Biology*, 2008 59, 595-624.

May, MJ; Vernoux, T; Leaver, C; Van Montagu, M; Inzé, D. Glutathione homeostasis in plants: implications for environmental sensing and plant development. *Journal of Experimental Botany*, 1998 49, 649-667.

Miller, G; Suzuki, N; Ciftci-Yilmaz, S; Mittler, R. Reactive oxygen species homeostasis and signalling during drought and salinity stresses. *Plant, Cell and Environment*, 2010 33, 453-467.

Niu, SL; Jiang, GM; Li, YG; Gao, LM; Liu, MZ; Peng, Y; Ding, L. Comparison of photosynthetic traits between two typical shrubs: legume and non-legume in Hunshandak Sandland. *Photosynthetica*, 2003 41, 111-116.

Parniske, M. Arbuscular mycorrhiza: the mother of plant root endosymbioses. *Nature Reviews Microbiology*, 2008 6, 763-775.

Pinior, A; Grunewaldt-Stocker, G; Strasser, RJ. Mycorrhizal impact on drought stress tolerance of rose plants probed by chlorophyll a fluorescence, praline content and visual scoring. *Mycorrhiza*, 2005 15, 596-605.

Porcel, R; Ruiz-Lozano, JM. Arbuscular mycorrhizal influence on leaf water potential, solute accumulation, and oxidative stress in soybean plants subjected to drought stress. *Journal of Experimental Botany*, 2004 55, 1743-1750.

Porcel, R; Aroca, R; Azcón, R; Ruiz-Lozano, JM. PIP aquaporin gene expression in arbuscular mycorrhizal *Glycine max* and *Lactuca sativa* plants in relation to drought stress tolerance. *Plant Molecular Biology*, 2006 60, 389-404.

Postaire, O; Tournaire-Roux, C; Grondin, A; Boursiac, Y; Morillon, R; Schäffner, AR; Maurel, C. A PIP1 aquaporin contributes to hydrostatic pressure-induced water transport in both root and rosette of Arabidopsis. *Plant Physiology*, 2010 152, 1418-1430.

Querejeta, JI; Allen, MF; Caravaca, F; Roldán, A. Differential modulation of host plant $\delta^{13}C$ and $\delta^{18}O$ by native and nonnative arbuscular mycorrhiazl fungi in semiarid environment. *New Phytologist*, 2006 169, 379-387.

Ruiz-Lozano, JM; Azcón, R. Hyphal contribution to water uptake in mycorrhizal plants as affected by fungal species and water status. *Physiologia Plantarum*, 1995 95, 472-478.

Ruiz-Lozano, JM; Azcón, R; Gómez, M. Effects of arbuscular-mycorrhizal *Glomus* species on drought tolerance: physiological and nutritional plant responses. *Applied and Environmental Microbiology*, 1995 61, 456-460.

Ruiz-Lozano, JM; Alguacil, MM; Bárzana, G; Vernieri, P; Aroca, R. Exogenous ABA accentuates the differences in root hydraulic properties between mycorrhizal and non mycorrhizal maize plants through regulation of PIP aquaporins. *Plant Molecular Biology*, 2009 70, 565-579.

Ruiz-Sánchez, M; Aroca, R; Muñoz, Y; Polón, R; Ruiz-Lozano, JM. The arbuscular mycorrhizal symbiosis enhances the photosynthetic efficiency and the antioxidative response of rice plants subjected to drought stress. *Journal of Plant Physiology*, 2010 167, 862-869.

Sánchez-Blanco, MJ; Ferrández, T; Morales, MA; Morte, A; Alarcón, JJ. Variations in water status, gas exchange, and growth in *Rosmarinus officinalis* plants infected with *Glomus deserticola* under drought stress. *Journal of Plant Physiology*, 2004 161, 675-682.

Steudle, E; Peterson, CA. How does water get through roots? *Journal of Experimental Botany*, 1998 49, 755-788.

Subramanian, KS; Santhanakrishnan, P; Balasubramanian, P. Responses of field grown tomato plants to arbuscular mycorrhizal fungal colonization under varying intensities of drought stress. *Scientia Horticulturae*, 2006 107, 245-253.

Tausz, M; Sircelj, H; Grill, D. The glutathione system as a stress marker in plant ecophysiology: is a stress-response concept valid? *Journal of Experimental Botany*, 2004 55, 1955-1962.

Valentine, AJ; Mortimer, PE; Lintnaar, A; Borgo, R. Drought responses of arbuscular mycorrhizal grapevines. *Symbiosis*, 2008 41, 127-133.

Vanrensburg, L; Kruger, GHJ. Evaluation of components of oxidative stress metabolism for use in selection of drought tolerant cultivars in *Nicotiana tabacum* L. *Journal of Plant Physiology*, 1994 143, 730-737.

Wang, Y; Chen, X; Xiang, CB. Stomatal density and bio-water saving. *Journal of Integrative Plant Biology*, 2007 49, 1435-1444.

Weber, E; Saxena, MC, George, E; Marschener, H. Effect of vesicular arbuscular mycorrhiza on vegetative growth and harvest index of chickpea grown in northern Syria. *Field Crop Research*, 1993 32, 115-128.

Wu, QS; Zou, YN. Mycorrhiza has a direct effect on reactive oxygen metabolism of drought-stressed citrus. *Plant, Soil and Environment*, 2009 55, 436-442.

Wu, QS; Zou, YN; Xia, RX. Effects of water stress and arbuscular mycorrhizal fungi on reactive oxygen metabolism and antioxidant production by citrus (*Citrus tangerine*) roots. *European Journal of Soil Biology*, 2006a 42, 166-172.

Wu, QS; Xia, RX; Zou, YN. Reactive oxygen metabolism in mycorrhizal and non-mycorrhizal citrus (*Poncirus trifoliate*) seedlings subjected to water stress. *Journal of Plant Physiology*, 2006b 163, 1101-1110.

In: Mycorrhizal Fungi
Editor: Susanne M. Fulton
ISBN 978-1-61122-659-1
© 2011 Nova Science Publishers, Inc.

Chapter 12

TOLERANCE TO HIGH TEMPERATURE STRESS AND ANTIOXIDANT RESPONSE IN MYCORRHIZAL STRAWBERRY PLANTS RAISED IN SUMMER

***Youhong Li*[1,2] *and Yoichi Matsubara*[2,*]**
[1]Southwest Forestry University, Kunming 650224, China
[2]Faculty of Applied Biological Sciences, Gifu University, Gifu 501-1193, Japan

ABSTRACT

High temperature in summer often restricts rooting, root growth and induces browning of roots and shoots during nursery production resulting in withering and causing great losses of plants in strawberry cultivation. Then, it is necessary to establish stable technique for nursery production in strawberry cultivation during the high temperature period in summer including the problems of global warming. Arbuscular mycorrhizal fungus (AMF) has the effect of promoting host plant growth mainly by enhancing phosphorus uptake through symbiosis. On the other hand, promotion of active oxygen synthesis in plants occurs under several stress conditions including temperature stress, and antioxidative ability has relation with the resolution. In this study, the changes in activities of antioxidative ability through symbiosis with AMF in strawberry plants under natural high temperature stress were investigated. The strawberry (*Fragaria*×*ananassa* Duch., cv. Nohime) runner plants were treated with split root system method, and plants were raised in greenhouse during summer period. Dry weight of shoots and roots became greater in AMF plants than in non-AMF, resulting that growth promotion appeared in mycorrhizal plants under high temperature stress in summer period. The severity of browning in roots caused by high temperature stress was eased in AMF plants than in non-AMF. In this case, the alleviation of root browning appeared in both AMF- and AMF+ root system, so that induced tolerance to high temperature stress was recognized in mycorrhizal plants. AMF plots showed higher activities in SOD and APX than non-AMF; the effect differs with plant portions and

[*] Corresponding author: ymatsu@gifu-u.ac.jp

investigated periods. In addition, some AMF plots showed higher values in followed items; chlorophyll content, ascorbic acid content and DPPH radical-scavenging activity. From these findings, tolerance to high temperature stress and the increase in antioxidative ability occurred in mycorrhizal strawberry plants, so that the activation of antioxidative ability is supposed to be closely associated with the high temperature stress tolerance in summer.

INTRODUCTION

In strawberry forcing cultivation, high temperature in summer often restricts rooting, root growth and induces browning of shoots and roots during nursery production resulting in withering and causing great losses of plants, especially in the warm climate regions in Japan (Saito et al., 2003). Then, it is necessary to establish stable technique for nursery production in strawberry cultivation during the high temperature period in summer including ever-bearing strawberry cultivation. However, characteristics of high temperature tolerance in strawberry plants remain unclear. Recently, Tsukiji et al. (2002) reported the evaluation method of high temperature tolerance in strawberry plants using leaf disc thermostability, and high temperature acclimation might be effective for affording high temperature tolerance in strawberry plants. On the other hand, cooling medium system by heat of vapourization in strawberry bench culture was tried (Takaichi, 2004), but some problems such as high cost are remained.

Arbuscular mycorrhizal fungus (AMF) has the effect of promoting host plant growth mainly by enhancing phosphorus uptake through symbiosis (Marschner and Dell, 1994). As for strawberry plants, growth enhancement by AMF inoculation was reported in several combinations of AMF species and strawberry cultivars (Niemi and Vestberg, 1992; Varma and Schuepp, 1994). In addition, AMF increased yields in strawberry plants cultivated in elevated bench culture method (Li et al., 2009). Schenck and Smith (1982) and Haugen and Smith (1992) also reported that AMF plants produced more dry weight of shoots and roots than non-inoculated ones under high soil temperature in soybean and mung bean, respectively. However, it has been unclear tolerance to natural high temperature stress during summer period in mycorrhizal strawberry plants.

On the other hand, promotion of active oxygen synthesis in plants occurs under environmental stresses, such as temperature, drought and plant disease, and the antioxidative abilities have relation with the detoxification (Fridovich, 1986; Sahoo et al., 2007). As for mycorrhizal plants, disease tolerance and the increase in superoxide dismutase (SOD) activity occurred in tomato (Pozo et al., 2002) and pepper (Garmendia et al., 2006), and drought tolerance and SOD increase appeared in lettuce (Lozano et al., 1996). However, it remains unclear the relationship between high temperature stress tolerance and antioxidative abilities in mycorrhizal plants.

In this study, the influence of AMF colonization on tolerance to natural high temperature stress in summer period and antioxidative responses in strawberry plants were investigated.

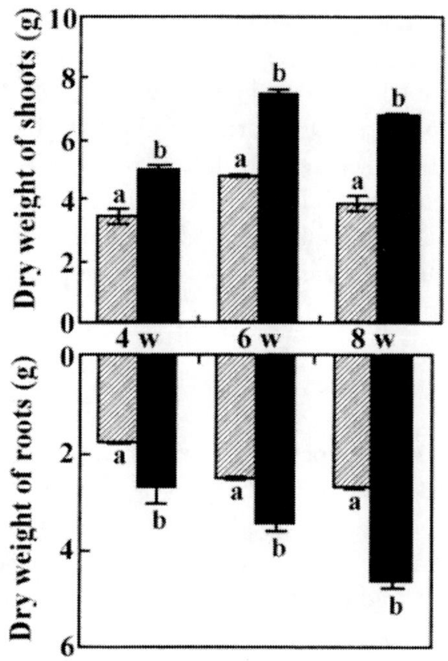

Figure 1. Influence of high temperature stress on dry weight of strawberry plants. ▨, control, ■, AMF. Bars represent standard errors (n=5). Different letters represent significant difference among the treatments according to Student's t-test (P<0.05).

MATERIALS AND METHODS

Inoculation of AMF

The strawberry (*Fragaria×ananassa* Duch., cv. Nohime) runner plants were treated with split root system method. Plant root system was separated into halves and each was placed in a different pot (10.5 cm in diameter) with bedding soil (autoclaved at 1.2 kg · cm^{-2} and 121°C for 1 h), and inoculated with *Glomus mosseae* (supplied by Idemitsukosan Co. Ltd.), according to Matsubara et al. (2004). Two weeks after AMF inoculation, the inoculated plants and the non-inoculated control plants were administered by mixed fertilizer (Long total 70 type, Asahi Kasei Co. Ltd.; N: P: K=13: 11: 13, 1 g per pot). Forty plants per plot with no replication, were irrigated by capillary watering as regularly in elevated bench culture. Plants were raised in greenhouse under daily fluctuating temperature ranged from 27±2 to 35±3 °C, natural light and daylength during summer period.

Estimation of Browning in Shoots and Roots

4, 6 and 8 weeks after AMF inoculation, the ratio of browning symptoms in roots of ten plants per plot were investigated.

Evaluation of AMF Colonization Level

Roots of AMF plants were harvested 4, 6 and 8 weeks after AMF inoculation. The roots were stained to observe the colonization according to Phillips and Hayman (1970). Then, the rate of AMF colonization in 1-cm segments of lateral roots (RFCSL) was calculated. Hence, RFCSL expresses the percentage of 1-cm AMF-colonized segments to the total 1-cm segments of all the lateral roots; the number of total segments was approx. 30 per a plant. The average colonization was calculated from the values of three plants.

Analysis of Antioxidative Abilities

4, 6 and 8 weeks after AMF inoculation, plants were sampled and partitioned into compound leaves, crowns and main roots characterized by no colonization and freezed in liquid nitrogen. Chlorophyll content in compound leaves was estimated by Porra et al. (1989). Analysis of superoxide dismutase (SOD) activity, ascorbic peroxidase (APX) activity, ascorbic acid content and DPPH radical-scavenging activity were carried out according to the methods of Beauchamp and Fridovich (1971), Li et al. (2004), Suzuki (2000), Suda (2000), respectively.

Results and Discussion

During the experimental periods, AMF plants had higher dry weight of shoots and roots than control (Figure 1), suggesting that growth promotion effect appeared in mycorrhizal strawberry plants under natural high temperature stress in summer. AMF colonization levels reached around 40% in all the sampling periods (Figure 2). The severity of browning in roots caused by high temperature became lower in AMF plants than in control plants 6 and 8 weeks after inoculation (Figure 3). In this case, severity of symptoms was alleviated in both AMF+ and AMF- plots than in control, thus, AMF- roots also showed high temperature tolerance. From these findings, tolerance to natural high temperature stress and induced tolerance to high temperature were recognized in mycorrhizal strawberry plants in summer period. Schenck and Smith (1982) reported that AMF plants produced more dry weight of shoots and roots than non-inoculated ones under high soil temperature in soybean, and the effect differed with fungal species. Shrestha et al. (1995) demonstrated that satsuma mandarin trees colonized with AMF showed greater growth and photosynthetic activity compared with non-inoculated ones under high temperature in August, though the effect differed with fungal species. On the other hand, Pozo et al. (2002) reported that in tomato plants with split root system, tolerance to *Phytophthora parasitica* appeared in both non-AMF inoculated roots and inoculated roots in AMF plants, so that induced tolerance to plant disease was recognized. The results in this study resembled with their findings.

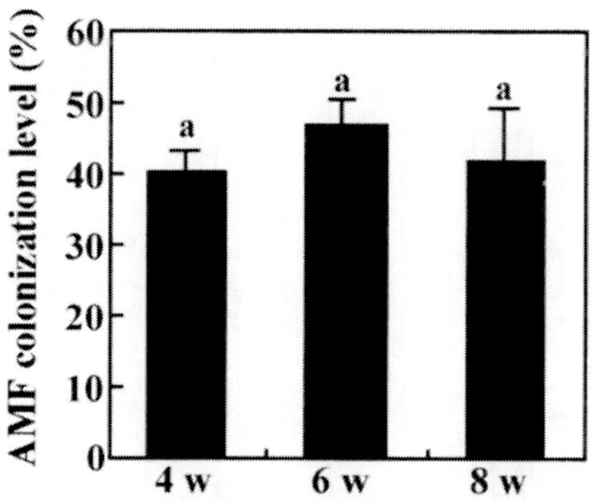

Figure 2. AMF colonization level (RFCSL) in strawberry plants. Different letters represent significant difference among the treatments according to Tukey's test (P<0.05).

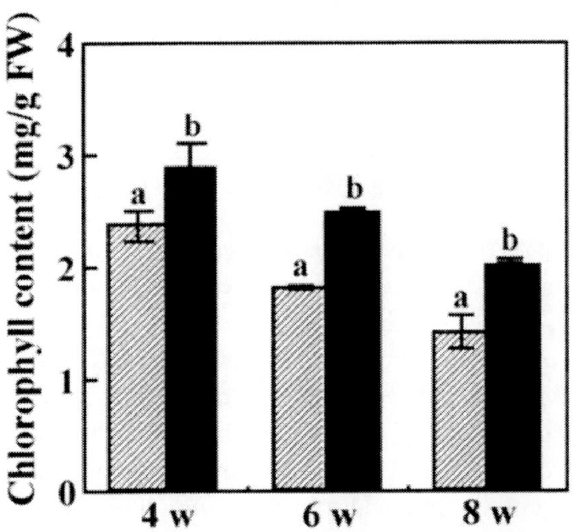

Figure 3. Chlorophyll content in compound leaves of strawberry plants. ▨, control, ■, AMF.

In this study, chlorophyll content became higher in AMF plots than in control among all of the investigated periods (Figure 4). As for antioxidative abilities, SOD activity became higher in crown and main roots (AMF+, AMF-) of AMF plots especially 6 and 8 weeks after inoculation except compound leaves (Figure 5). AMF plots showed higher values in APX activity in compound leaves 4w after inoculation and main roots 6w after inoculation. Ascorbic acid content increased in crown and main roots (AMF+, AMF-) of AMF plots 6w after inoculation. DPPH radical-scavenging activity increased in main roots of AMF plots 4 and 6w after inoculation. From these findings, tolerance to natural high temperature stress in summer and the increase in antioxidative abilities occurred in mycorrhizal strawberry plants, though the changes in antioxidative abilities varied depending on plant portions, so that the activation of antioxidative abilities supposed to be associated with the temperature stress tolerance.

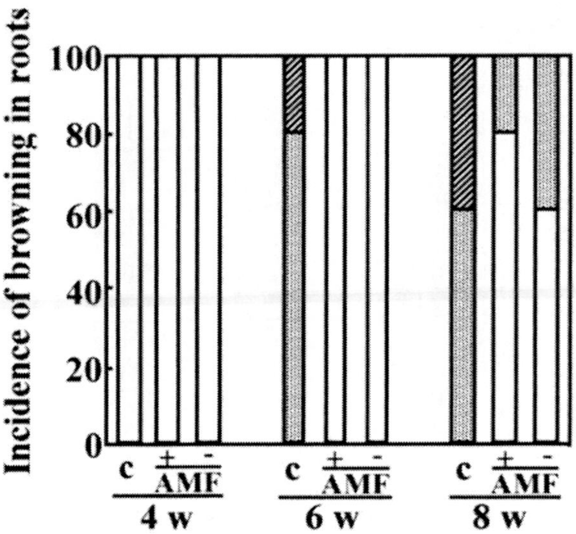

Ratio of browning in roots:
□ , 0-20; ▨ , 20-40; ▩ , 40-60; ▧ , 60-80; ■ , 80-100 (%).
c, control; +, AMF-inoculated roots; - , AMF-noninoculated roots.

Figure 4. Incidence of browning in roots of strawberry plants.

SOD play a primary role in defensive reactions and detoxify superoxide (O_2^-) among the antioxidative enzymes, thus, SOD activity is considered as the most important key enzyme in antioxidative abilities in plants (Fridovich, 1986). Pozo et al. (2002) demonstrated that tolerance to *Phytophthora parasitica* appeared in mycorrhizal tomato plants and SOD activity increased in both non-AMF inoculated and AMF-inoculated roots in split root system. In this study, non-AMF inoculated roots showed tolerance to high temperature stress, and the increase in SOD activity occurred in AMF- roots, supposing that antioxidative abilities might be also some association with the induced tolerance to high temperature stress in mycorrhizal strawberry plants.

Some reports described that mycorrhizal colonization itself induced temporal increase in antioxidative enzymes (APX, CAT), H_2O_2, flavonoid content, suggesting that colonization might become temporal stress for host plants (Volpin et al., 1995; Salzer et al., 1999; Blilou et al., 2000). In this study, no characteristics concerning colonization were found in the changes in antioxidative abilities. However, most of the antioxidative factors became higher levels in mycorrhizal plants during the experimental period. Hence, colonization itself induced little changes in mycorrhizal strawberry plants in this study.

The results in this study reveal that the symbiotic relationship with AMF provides vigorous shoot and root growth under uncertainly fluctuating temperatures in summer period, resulting in fast growing, vigorous plants with a high yield potential in strawberry.

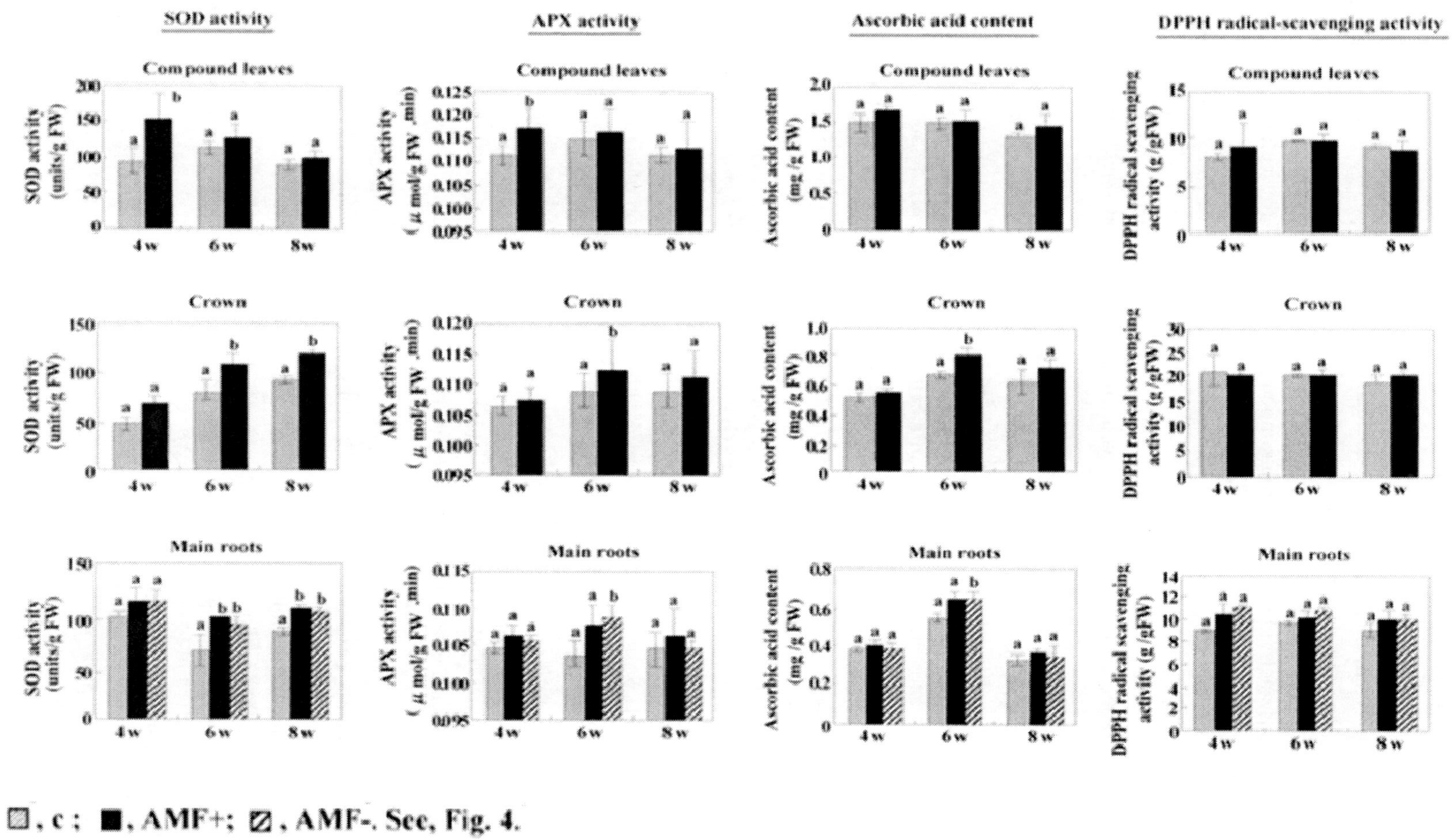

Figure 5. Antioxidative ability in strawberry plants under high temperature stress.

REFERENCES

Beauchamp, C. & Fridovich, I. (1971). Superoxide dismutase: Improved assays and an assay applicable to acrylamide gels. *Anal. Biochem.,* 44, 276-287.

Blilou, I., Bueno, P., Ocampo, J. A. & Garrido, J. M. (2000). Induction of catalase and ascorbate peroxidase activities tobacco roots inoculated with the arbuscular mycorrhizal *Glomus mosseae*. *Mycol. Res.,* 104, 722-725.

Fridovich, I. (1986). Biological effects of the superoxide radicals. *Arch. Biochem. Biophys.,* 247, 1-11.

Garmendia, I., Aguirreolea, J. & Goicoechea, N. (2006). Defence-related enzymes in pepper roots during interactions with arbuscular mycorrhizal fungi and/or *Verticillium dahliae*. *Biocont.,* 51, 293-310.

Haugen, L. M. & Smith, S. E. (1992). The effect of high temperature and fallow period on infection of mung bean and cashew roots by the vesicular-arbuscular mycorrhizal fungus *Glomus intraradices*. *Plant Soil,* 145, 71-80.

Li, Y., Imai, K., Ohno, H. & Matsui, S. (2004). Effects of acclimation temperatures on antioxidant enzyme activities in mericlones of a cattleya hybrid. *J. Japan. Soc. Hort. Sci.,* 73, 386-392 (In Japanese with English abstract).

Li, Y., Sassa, D., Matsubara, Y. & Koshikawa, K. (2009). Yield in mycorrhizal strawberry plants by elevated bench culture using fabric pots. *Hort. Res.,* 8, 215-219 (In Japanese with English abstract).

Lozano, J. M., Azcon, R. & Palma, J. M. (1996). Superoxide dismutase activity in arbuscular mycorrhizal *Lactuca sativa* plants subjected to drought stress. *New Phytol.,* 134, 327-333.

Marschner, H. & Dell, B. (1994). Nutrient uptake in mycorrhizal symbiosis. *Plant Soil,* 159, 89-102.

Matsubara, Y., Hirano, I., Sassa, D. & Koshikawa, K. (2004). Alleviation of high temperature stress in strawberry plants infected with arbuscular mycorrhizal fungi. *Environ. Control Biol.,* 42, 105-111.

Niemi, M. & Vestberg, M. (1992). Inoculation of commercially grown strawberry with VA mycorrhizal fungi. *Plant Soil,* 144, 133-142.

Phillips, J. M. & Hayman, D. S. (1970). Improved procedures for clearing roots and staining parasitic and vesicular-arbuscular mycorrhizal fungi for rapid assessment of infection.*Trans. Br. Mycol. Soc.,* 55, 158-163.

Porra, R. J., Thompson, W. A. & Kriedemann, P. E. (1989). Determination of accurate extinction coefficients and simultaneous equations for assaying chlorophylls a and b extracted with four different solvents: verification of the concentration of chlorophyll standards by atomic absorption spectroscopy. *Biochem. Biophys. Acta,* 975, 384-394.

Pozo, M. J., Cordier, C., Gaudot, E. D., Barea, J. M. & Aguilar, C. A. (2002). Localized versus systemic effect of arbuscular mycorrhizal fungi on defence responses to *Phytophthora* infection in tomato plants. *J. Exp. Bot.,* 53, 525-534.

Sahoo, M. R., Dasgupta, A. M., Paresh, A., Kole, C., Jayant, A., Bhat, S. & Mukherjee, A. (2007). Antioxidative enzymes and isozymes analysis of taro genotypes and their implications in Phytophthora blight disease resistance. *Mycopathol.,* 163, 241-248.

Saito, Y., Nagasaka, Y. & Imagawa, M. (2003). Repression of rooting in high temperature and effect of the pretreatment of short time low temperature storage in strawberry runner cutting. *J. Japan. Soc. Hort. Sci.,* 72 (suppl. 1), 244 (In Japanese).

Salzer, P., Corbiere, H. & Boller, T. (1999). Hydrogwn peroxidase accumulation in *Medicago truncatula* roots colonized by the arbuscular mycorrhiza-forming fungus *Glomus intraradices*. *Planta,* 208, 319-325.

Schenck, N. C. & Smith, G. S. (1982). Responses of six species of vesicular-arbuscular mycorrhizal fungi and their effects on soybean at four soil temperatures. *New Phytol.,* 92, 193-201.

Shrestha, Y. Ishii, T. & Kadoya, K. (1995). Effect of vesicular-arbuscular mycorrhizal fungi on the growth, photosynthesis, transpiration and the distribution of photosynthates of bearing satsuma mandarin trees. *J. Japan. Soc. Hort. Sci.,* 64, 517-525.

Suda, I. (2000). Estimation of DPPH radical scavenging activity. In: K. Shinohara, T. Suzuki and S. Kaminogawa (eds.), *Study methods of food function:* pp. 218-223. Tokyo, Korin (In Japanese).

Suzuki, T. (2000). Analysis of ascorbic acid content. In: T. Sugawara and A. Maekawa (eds.), *New handbook of food analysis:* pp. 233-239. Tokyo, Kenpakusya (In Japanese).

Takaichi, M. (2004). Medium cooling system by heat of vapourization in strawberry bench culture. *Agr. Hort.,* 59, 55-57 (In Japanese).

Tsukiji, K., Imada, S. & Dan, K. (2002). Evaluation of high temperature acclimation in strawberry by leaf discs thermostability. *J. Japan. Soc. Hort. Sci.,* 71 (suppl. 2), 373 (In Japanese).

Varma, A. & Schuepp, H. (1994). Infectivity and effectiveness of *Glomus intraradices* on micropropagated plants. *Mycorrhiza,* 5, 29-37.

Volpin, H., Phillips, D. A., Okon, Y. & Kapulnik, Y. (1995). Suppression of an isoflavonoid phytoalexin defense response on mycorrhizal alfalfa roots. *Plant Physiol.,* 108, 1449-1454.

INDEX

A

B

C

D

E

F

G

H

I

J

K

L

M

N

O

P

Q

R

S

T

U